Reactive Oxygen Species

Scrivener Publishing
100 Cummings Center, Suite 541J
Beverly, MA 01915-6106

Publishers at Scrivener
Martin Scrivener (martin@scrivenerpublishing.com)
Phillip Carmical (pcarmical@scrivenerpublishing.com)

Reactive Oxygen Species

Signaling Between Hierarchical Levels in Plants

Edited by

Franz-Josef Schmitt
and Suleyman I. Allakhverdiev

WILEY

This edition first published 2017 by John Wiley & Sons, Inc., 111 River Street, Hoboken, NJ 07030, USA and Scrivener Publishing LLC, 100 Cummings Center, Suite 541J, Beverly, MA 01915, USA

For more information about Scrivener publications please visit www.scrivenerpublishing.com.

Wiley Global Headquarters
111 River Street, Hoboken, NJ 07030, USA

For details of our global editorial offices, customer services, and more information about Wiley products visit us at www.wiley.com.

Library of Congress Cataloging-in-Publication Data
ISBN 978-1-119-18488-1

Cover image courtesy of Suleyman I. Allakhverdiev
Cover design by: Kris Hackerott

Set in size of 11pt and Minion Pro by Exeter Premedia Services Private Ltd., Chennai, India

10 9 8 7 6 5 4 3 2 1

Contents

Abstract ix

Foreward 1 xi

Foreward 2 xiii

Preface xv

1 Multiscale Hierarchical Processes 1
- 1.1 Coupled Systems, Hierarchy and Emergence 2
- 1.2 Principles of Synergetics 12
- 1.3 Axiomatic Motivation of Rate Equations 15
- 1.4 Rate Equations in Photosynthesis 19
- 1.5 Top down and Bottom up Signaling 23

2 Photophysics, Photobiology and Photosynthesis 27
- 2.1 Light Induced State Dynamics 27
  - 2.1.1 Light Induced Transition Probabilities and Rate Equations 32
  - 2.1.2 Absorption and Emission of Light 33
  - 2.1.3 Relaxation Processes and Fluorescence Dynamics 35
  - 2.1.4 Decay Associated Spectra (DAS) 39
- 2.2 Rate Equations and Excited State Dynamics in Coupled Systems 41
  - 2.2.1 Simulation of Decay-Associated Spectra 47
  - 2.2.2 Excited States in Coupled Pigments 52
  - 2.2.3 Förster Resonance Energy Transfer (FRET) 55
- 2.3 Light-Harvesting, Energy and Charge Transfer and Primary Processes of Photosynthesis 64
- 2.4 Antenna Complexes in Photosynthetic Systems 70
  - 2.4.1 The Light-Harvesting Complex of PS II (LHCII) of Higher Plants 72
  - 2.4.2 The LH1 and LH2 of Purple Bacteria 75

2.4.3 The Fenna-Matthews-Olson (FMO) Complex of Green Sulfur Bacteria 79
2.4.4 Phycobilisomes in Cyanobacteria 80
2.4.5 Antenna Structures and Core Complexes of *A.marina* 86
2.5 Fluorescence Emission as a Tool for Monitoring PS II Function 91
2.6 Excitation Energy Transfer and Electron Transfer Steps in Cyanobacteria Modeled with Rate Equations 93
2.7 Excitation Energy and Electron Transfer in Higher Plants Modeled with Rate Equations 105
2.8 Nonphotochemical Quenching in Plants and Cyanobacteria 114
2.9 Hierarchical Architecture of Plants 118

3 Formation and Functional Role of Reactive Oxygen Species (ROS) 123
3.1 Generation, Decay and Deleterious Action of ROS 125
3.1.1 Direct $^1\Delta_gO_2$ Generation by Triplet-triplet Interaction 126
3.1.2 The $O_2^{-\bullet}/H_2O_2$ System 130
3.1.3 H_2O_2 and Formation of $^1\Delta_gO_2$ and Other Reactive Species like $HO^\bullet$ 134
3.1.4 The $HO^\bullet$ Radical 136
3.2 Monitoring of ROS 137
3.2.1 Exogenic ROS Sensors 138
3.2.2 Spin Traps 145
3.2.3 Genetically Encoded ROS Sensors 146
3.2.4 Electrochemical Biosensors 150
3.3 Signaling Role of ROS 151

4 ROS Signaling in Coupled Nonlinear Systems 157
4.1 Signaling by Superoxide and Hydrogen Peroxide in Cyanobacteria 158
4.2 Signaling by Singlet Oxygen and Hydrogen Peroxide in Eukaryotic Cells and Plants 163
4.3 ROS and Cell Redox Control and Interaction with the Nuclear Gene Expression 167
4.4 ROS as Top down and Bottom up Messengers 174
4.4.1 Stoichiometric and Energetic Considerations and the Role of Entropy 179
4.4.2 The Entropy in the Ensemble of Coupled Pigments 186

4.5 Second Messengers and Signaling Molecules in H_2O_2 Signaling Chains and (Nonlinear) Networking 191
4.6 ROS-Waves and Prey-Predator Models 192
4.7 Open Questions on ROS Coupling in Nonlinear Systems 196

5 The Role of ROS in Evolution 199
5.1 The Big Bang of the Ecosphere 200
5.2 Complicated Patterns Result from Simple Rules but Only the Useful Patterns are Stable 201
5.3 Genetic Diversity and Selection Pressure as Driving Forces for Evolution 205

6 Outlook: Control and Feedback in Hierarchical Systems in Society, Politics and Economics 209

Bibliography 213

Appendix 249

Index 259

Abstract

Reactive oxygen species (ROS) play different roles in oxidative degradation and signal transduction in photosynthetic organisms. This book introduces basic principles of light-matter interaction, photophysics and photosynthesis to elucidate complex signaling networks that enable spatiotemporally directed macroscopic processes. Reaction schemes are presented for the formation and monitoring of ROS and their participation in stress signal transduction pathways within prokaryotic cyanobacteria as well as from chloroplasts to the nuclear genome in plants. Redoxregulated systems, mitogen-activated protein kinase cascades and transcription factor networks play a key role in ROS-dependent signaling systems in plant cells and for their spatiotemporal morphology and lifespan. ROS are understood as bottom up messengers and as targets of top down communication in plants. The role of the chemical environment in introducing genetic diversity as a prerequisite for efficient adaption is elucidated. Finally, it is suggested how the presented concepts can be used to describe other biological principles and multiscale hierarchical systems in society, politics and economics.

We hope that this book will be an invaluable reference source for university and research institute libraries, as well as for individual research groups and scientists working in the fields. It will be helpful not only for photo–biophysicists, biochemists, and plant physiologists, but also for a wider group of physicists and biologists. Lastly, and most importantly, it will serve to educate undergraduate, graduate and post-graduate students around the world.

Foreword 1

This book introduces basic principles of light-matter interaction, photophysics and photosynthesis in a brief manner in order to elucidate complex signaling networks that enable spatiotemporally directed macroscopic processes. The roles of reactive oxygen species (ROS) are explained both as bottom up messengers and as targets of top down communication. Starting with Hermann Haken's Principles of Synergetics, the role of ROS is investigated in order to explain emergent phenomena that occur during light-driven chemical reactions. Rate equations are used to describe energy transfer processes in photosynthesis, nonlinear phenomena that fill up energy reservoirs, and drive forces that are able to control macroscopic dynamics. The second chapter describes the basic principles of the light reaction in photosynthesis from light absorption to the storage of free Gibbs energy in the form of energy rich chemical compounds. The roles of different processes that support light energy transduction, on the one hand, and non-photochemical quenching, on the other hand, are elucidated with special focus on the context of a highly adapted system that has developed as an advanced structure for energy conversion during evolution in its local environment. Rate equations are not only used to understand optical transitions, excitation energy and electron transfer processes, and chemical reactions but are used to much more generally describe information processing in complex networks.

In the third and fourth chapters, the messenger role of ROS is described including formation, decay, monitoring and functional role of ROS – mainly H_2O_2, $^1O_2$, $O^{\bullet}{}_2^-$ – in both oxidative degradation and signal transduction during exposure of oxygen-evolving photosynthetic organisms to oxidative stress. These chapters focus on phenomena and mechanisms of ROS signaling. Reaction schemes are presented for the for mation and monitoring of ROS and their participation in stress signal transduction pathways, both within prokaryotic cyanobacteria as well as from chloroplasts to the nuclear genome in plants. It is suggested that redox-regulated systems, mitogen-activated protein kinase cascades, and transcription factor networks play a key role in the ROS-dependent signaling systems in plant cells and for their spatiotemporal morphology and lifespan.

Vladimir A. Shuvalov,
Academician (Russia)

Foreword 2

Reactive oxygen species (ROS) play different roles in oxidative degradation and signal transduction in photosynthetic organisms. Since modern microscopic, genetic, and chemical techniques for ROS detection and controlled generation have improved significantly in recent years, our knowledge of the complex interaction patterns of ROS has significantly increased. This book introduces basic principles of light-matter interaction, photophysics and photosynthesis in a brief manner in order to generally elucidate complex signaling networks that enable spatiotemporally directed macroscopic processes. ROS are understood as bottom up messengers and as targets of top down communication. Reaction schemes are presented for the formation and monitoring of ROS and their participation in stress signal transduction pathways, both within prokaryotic cyanobacteria as well as from chloroplasts to the nuclear genome in plants. It is suggested that redox-regulated systems, mitogen-activated protein kinase cascades, and transcription factor networks play a key role in the ROS-dependent signaling systems in plant cells and for their spatiotemporal morphology and lifespan.

Rate equations are used to explain the dynamics of excitation energy and electron transfer during light-driven reactions. Energy transfer processes and subsequent chemical reactions in photosynthesis are nonlinearly coupled. They fill up energy reservoirs and drive forces able to control macroscopic dynamics.

Photosynthetic organisms (as all organisms) represent highly adapted systems that have developed advanced structures for energy conversion during evolution in their local environment. Basic principles of evolution as the continuous adaption to environmental constraints are derived from considerations based on state transitions as basic theory and ROS as an example for a chemical reaction partner that contributes to selection and mutation. It will be shown that the structure of the environment mainly allows for genetic diversity as a prerequisite for efficient adaption and that mutations are of minor relevance in that context.

Finally, it is suggested how the presented concepts can be used to describe other biological principles and multiscale hierarchical systems in society, politics and economics.

This book is intended for a broad range of researchers and students, and all who are interested in learning more about the most important global process on our planet – the process of photosynthesis.

Tatsuya Tomo,
Professor (Japan)

Preface

To my family and my friends
Franz-Josef Schmitt

To my mother, my wife and my son
Suleyman I. Allakhverdiev

This book introduces basic principles of light-matter interaction, photophysics, and photosynthesis in a brief manner in order to elucidate principles of complex signaling networks that enable spatiotemporally directed macroscopic processes. We will start with random walk processes typically used to describe excitation energy transfer in plant lightharvesting complexes and later focus on coupled systems such as the communication network of reactive oxygen species (ROS) which are embedded into the plant metabolism and enable an information transfer from molecules to the overall organism as a bottom up process. In addition the microscale dynamics always accepts signals on the bottom level that can be understood as a top down communication. One example is the activation of genes by ROS caused by macroscopic events like strong sunlight or atmospheric variations that activate a change in composition of the microscopic environment by producing new proteins.

Typical bottom up processes are cascades that start with gene activation and protein translation regulating plant growth and morphology. Top down messengers triggered by macroscopic actuators like sunlight, gravity, environment or any forms of stress, on the other hand, activate gene regulation on the molecular level and therefore concern the dynamics of single molecules under the constraints of macroscopic factors. In this book the generation and monitoring as well as the role of ROS in photosynthetic organisms as typical messengers in complex networks are primarily treated in a scientific manner. All findings are supported by our own research results and recent publications. Additionally, the principles of top down and bottom up messaging are presented in the form of a philosophical discussion.

The first chapter focuses on a theoretical approach according to the Principle of Synergetics (Haken, 1990) to understand coupling, networks, and emergence of unpredictable phenomena. The approach is used to model light absorption, electron transfer and membrane dynamics in plants. Special focus will be placed on nonlinear processes that form the basic principle for the accumulation of energy reservoirs and on the formation of forces that are able to control the dynamics of macroscopic devices.

The formalism of rate equations is presented as a general scheme to formulate dynamical equations for arbitrarily complex systems. Key is the definition of "states" as an intensity level or a pattern that carries a certain amount of information, and their dynamics which are assessed by evaluating the probability of transfer from one state to another. In fact, rate equations are not only used to describe energy transfer processes in photosynthesis but in many systems, for instance, optical transitions, particle reactions, complex chemical reactions and more general information processing in complex networks. Rate equations are also used to describe complex systems such as sociological networks.

The second chapter describes the basic principles of the light reaction in photosynthesis from absorption to the storage of free Gibbs energy in the form of energy rich chemical compounds. The roles of different processes that support light energy transduction, on the one hand, and non-photochemical quenching, on the other hand, are elucidated with special focus on the context of a highly adapted system that has developed as an advanced structure for energy conversion during evolution in its local environment.

In the third chapter the formation, decay, monitoring, and the functional role of ROS – mainly H_2O_2, $^1O_2$, $O^{\bullet}{}_2^-$ are described and the fourth chapter especially focuses on the messenger role of ROS. The ambivalent picture of oxidative degradation and signal transduction during exposure of oxygen-evolving, photosynthetic organisms to oxidative stress isteluci-dated. Both degradation and activation are important mechanisms of ROS signaling. Reaction schemes are presented for the formation and monitoring of ROS and their participation in stress signal transduction pathways both within prokaryotic cyanobacteria as well as from chloroplasts to the nuclear genome in plants. It is suggested that redoxregulated systems, mitogen-activated protein kinase cascades and transcription factor networks play a key role in the ROS-dependent signaling systems in plant cells and for their spatiotemporal morphology and lifespan.

Chapter five focuses on evolution. It is emphasized that mainly the local environment of evolving organisms enforces directed evolution resulting in quick changes of the phenotype if the genetic diversity of the organisms is large enough. In that sense leaps in evolution necessarily follow volatile changes of the environment. It might be possible that ROS are the most important driving forces in evolution.

The last chapter finally offers a glance at how the described concepts can be used to describe other biological principles and multiscale hierarchical systems in society, politics and economics. The book is intended for a broad range of researchers and students, and everyone who is interested in learning about the most important global process on our planet – the process of photosynthesis. We would like to believe that this book will stimulate future researchers of photosynthesis and lead to progress in our understanding of the mechanisms of photosynthesis and their practical use in biotechnology and in human life.

We express our sincere gratitude to the two referees: the Academician of the Russian Academy of Sciences (RAS) Prof. V.A. Shuvalov and Prof. T. Tomo of Tokyo University of Science, Tokyo, Japan. We are extremely grateful to Corresponding Member of RAS Vl.V. Kuznetsov, Corresponding Member of RAS A.B. Rubin, and Professors D.A. Los, A.M. Nosov, V.Z. Paschenko, A.N. Tikhonov, G.V. Maksimov, V.V. Klimov, A.A. Tsygankov, and Drs. V.D. Kreslavski, S.K. Zharmukhamedov, I.R. Fomina, J. Karakeyan for their permanent help and fruitful advice. We are also indebted to Professors T. Friedrich, N. Budisa, P. Hildebrandt, L. Kroh, H.J. Eichler, Drs. M. Vitali, V. Tejwani, C. Junghans, N. Tavraz, J. Laufer, and J. Märk from TU Berlin and Drs. E.G. Maksimov and N. Belyaeva from Moscow State University, Prof. J. Pieper from University of Tartu, Prof. H. Paulsen from Johannes Gutenberg-Universität Mainz, Prof. F. Zappa and Dr. D. Bronzi from Politecnico di Milano, and Prof. R. Rigler, Drs. J. Jarvet, and V. Vukojević from the Karolinska Institute in Stockholm.

We express our deepest gratitude to Russian Science Foundation (№ 14-04-00039) and the German Research Foundation DFG (cluster of excellence "Unifying Concepts in Catalysis") and the Federal Ministry of Education and Research for funding bilateral cooperation between Germany and Russia (RUS 10/026 and 11/014). We acknowledge COST for financial support in the framework of COST action MP1205. We thank F. Schmitt for preparing Figs. 4, 5, 7, 10, 11, 18, 27A, 50, 51, 71, 75, 79 and 80. Further gratitude belongs to M. Nabugodi and J.M. Zinn for proofreading of the manuscript. F.-J. Schmitt especially thanks Joachim Herz

Stiftung and Stifterverband für die Deutsche Wissenschaft for the fellowship IGT-educationTUB.

We are grateful to **Scrivener Publishing, Wiley and Izhevsk Institute of Computer Sciences** for their cooperation in producing this book.

Franz-Josef Schmitt
Institute of Physical Chemistry,
Max-Volmer-Laboratory for Biophysical Chemistry,
Technische Universität Berlin
(e-mail: schmitt@physik.tu-berlin.de)

Suleyman I. Allakhverdiev
Controlled Photobiosynthesis Laboratory,
Institute of Plant Physiology, Russian Academy of Sciences, Moscow;
Institute of Basic Biological Problems,
Russian Academy of Sciences, Pushchino, Moscow Region;
Department of Plant Physiology, Faculty of Biology,
M.V. Lomonosov Moscow State University;
Department of Biological and Medical Physics,
Moscow Institute of Physics and Technology (State University),
Moscow, Russia;
Bionanotechnology Laboratory, Institute of Molecular Biology and
Biotechnology, Azerbaijan National Academy of Sciences,
Baku, Azerbaijan;
Department of New Biology, Daegu Gyeongbuk Institute
of Science & Technology (DGIST), Daegu, Republic of Korea
(e-mail: suleyman.allakhverdiev@gmail.com)

Franz-Josef Schmitt is a research assistant at the Institute of Physical Chemistry, Technische Universität Berlin (TU Berlin). He finished his doctoral thesis with "summa cum laude" in physics in 2011 in Prof. Hans-Joachim Eichler's working group on Laser Physics under the supervision of Prof. Gernot Renger. The thesis is entitled "Picobiophotonics for the investigation of Pigment-Pigment and Pigment-Protein interactions in Photosynthetic Complexes". Dr. Schmitt holds a series of scientific awards including the Chorafas award for extraordinary scientific results (2009) and the Stifterverb and Fellowship for excellence in teaching (2014) awarded by Joachim Herz Stiftung for his teaching project IGT-education TUB. He was recently awarded twice with a young talents award (2013) and received best poster awards (2014) for his invited presentations at the conference on Photosynthesis research for Sustainability in Baku, Azerbaijan and Moscow, Russia, respectively. More than 80 research papers, two patents, one book chapter and 200 partially invited talks on international conferences summarize his research in Photosynthesis, Nanobiophotonics, Environmental Spectroscopy and didactics. Dr. Schmitt was group leader of the reform project "educationZEN" that developed new teaching formats for Mathematics education and internships in the Nature Sciences and Engineering. His research fields comprise protein dynamics and protein-cofactor interaction, energy transfer and conversion in organic systems, information processing in complex networks and general biophysics and biotechnology. Dr. Schmitt is coordinating editor of the topical issue on "Optofluidics and Biological Materials" of the open access journal "Optofluidics, Microfluidics and Nanofluidics" (de Gruyter open) and guest editor of the open access journal SOAJ NanoPhotoBioSciences. He was substitutional Management Committee member of COST action MP 1205 and member of the Faculty Council of the Department of Mathematics and Nature Sciences (2007–2017). He is a member of the Academic Senate (since 2010), and Chairman of the Extended Academic Senate of TU Berlin (since 2015).

Suleyman I. Allakhverdiev is the Head of the Controlled Photobiosynthesis Laboratory at the Institute of Plant Physiology of the Russian Academy of Sciences (RAS), Moscow; Chief Research Scientist at the Institute of Basic Biological Problems RAS, Pushchino, Moscow Region; Professor at M.V. Lomonosov Moscow State University, Moscow; Professor at Moscow Institute of Physics and Technology (State University), Moscow, Russia; Head of Bionanotechnology Laboratory at the Institute of Molecular Biology and Biotechnology of the Azerbaijan National Academy of Sciences, Baku, Azerbaijan, and Invited-Adjunct Professor at the Department of New Biology, Daegu Gyeongbuk Institute of Science & Technology (DGIST), Daegu, Republic of Korea. He is originally from Chaykend (Karagoyunly/Dilichanderesi), Armenia, and he obtained both his B.S. and M.S. in Physics from the Department of Physics, Azerbaijan State University, Baku. He obtained his Dr.Sci. degree (highest/top degree in science) in Plant Physiology and Photobiochemistry from the Institute of Plant Physiology, RAS (2002, Moscow), and Ph.D. in Physics and Mathematics (Biophysics), from the Institute of Biophysics, USSR (1984, Pushchino). His Ph.D. advisors were Academician Alexander A. Krasnovsky and Dr. Sci. Vyacheslav V. Klimov. He worked for many years (1990–2007) as visiting scientist at the National Institute for Basic Biology (with Prof. Norio Murata), Okazaki, Japan, and in the Department de Chimie-Biologie, Universite du Quebec at Trois Rivieres (with Prof. Robert Carpentier), Quebec, Canada (1988–1990). He has been a guest editor of more than thirty special issues in international peer-reviewed journals. At present, he is a member of the Editorial Board of more than fifteen international journals. Besides being editor-in-chief of SOAJ NanoPhotoBioSciences, associate editor of the International Journal of Hydrogen Energy, section editor of the BBA Bioenergetics, associate editor of the Photosynthesis Research, associate editor of the Functional Plant Biology, and associate editor of the Photosynthetica, he also acts as a referee for major international journals and grant proposals. He has authored (or co-authored) more than 400 research papers, six patents and eight books. He has organized more than ten international conferences on photosynthesis. His research interests include the structure and function of photosystem II, water-oxidizing complex, artificial photosynthesis, hydrogen photoproduction, catalytic conversion of solar energy, plants under environmental stress, and photoreceptor signaling.

1 Multiscale Hierarchical Processes

Träumen wir uns für einen Moment in einen zukünftigen Zustand der Naturwissenschaft, in dem die Biologie ebenso vollständig mit Physik und Chemie verschmolzen sein wird, wie in der heutigen Quantenmechanik Physik und Chemie miteinander verschmolzen sind. Glaubst du, dass die Naturgesetze in dieser gesamten Wissenschaft dann einfach die Gesetze der Quantenmechanik sein werden, denen man noch biologische Begriffe zugeordnet hat, so wie man den Gesetzen der Newtonschen Mechanik noch statistische Begriffe wie Temperatur und Entropie zuordnen kann; oder meinst du, in dieser einheitlichen Naturwissenschaft gelten dann umfassendere Naturgesetze, von denen aus die Quantenmechanik nur als ein spezieller Grenzfall erscheint, so wie die Newtonsche Mechanik als Grenzfall der Quantenmechanik betrachtet werden kann ?

Werner Heisenberg (1901–1976),
Deutscher Physiker (Heisenberg, 1986)

Multiscale hierarchical processes are understood as information transduction in networks which are hierarchically structured. The most simple assumption might be a house which is structured into rooms, rooms are structured into furnishings but also people that move from one room to another. Of course cupboards and chairs, computers and TVs as well as human beings are hierarchically structured in a somehow comparable way. We would call our house a hierarchically structured system. If information flows from one room to another – and everyone would agree that this is the case when people live in that house and move objects or direct information – we can speak about hierarchical processes. These processes might comprise the information conveyed by the parents that the food is prepared which leads to a movement of the children towards the kitchen and the covering of the table by dishes, not to mention all the processes that are correlated with eating and enjoying the wonderful meal.

If we agree that the type of hierarchical structure might additionally vary if elucidated from different aspects we speak about multiscale hierarchical processes. Such different aspects can be aspects of spatial organization as it is the case in our example, the house. But in addition also other, for example temporal, organization principles are possible. To summarize all these organization principles we generalize the hierarchical systems to multiscale hierarchical systems housing the dynamics of multiscale hierarchical processes.

Living systems are always spatially hierarchically organized: in this case molecules are the basic entities that form genes and proteins as an intermediate structure on a mesoscale. The proteins aggregate in a quaternary structure to form higher ordered systems that do not necessarily need to be stable in time. The network of interacting proteins is in its turn forming a metastructure that can be understood as a network formed from single proteins. However, also on the temporal scale multiscale hierarchical processes arise. For example, a reaction scheme may represent the dynamics of the chemical reaction of two compounds on the temporal microscale. However, if a certain threshold of concentration of its output is present, another chemical reaction may start and is therefore triggered by the first reaction scheme. Long-term effects like the active movement of our extremities, circadian rhythms, the growth of an organism and senescence are typical examples of processes that change their appearance over time and are therefore a hierarchical metastructure that arises on the network of microscale processes.

1.1 Coupled Systems, Hierarchy and Emergence

Naturally, coupled systems are generally nonlinear. That is always the case if two compounds form a special reaction pathway that is not possible if only a single compound is present. If two molecules of two different compounds interact, then the reaction is bilinear or bimolecular, which is a nonlinearity. If two molecules (or two photons) of the same kind are able to reach a state that is not accessible by a single molecule (or photon) then the outcome is typically in a nonlinear dependence on the input. One prominent example of nonlinear optics in physics is the two photon absorption where the absorbing state is reached by interaction of two photons with the ground state within a certain time interval. If the photon density is too low for that to happen then the output is zero, but when the photon density increases the probability for the two photon absorption increases with the square of the photon density. Therefore two (or more) photons are needed to activate a state transition. Only spontaneous population and depopulation of certain states, which is not the typical situation for characteristic biochemical reactions, are truly linear. If several molecules of one or several compounds have to interact within a certain time interval, then the reaction scheme is nonlinear and characterized by the typical mathematical problems and challenges of nonlinear systems.

Photosynthesis is a truly nonlinear reaction as at least eight photons have to be absorbed by two different photosystems to split two molecules of water and release one molecule of oxygen. Photon absorption drives an electron transfer in photosynthesis. However, the involved molecules are reduced and/or oxidized by more than one electron and the coupled proton transfer again forms ATP from ADP and phosphate in a nonlinear process. Biochemistry is truly a hierarchy of nonlinear processes.

The "cycles" of nonlinearities that form the overall, hierarchical structure also include loss processes. After photon absorption excitation-energy can be lost and the following electron transfer processes are likewise restricted by loss processes that limit the production of one molecule of oxygen with a demand of at least 11–12 photons. Other sources report 60 photons per molecule glucose (Häder, 1999; Campbell and Reece, 2009) which would equal 10 photons per molecule glucose according to the basic equation of photosynthesis understood as the light-induced chemical reaction of water with carbon dioxide to glucose:

$$12\,H_2O + 6\,CO_2 \xrightarrow{h \cdot \nu} C_6H_{12}O_6 + 6\,O_2 + 6\,H_2O \qquad \textbf{chemical equation 1}$$

The energetic stoichiometry of light and dark reactions in photosynthesis are again discussed in chap. 4.4.1 as well as in the literature (Häder, 1999). This also features discussions of coupled reaction schemes like the proton assisted electron transfer (Renger, 2008, 2012; Renger and Ludwig, 2011). This question shall therefore not be discussed here in more detail. The basic principles of the photosynthetic light reaction are presented in chapter two.

Here we primarily intend to elucidate the highly nonlinear character of photosynthesis. Indeed, the nonlinearity of photosynthesis goes far beyond this discussion. If one regards the hierarchy of the spatiotemporal order of a plant as an overall reaction system, one could ask how many photons are involved in the construction of a new leaf. Analyzing the biomass of a dried leaf, which is strongly dependent on the organism, we might look at, say, 100 mg and find that the fixation of an order of 10^{21} carbon molecules was necessary with a corresponding nonlinear response of a new "leaf" to more than 10^{22} absorbed photons. That means absorption of 10^{22} photons finally leads to the spontaneous appearance of a single "leaf". Of course these photons have to be absorbed within a certain time interval. If illumination stays under a certain threshold, nothing happens, but if bright sunlight, sufficient day length and adequate temperature trigger the mechanisms correctly in spring, leaves might appear proportional to "packages" of 10^{22} absorbed photons. In this sense, the process might seem to be a linear response, but it is surely not and stops completely after a short growth period when new priorities like the production and storage of biomass take over in summer.

Even if we understand this reaction as a subsequent construction which can be analyzed step by step, it might still be a matter for discussion whether this reductionism leads to the loss of information and prevents our overall understanding of the growth of a plant (Heisenberg, 1986). After all, we have the appearance of one single leaf after the absorption of 10^{22} photons if we work as a pure phycisist who did not learn the details of biochemistry and does not know anything from gene activation and proteomics.

Our first identified nonlinear system (the water splitting and oxygen evolution) forms the trigger or the "input" for processes that are highly nonlinear themselves since they require several molecules of glucose, ATP or NADH to drive the production of one single further unit like for example a whole cell. We have a complicated spatiotemporal network of nonlinear systems that are coupled to nonlinear networks on the next hierarchically higher "level". In this way, the complexity of the overall system, the plant, arises. However, if only bottom up processes from the molecular interaction on the single molecule level are taken into account, then reductionism will fail to explain the details of the plant's morphology and lifespan.

Therefore it might still be wondered, like Heisenberg did in his book *Der Teil und das Ganze* (Heisenberg, 1986), whether we can expect that a possible picture of a plant as an organism understood in full detail will still use the language of physics or whether this picture will require that we formulate its propositions with novel approaches. Of course, taking the view of a physicist, the formalism that enables a scientist to understand an organism in detail and therefore enables the possibility to simulate the response of the system to a parameter change will be understood as a novel formalism. Therefore, Heisenberg's skepticism might not necessarily lead to the termination of actual scientific approaches. It is more likely that in the future science will overcome actual limitations as it has always done in the past: by inventing new approaches, new languages and more computational power. Current novel, highly-funded activities in the field of synthetic biology give rise to the hope that at least the composition of a single working cell from its basic chemical compounds will soon be possible.

Outputs of highly nonlinear reaction schemes function as substrate for highly nonlinear processes – on a hierarchic order of the scaling. In fact, interaction partners, cofactors, substrates, a series of several molecules or whole reaction networks, and even time can be a nonlinearity or a nonlinear scaling factor if there exists a feedback parameter that influences the dynamics in such way that it is no longer dependent on the substrate in a purely linear manner. Mathematical and physical representations of quite simple nonlinear systems show that the dynamics of such systems can change extremely if one parameter only changes slightly. This is called "chaotic behavior" or simply "chaos".

In contrast to the fact that nonlinear systems typically respond with chaotic behaviour, such coupled nonlinear structures can also be stable. Stability can arise from chaos. The chaotic behaviour of simple reaction schemes and the complexity that arises as well as the simplicity that arises from a complex pattern were topics in a broad series of literature that flourished 15–20 years ago (Gell-Mann, 1994; Cohen and Stewart, 1997; Wolfram, 2002). These excellent research books show how simple rules lead to complex structures and/or mathematically analyze chaos in deep detail using a scientific procedure.

At this point we might go a step backwards and focus on our initially introduced question how complex phenomena can arise from simple processes and how a hierarchical organization might deliver patterns with novel properties. It helps also to understand principles of self-similarity and macroscopic structuring arising from simple rules for single (chemical) reactions. Stephen Wolfram´s book *A New Kind of Science* deals with such an approach to motivate rise and decay of complex macroscopic patterns from simple rules on the microscale.

He works with so called "cellular automata" that define rules how to color a square in dependency of neighbouring squares. Programming such rules he found that the computer starts to draw interesting patterns and complex networks that arise even if the underlying rule is the most possible simple one working with black and white squares only.

Stephen Wolfram's underlying aproach is clearly bottom up. The concept of cellular automata explains complex patterns and their self-similarity by basic rules that the smallest entities in the concept have to follow. From the simple rule that determines in which color neighbouring squares on a white sheet of paper have to be imaged he can principally generate any desired pattern while discussing complexity, spatial and temporal hierarchies in the structures and the phenomenon of emergence.

Figure 1 is a computation of a simple rule, called "rule 30" by Wolfram with the commercial program Mathematica® also invented by Wolfram and distributed by Wolfram research. Wolfram has invented the concept of cellular automata that propose certain rules for the generation of binary two dimensional (or also more complicated) patterns of elementary cells only from the information content of neighbouring cells. Some selected rules as published in (Wolfram, 2002) are shown in Figure 2.

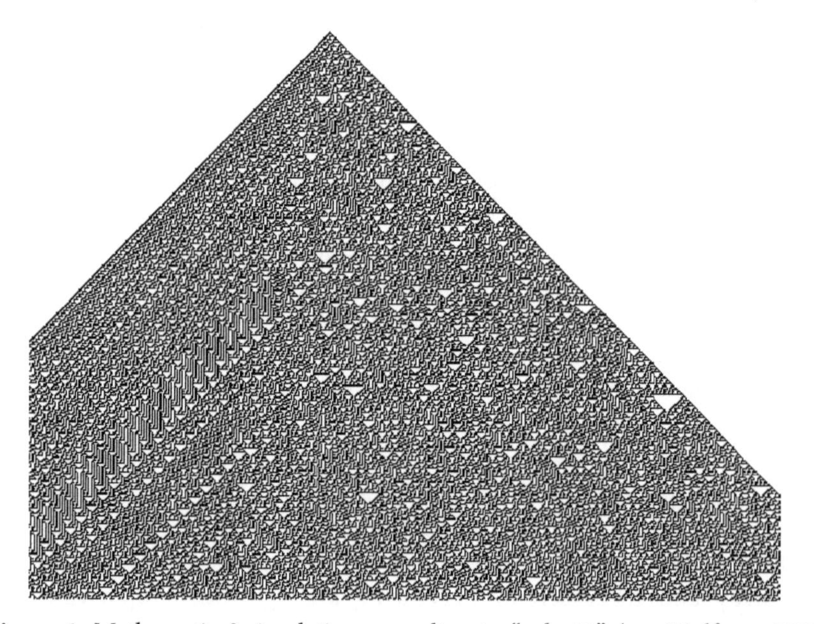

Figure 1. Mathematica® simulation according to "rule 30" (see Wolfram, 2002 and Figure 2) with 400 lines/iteration cycles. The image is cut asymmetrically at the left and right side.

Cellular automata generate patterns starting with a single black elementary cell in the middle of the first line on a white sheet of paper. The rule delivers the information how the neighbouring elementary cells in the next line have to be colored. The output of the rules indicated in

Figure 2 is shown in Figure 3. Some of these rules generate quite boring patterns like rule 222 which forms a black pyramid and rule 250 forming a chess board pattern (see Figure 2 and Figure 3). However, there exist rules that generate complicated patterns with typical properties of self-similarity. For example rule 30 (see Figure 1, Figure 2 and Figure 3) iteratively generates white triangles standing on the top with varying size (albeit with a limited size distribution).

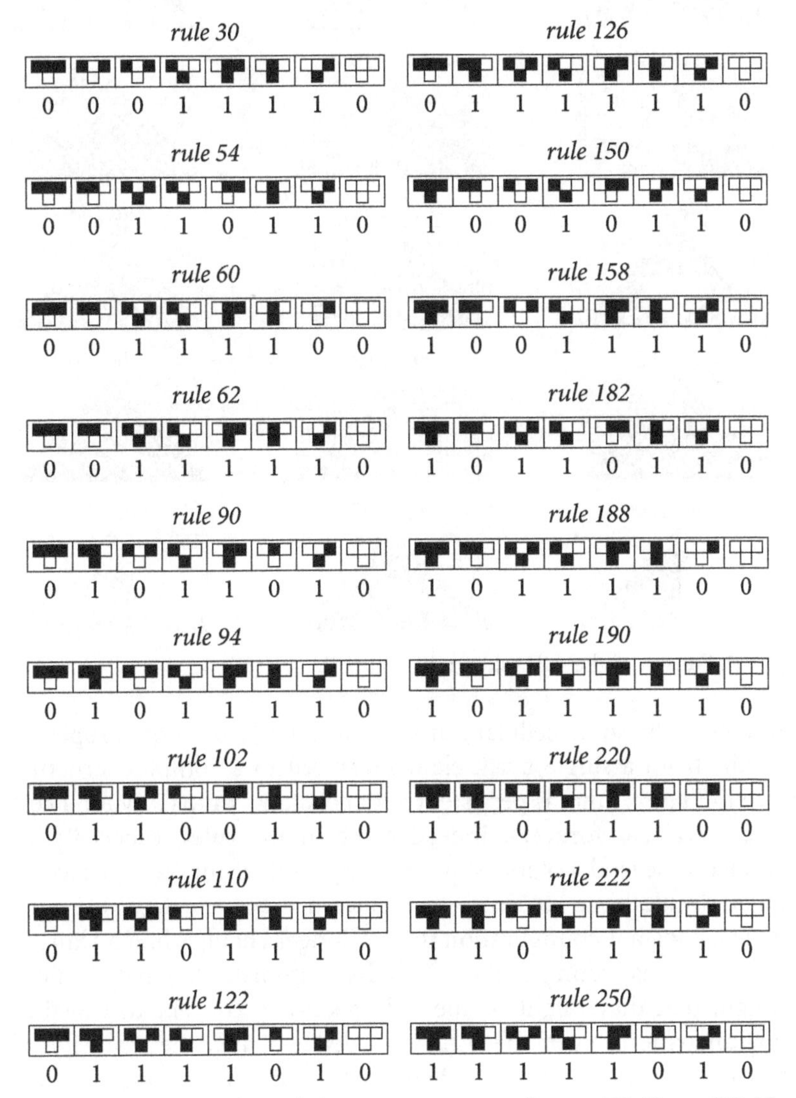

Figure 2. Typical rules for cellular automata according to Wolfram (Wolfram, 2002). Image reproduced with permission.

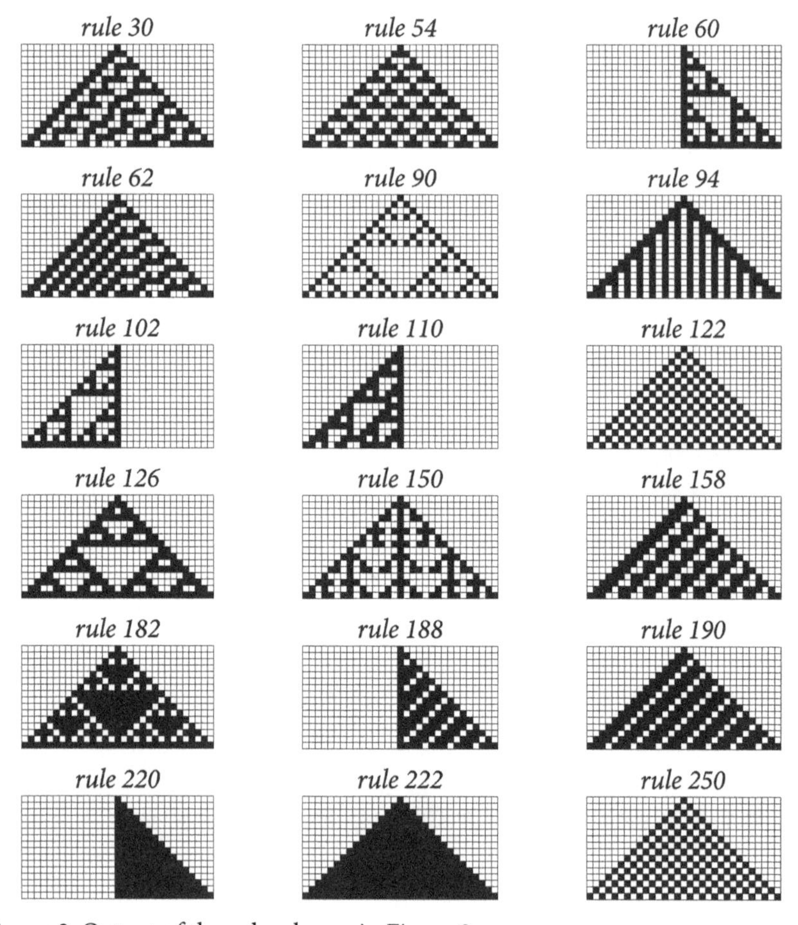

Figure 3. Output of the rules shown in Figure 2.

Interestingly, some cellular automata also typically show properties of "growth" from a single black elementary cell to a complex structure as shown in Figure 1 that is persistent with constant complexity for a certain time interval (or forever). Indeed some of the rules, especially when introducing several colors, show an apparent rising complexity from a rather simple basic structure to a pattern that then decays again and leads to a constant configuration that no longer changes line by line.

When I was younger these cellular automata fascinated me and I thought that they might be the only necessary concept to explain any biological pattern, even the birth and death of life, the diversity of its forms and general emergent phenomena like consciousness. Wolfram understands the unforeseen growth of novel structures on the patterns generated by cellular automata as emergence. And he refers to his automata and their possibility to derive complexity from a few simple rules directly to biology. In his index, the entry for "emergence, general concept" points to page 3 (Wolfram, 2002) where the word is not found a single time. Instead Wolfram states:

"It could have been, after all, that in the natural world we would mostly see forms like squares and circles that we consider simple. But in fact one of the most striking features of the natural world is that across a vast range of physical, biological and other systems we are continually confronted with what seems to be immense complexity. And indeed throughout most of history it has been taken almost for granted that such complexity – being so vastly greater than in the works of humans – could only be the work of a supernatural being. But my discovery that many very simple programs produce great complexity immediately suggests a rather different explanation. For all it takes is that systems in nature operate like typical programs and then it follows that their behavior will often be complex. And the reason that such complexity is not usually seen in human artifacts is just that in building these we tend in effect to use programs that are specially chosen to give only behavior simple enough for us to be able to see that it will achieve the purposes we want" (Wolfram, 2002, page 2f.).

Indeed, Wolfram here performs a stroke of genius in his proposition that the natural behavior of a machine following simple rules is the production of a complex output and that we therefore do not have to answer the question why nature is complex. Rather, complexity is natural and what we need to ask is why mankind thinks in such simple manner that everyone picks the few algorithms from the broad range of possible rules that Wolfram proposes in order to generate simple enough output. This output is not prone to complexity, rather it serves to avoid the possibility that the output might stress the architect's mind. This finding is highly plausible and there should not be any doubt or any restriction to the work of Wolfram. Nonetheless, some questions remain unanswered. These questions can be formulated as top down or bottom up, a bit like the general concept of thinking that should be kept in mind while reading this book. From bottom up we might ask: why do we have chemical reactions of higher order introducing such strong nonlinearities (such as the nonlinear optics of photosynthesis) assuming that a more simple pattern generator is already capable of simulating any natural behavior? Why does nature behave nonlinearily? When does it produce chaotic output? The output of Wolfram's cellular automata shows high complexity and self-similarity – however, these automata do not behave in a chaotic way. Or do they? If we change a rule "slightly" the output might change completely. If we shift a black square the output does not change at all. However, it does if we shift the position of the lines next to each other.

Chaos should generally be defined as the fact that the deviation of initial conditions leads to an inherent local inpredictability that rises exponentially over time. Even assuming chaos as deterministic, limited computational power always limits the predictability of an output that arises later in time. Does chaos arise from nonlinearities and is chaos theory a special point of view on nature that can be fully described by imaging all processes on a computer program? It might be possible and in the following we will try to elucidate and answer this question. The concept of computational irreducibility is the final reason for Wolfram to assume that computational models can sufficiently describe nature and especially the reason why it is absolutely necessary to take such computational models into account. Irreducibility is the reason for the generation of intrinsic randomness, which means that his cellular automata inherently generate randomness and he even proposes that neither chaos theory (which is deterministic) nor stochastic perturbations nor quantum theory are necessary for randomness.

So there might be some "features" of natural systems already on the molecular scale that are not implemented in cellular automata. From the other perspective, a top down argument might exist that calls into question whether cellular automata are sufficient to describe any observed natural entity and cover the full range of all existing emergent phenomena. For example, the existence of emotions, or, to put it in a wider sense, consciousness, might be an emergent phenomenon that does not necessarily arise from a computational approach that aims to explain the dynamics of the universe on the scale of mankind employing cellular automata. We cannot answer this question at this point. It is an interesting phenomenon that living systems develop consciousness.

Some authors state that already the unpredictable behavior of irreducible algorithms is sufficient to explain consciousness. They take an external point of view on the program to argue that it has a free will; free will is generally considered to be the most powerful argument against full determinism in nature. However, it might be a phenomenon that emerges from basic simple deterministic rules. At least Wolfram seems to assume so when he states:

"One might have thought that with all their successes over the past few centuries the existing sciences would long ago have managed to address the issue of complexity. But in fact they have not. And indeed for the most part they have specifically defined their scope in order to avoid direct contact with it. For while their basic idea of describing behavior in terms of mathematical equations works well in cases like planetary motion where the behavior is fairly simple, it almost inevitably fails whenever the behavior is more complex. And more or less the same is true of descriptions based on ideas like natural selection in biology. But by thinking in terms of programs the new kind of science that I develop in this book is for the first time able to make meaningful statements about even immensely complex behavior" (Wolfram, 2002).

This is certainly a self-assured stance.

Luckily, since it is still open to question whether plants develop consciousness, we do not have to determine whether consciousness can arise from a computational approach to describing nature for the purposes of this book. The book rather wants to elucidate the complex interplay of nonlinear systems with feedback loops on a spatiotemporal hierarchy of scales based on the example of ROS networking in plants.

Life couples nonlinearities to each other. Evolution means stability in time due to the adaption to a given environment under competition for resources and evolution describes change if the environment changes or if the genetic diversity changes. Nature selects the structures that are observable and therefore there exists a natural selection mechanism that prints the few stable structures emerging by cellular automata or chaotic nonlinear systems into reality. These aspects of nonlinear feedback in evolution and possible meanings for top down signalling are elucidated in chapter five. Thereby we will show that pure Darwinism is sufficient to explain both bottom up as well as the "type" of top down phenomena that we investigate in this book. There is absolutely no need to give space for an intelligent designer of any kind. In this way, we are in full agreement with Wolfram and most recent researchers on the topic.

We do not know whether the emergence of emotions, consciousness, or the fundamental and rather complicated question of the existence of a free will is somehow accessible by concepts like cellular automata based on rate equations to describe their dynamics. Therefore we choose this theoretical approach to describe the most complex (biological) system that is typically described by rate equations in the literature, but that does not bring us into the complication of accounting for striking behavior that is dominated by an emerging ununderstandable effect like consciousness, namely plants.

A top down control feedback in plants is for example represented by the feedback given by a certain concentration of ROS generated by macroscopic events like stress. Stress is a macroscopic event that can even be activated based on free will running on the consciousness, but it does not necessarily need to. It can just arise from environmental conditions. On the other hand, consciousness emerges from the molecular reactions.

Current books on the market elucidate the novel topics in the field of primary processes of photosynthesis (Renger, 2008, 2012; Renger and Ludwig, 2011), nonphotochemical quenching (Demmig-Adams et al., 2014) and, more specifically, the generation and decay of reactive oxygen species (Del Rio and Puppo, 2009). Some authors state that the actual overview on ROS is missing a generalized link to plant morphology and evolution. Even if it might be impossible to achieve such a link without a certain degree of speculation, it is necessary to deal with the topic in a fundamentally scientific manner in order to avoid the postulation of novel unexplainable and experimentally inaccessible entities such as Rupert Sheldrake's proposition of a "morphogenetic field" (Sheldrake, 2008) or, much worse, theories of an "intelligent designer" that not only compete with recent scientific findings but even actively reject them and support finally fundamentalists and rabble-rousers. We always pay careful respect to the first, basic property of a good scientific theory: the theory must be falsifiable. Everyone who demands inerrability is surely wrong.

1.2 Principles of Synergetics

Since Hermann Haken wrote his book *Synergetics* several (successful) attempts have been made to apply his formalism to systems that can be described by rate equations. Haken himself applied his concept mainly to laser theory and proceeded to show the general capability of his approach to describe any system that is characterized by physically distinguishable states and natural transition probabilities between these states as well as a control parameter that can induce such transitions (Haken, 1990). Rate equations are the next step to transfer the concept of cellular automata from Wolfram (2002) (see chapter 1.1) into a mathematical formalism. We introduce time and probability to generate dynamics from the single elementary cells that ultimately form the pattern.

In this way it is inferred that the concept of cellular automata can describe biological systems by computational power as stated by Wolfram. A computer that is equipped with the capability of calculating rate equations and modelling states according to a given algorithm might eventually compute the evolution of our biosphere by iterative generation of each single step in this process. This simulation can be done with rate equations modeling the development of cellular automata.

For a simulation of the biosphere's evolution, formal constraints might be added to that concept, such as the fact that organisms need a special environment and food supply and that their living space is limited. However, on the molecular level these constraints, which appear according to the competition in the biosphere, might arise as a product of the calculation itself as growth leads to limitation.

Such approaches to simulate complex systems generally deliver qualitative results as the outcome depends crucially on the exact values of single transfer steps and the initial conditions. Both cannot be computed with absolute accuracy but deviate in such ways that the system becomes inpredictable after a few cycles. Such models can make complexity understandable in general or they can show the development of self-similarity. However, any attempt is far from a quantitative description of evolution as it is observed in our biosphere. One example is the well known "game of life". Here the constraints of the system – typically the size of the system, the growth dynamics that depend on size, possible food supply factors or predators and their interaction strength – determine the final "phenotype" of the structure.

The limited space but more likely the structure and the composition of the earth at the beginning of this "game of life" should of course, set the initial conditions.

This concept, the concept of the "game of life", is well known and therefore represents a generalized description formalism that can be used to imitate the generation and decay of reactive oxygen species (ROS) in plants, the interaction of ROS with other chemical compounds, ROS networking and its impact on the resulting plant morphology. The approach is straightforward. Therefore it appears surprising that only few attempts have been made to use time dependent "cellular automata", which are able to describe the dynamical development of any system, to describe and model biological systems. We provide examples for the general application of rate equations to biological systems: the simulation of the PS II dynamics after light absorption is shown in chapter 2.6 for excitation energy transfer (EET) in light-harvesting complexes, and in chapter 2.7 for electron transfer (ET) processes in the thylakoid membrane.

Haken pointed out how novel and synergetic approaches can be used to better understand the function of the human brain. This approach is useful to find the link that connects complex systems to the general interaction of single compounds and to entities in macroscopic systems, a link that is still not completely understood. We intend to shift the focus to novel ideas and approaches and believe that ROS in photosynthetic organisms paves the road for further concepts that in their turn might help develop new experiments that are suitable to elucidate currently unknown topics, such as the missing links between ROS driven communicating networks. It might be of higher relevance to understand such links between communicating networks than the single steps inside the networks. The accurate quantitative output of a product like this is dependent on exact parameter settings and simply cannot be predicted with absolute accuracy.

This book transfers the very general concepts developed in Haken's book *Synergetics* to the specific application of its formalism to signaling networks driven by ROS. It elucidates some mechanisms of coupled top down and bottom up signaling. These ideas are embedded into an overview of basic principles of photophysics and photobiology, light-harvesting, charge transfer and primary processes of photosynthesis and nonphotochemical quenching in plants and cyanobacteria. Therefore, the reader can identify and compare the proposed ideas with a large review of the current state of knowledge in the discipline. The recent knowledge of ROS monitoring, generation, decay and signaling in plant cells and cyanobacteria is presented in the following chapters alongside a number of novel reaction schemes for the production of secondary ROS and inter-organelle signaling.

Understanding the role of reactive oxygen species is a key factor for efficient strategies to target basic research on primary plant metabolism and crop enhancement. Furthermore, the understanding of deleterious effects and protection from ROS is of high interest in diagnostics and therapy of many diseases and can have further biomedical applications, for example ROS play a distinct role in carcinogenesis and ROS form the main therapeutic substrates in photodynamic therapy. Cells control ROS levels by tuning generation and scavenging mechanisms. It was recently shown that cancer cells seem to be characterized by even more sensitive fine adjustment of ROS levels as they contain higher contents of antioxidants and a more stable ROS concentration (Liou and Storz, 2010). Therefore control of ROS is an attractive target for therapeutic concepts in cancer treatment.

Underlying mechanisms such as enhanced metabolic action as well as changes in cellular signaling are key factors in the development of novel therapeutic concepts. While findings on plant cells are not easily transferable to mammalian cells, the suggested strategies are generalized to emphasize medical applications. The book focuses on the stimulation of new ideas for research strategies and experiments that help develop biological and medical progress.

1.3 Axiomatic Motivation of Rate Equations

Most calculations that are presented in this work were done with rate equations. The concept of rate equations is presented together with the way of thinking that underlies the concept of cellular automata. This concept was presented as a general and basic theory that is suitable to predict the dynamics of any system that can be described by a mathematical expression that denotes information about the system by referring to its "state" and the probability for transitions between these states. In fact, rate equations are not only used to describe EET and ET processes but many systems, including optical transitions, particle reactions, complex chemical reactions and more general information processing in complex networks. Rate equations can even be used to describe complex systems such as sociological networks (Haken, 1990).

To understand the very basic concept that underlies the rate equation concept we present an axiomatic fundament that is suitable to motivate rate equations as a general mathematical framework that does not restrict the modeling with any constraints except basic axioms. We will speak about "systems" in the following. As systems we understand the objects that we want to investigate. These systems could be a piece of semiconductor, a cell, a plant, or the universe. It should be pointed out that the structure of the book allows for an omission of mathematical details to target the readers who are interested in biological background but not in the mathematical details. These readers are encouraged to skip the following parts and directly proceed with the next chapter.

As presented in (Schmitt, 2011), one can start from two very basic axiomatic prepositions to motivate rate equations:

1) The assumption of "states" that define the system in form of information strings which could be numbers or any other type of coding. These strings contain all information that is experimentally accessible in a given system. One could talk about several states or substates of a system that define the "overall state".

2) The substates change over time. The time evolution can be described by differential equations. That means that we agree that there exists a transition in these systems that can be described by accurate mathematical treatment. The choice of differential equations is a specialization, however, it seems to be the most straightforward assumption that can define an equation for the "change" of the system in time.

Axiom 1) suggests that each state might be measurable in form of a mathematical object N_i with i forming a tupel of numbers or parameters that completely characterize the state.

According to 2) $\frac{dN_i}{dt} \neq 0$ and therefore there must exist a function f with $\frac{dN_i}{dt} = f_i\left(x_j, N_k, t\right)$. f could depend on time t, any linear combination of the substates N_k themselves and other variables x_j:

$$f_i\left(x_j, N_k, t\right) = \sum_k T_{ik}\left(x_j, N_j, t\right) N_k\left(t\right). \tag{1}$$

The so-called transfer matrix T_{ik} contains all the factors for the linear combination of all substates k (taken together, they form the "state" of the system) and possible additional mathematical operators that describe the time evolution of substate i. Even though $f_i\left(x_j, N_k, t\right)$ could be a nonlinear function in the substates, for the sake of simplicity we assume that $f_i\left(x_j, N_k, t\right)$ is a linear function in the substates as a kind of first approximation. Since x_j are the variables that encode the information about the substates, these states (we might refrain from mentioning "substates" and call them just "states") themselves should contain these variables as they contain all information that is experimentally accessible. As a consequence we can simplify eq. 1:

$$f_i\left(N_k, t\right) = \sum_k T_{ik}\left(t\right) N_k\left(t\right). \tag{2}$$

In a further simplification, we can assume that the dynamical behaviour may not change over time. The states themselves will of course change. Only the underlying dynamical concept is constant. In that case the proportionality factors $T_{ik}\left(t\right)$ do not depend on the time and we get

$$f_i\left(N_k, t\right) = \sum_k T_{ik} N_k\left(t\right). \tag{3}$$

As we will see in chapter 1.4 (Rate Equations in Photosynthesis), the (protein) environment of a molecule can be assumed to introduce additional states (e.g. "relaxed" or "unrelaxed" states of the environment of the system). Therefore, the treatment of a closed system is in principle suitable to analyze even open systems if the states of the environment are known and incorporated into the theoretical description of the system.

From our simple axioms that there exist states and dynamics a kind of classical master equation formalism directly follows, as is visible in eq. 3. Eq. 3 is directly derived from the axiomatic postulate that ensembles are described by states and dynamic changes lead to transitions between these states. The number of transitions from one state to another is proportional to the population of this state and equal to the product of the transfer probability and the population of the states.

These axioms deliver a dynamical formulation of the temporal change of our states:

$$\frac{dN_i}{dt} = \sum_k T_{ik} N_k(t) \tag{4}$$

with the transfer matrix $T_{ik} = \bar{\bar{T}}$. If we separate the supplying processes and the emptying processes (i.e. we separate the sum avoiding summation over $k=i$) we get:

$$\dot{N}_i = \sum_{k \neq i} T_{ik} N_k(t) - \sum_{k \neq i} T_{ki} N_i(t). \tag{5}$$

Eq. 5 is a master equation for the evolution of population probabilities. For an excellent depiction of rate equations, the master equation and their applications see (Haken, 1990). The probability to find a certain state populated denotes to

$$P_i = \frac{N_i}{\sum_j N_j(t)}. \tag{6}$$

If the set is complete, i.e. $\frac{d}{dt}\sum_j N_j(t) = 0$ then this formulation delivers the general master equation:

$$\begin{aligned} \dot{P}_i &= \frac{\dot{N}_i}{\sum_j N_j(t)} = \sum_{k \neq i} T_{ik} \frac{N_k(t)}{\sum_j N_j(t)} - \sum_{k \neq i} T_{ki} \frac{N_i(t)}{\sum_j N_j(t)} \\ &= \sum_{k \neq i} \left(T_{ik} P_k(t) - T_{ki} P_i(t) \right). \end{aligned} \tag{7}$$

We have to keep in mind that eq. 5 is a simplified and linearized form of master equation. The full general ansatz is found in eq. 1. For example, eq. 5 does not describe biomolecular processes or processes that amplify or inhibit themselves (e.g. a cell division process or an allosteric protein regulation). But from eq. 1 a description for any process that is thinkable in the context of states and their change can be derived and solved numerically.

The time reversible/equilibrium case is equivalent to the existence of equilibrium probabilities P_k and P_i in such a manner that each term on the right side of eq. 7 vanishes separately, i.e. $T_{ik}P_k = T_{ki}P_i$. This means that the number of transfers from a substate N_k (which is found with probability P_k) to a substate N_i is the same as the backward reaction from state N_i to N_k.

In that case, the master equation (7) is called to be at a detailed balance. From eq. 7 we can see that due to this detailed balance condition the population probabilities are reciprocal to the transition matrix elements (determining the transition probabilities) for all times: $\frac{N_k}{N_i} = \frac{T_{ki}}{T_{ik}}$. This means that the equilibrium is reached if there is no time dependency in the population densities of all substates (i.e. the population densities do not change over time).

As the number of transitions from a single state to another is same for forward and backward reaction in the equilibrium, the transition probabilities for forward and backward reaction are proportional to the relation between the populated states. In a microcanonical equilibrium, the Boltzmann distribution determines the population of states, i.e. forward and backward transfer of two coupled states are found to satisfy the Boltzmann relation

$$\frac{T_{ki}}{T_{ik}} = e^{\frac{(E_i - E_k)}{k_B T}}, \tag{8}$$

where E_i denotes the energy of the state i and $k_B T$ the thermal energy at temperature T. As we would expect from a single electron spreading along empty states, the overall equilibrium population is (see Haken, 1990).

$$\frac{N_k}{N_i} = e^{\frac{(E_i - E_k)}{k_B T}}. \tag{9}$$

For degenerated states eq. 9 has to be extended to (see Haken, 1990)

$$\frac{N_k}{N_i} = \frac{g_k}{g_i} e^{\frac{(E_i - E_k)}{k_B T}} \tag{10}$$

if N_k is g_k-fold degenerated and N_i is g_i-fold degenerated.

The formalism of rate equations as presented here has several advantages:

1) The theory contains all information about the population of the states in the system.

2) Only two basic axioms and the linearity condition are sufficient to motivate the basic theory.
3) The equations are rather simple (DGL 1st order).
4) The theory allows the incorporation of further aspects by extending the space of states (see for example chapters 2.6 and 2.7 for concrete applications on EET and ET processes in plants).
5) There exist common algorithms for solving the rate equations.

1.4 Rate Equations in Photosynthesis

The mathematical concept of rate equations is suitable to analyze the synergetic behaviour of a complex coupled network if the coupling strength and the transition probabilities between single states are known, or if measurement results for the fluorescence dynamics or other experimental results that deliver information on the transfer probabilities exist. We will now have a look at the underlying concept of the theory of coupled pigments and the derivation of the rate constants for excitation energy (EET) and electron transfer (ET) and proton transfer (PT) processes in photosynthesis from experimental results. As one example the theory of Förster Resonance Energy Transfer is elaborated (see chap. 2.2.3) while other concepts like the famous Marcus theory are omitted for the sake of a compact description.

In earlier works we performed a theoretical description of the EET dynamics and ET in systems with pigment-pigment and pigment-protein interaction using rate equations that were applied to structures with increasing hierarchical complexity (Schmitt, 2011). The study started with a system consisting of two excitonically coupled Chl molecules in a tetrameric protein environment represented by the recombinant water soluble Chl binding protein (WSCP) of type IIa (Theiss et al., 2007b; Schmitt et al., 2008; Renger et al., 2009, 2011). It was completed with a study of the photosystem II (PSII) dynamics in whole cells of *Chlorella pyrenoidosa* Chick and whole leaves of the higher plant *Arabidopsis thaliana* (Belyaeva et al., 2008, 2011, 2014, 2015). In this way a quantification of dissipative excited state relaxation processes as a function of increasing excitation light intensity was achieved. The approach permits the determination of selected parameter values, their probability and stability in dynamical systems. A way to calculate thermodynamic quantities (e.g. entropy) under nonequilibrium conditions from rate equations was proposed (Schmitt, 2011) and is presented in chapter 4.4.

Photosynthetic EET and ET steps are predestined for a description by classical rate equations as it is assumed that (incoherent) excitons or electrons are transferred from one pigment protein complex (PC) to another one with given rate constant (probability per time unit). Coherence effects as studied in quantum mechanical approaches are omitted here since the description focuses on the approach to describe the behaviour of individual elements on timescales where coherence effects seem to be of minor relevance. Polarizations and coherence can be added into the formalism and extend it to optical Bloch equations and quantum mechanical approaches. Since the typical coherence times for the quantum mechanical states that are excited by light or due to energy or particle transfer are short in comparison to ET and most EET steps, the rate equations are treated classically and the formalism as presented by eq. 1 and eq. 7 represents an useful approach. This is especially assumed for ROS signaling in a thermalized environment. For fast EET it is known that coherence effects might play a role for efficient transfer (Calhoun et al., 2009). The photochemically active chlorophyll containing photosynthetic pigment protein complexes (PPCs) are classified into the photosystems (PS) PS I and PS II (see Figure 4). Electrons flow from PS II to PS I. Light absorption changes the redox properties of a single electron in each photosystem. The nomenclature has developed historically denoting the "first" PS as PS II which delivers electrons to the PS I (Byrdin, 1999).

It should be mentioned that in contrast to the simplified schemes shown in Figure 4, the distribution of different pigment-protein complexes with different optical properties is not homogeneous along the thylakoid membrane (Steffen, 2001) and that there exist PS II complexes with different antenna sizes, so called alpha centers and beta centers (Albertsson et al., 1990) which are inhomogeneously distributed along the thylakoid membrane.

The light energy is absorbed by chromophores bound to the photosynthetic complexes of PS I and PS II. These chromophores are mainly chlorophyll and carotenoid molecules. For a description of PS I see (Byrdin, 1999) and references therein.

The electronically excited singlet states formed by light absorption of the Chl molecules are not completely transformed into Gibbs free energy. A fraction is emitted as red fluorescence and the dynamics of the fluorescence emission of all samples containing PS I and PS II is mainly determined by the properties (organization and coupling) of the photosynthetic pigment-protein complexes of PS II (see Schatz et al., 1988). Therefore, fluorescence emission delivers a strong tool to monitor the excited state dynamics in photosynthetic complexes, especially PS II.

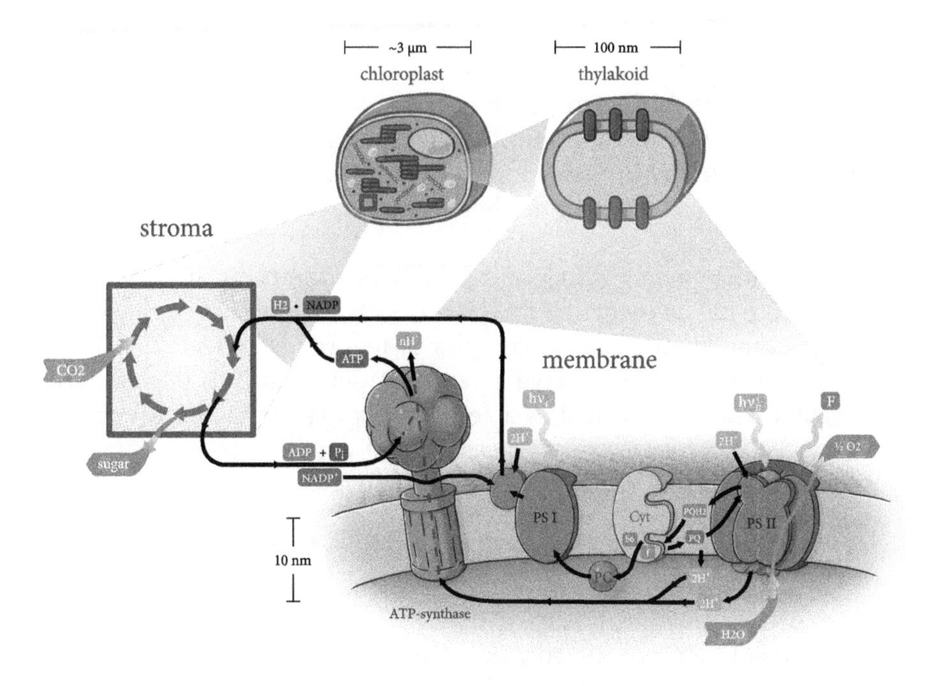

Figure 4. Membrane proteins inside the lipid double membrane of the thylakoids and light reaction in PS II, PS I and ATP-synthase.

The "light" reaction performs the exploitation of solar energy by highly functionalized PPCs. Solar energy represents the unique Gibbs free energy source of the earth's biosphere. The Gibbs free energy is converted into high energy chemical compounds via the process of photosynthesis. This is achieved perfectly by incorporation of suitable chromophores into protein matrices. The PPCs are optimized to energy absorption and transfer, thereby producing the high energetic compounds ATP and NADP H_2 (see Figure 4).

PS II and PS I are functionally connected by the Cytb$_6$f (Cyt) complex where plastoquinol PQH_2 formed at PS II is oxidized and the electrons are transferred to PS I via plastocyanin (PC) as mobile carrier (see Figure 4). This process is coupled with proton transport from the stroma to the lumen, thus increasing the proton concentration in the lumen. At PS I the light driven reaction leads to the reduction of $NADP^+$ to NADP H_2. The proton gradient provides the driving force for the ATP synthase where ATP is formed from ADP + P. NADP H_2 is used in the dark reactions for CO_2 reduction to produce glucose inside the chloroplast stroma.

The spatial separation between lumen and stroma, and the oriented arrangement of PS I, PS II and $Cytb_6f$ enables a directional electron flow coupled with the formation of a transmembrane electrochemical potential difference. This way the absorbed Gibbs free energy of the photons is partially and transiently "stored". For further thermodynamic considerations of the initial processes of photosynthesis, see chapter 2 and chapter 4.4.

A simplified model for EET and ET processes employs the concept of "compartments" as suggested by (Häder, 1999). According to this compartmentation model, a rather complex system can be treated in form of "compartments" if the energetic equilibration between the coupled molecules which are treated as a compartment is fast in comparison to the achievable resolution of the measurement setup. Then there generally occurs an incoherent EET between different compartments which can be understood as a single "transfer step".

The advantage of the compartment model is a reduction of the number of states that needs to be analyzed to characterize the system. For example, one can treat the light-harvesting antenna as a composition of compartments which are represented as intrinsic antenna systems or extrinsic antenna systems (see Figure 5). The compartments of the thylakoid membrane as shown in Figure 4 typically divide into the mentioned photosystems PS I and PS II or $Cytb_6f$ or the ATPase. Also lumen or stroma could be understood as compartments if one speaks about the transfer of single protons modeled during equilibration across the membrane. In this way single states do not necessarily demand a treatment of

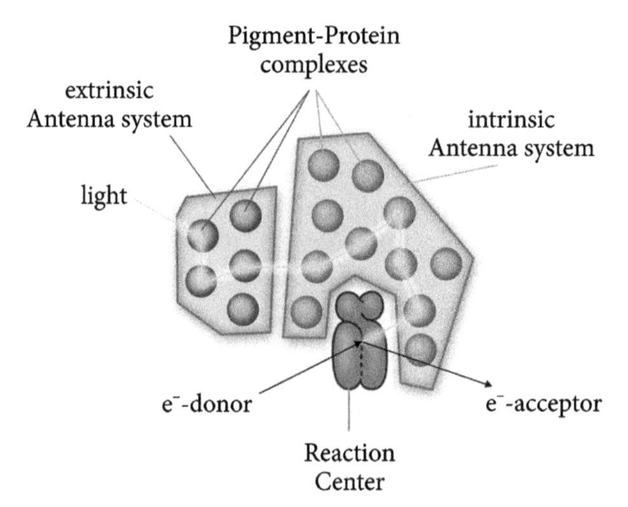

Figure 5. Compartment model according to (Häder, 1999).

each single molecule. The whole light-harvesting complex, including the core antenna and the Chl molecules in the reaction center (RC), might even be treated as one single compartment. In a more complex approach one can treat a whole cell as a single compartment. The question is what kind of dynamics we intend to describe and if there is need for the description of the intrinsic dynamics of the compartments. Such simplifications are absolutely necessary to achieve an unambiguous description of the biological structures and to have a chance to derive clear results from observed trajectories, such as time resolved fluorescence signals. As can be seen in the discussion on the description of the overall PS II dynamics from a set of trajectories of time resolved fluorescence emission presented in chapter 2.7, such schemes are calculated on a huge parameter space and a reduction of this parameter space is inevitably necessary to avoid overparametrization and concomitant arbitrariness in the gathered results. If there are too many parameters imbedded into a system one can use it to "fit an elephant" or even "let him wiggle his trunk" as stated in (Dyson, 2004). The phrase is attributed to John von Neumann who already claimed the possibility to fit an elephant from only four parameters, while five parameters are enough to let the elephant dance. Compartment models help to control the elephant's trunk and to draw clear conclusions from measurement results in the framework of the simplified system provided that the system is a suitable framework for the description of the results.

Figure 6 shows how EET conducts excitation energy between different compartments of cyanobacterial phycobiliprotein (PBP) antenna complexes. On the left panel EET inside the PBP antenna rod and between the PBP antenna and Chl *d* in *A.marina* is shown. At the top, the model scheme of the phycocyanin (PC) trimer with its bilin chromophores is shown. The right panel represents a model scheme and time constants for the EET inside the phycobilisomes of *Synechococcus* 6301, and from there to the RC. This gives a résumé of the extant literature on EET processes inside the PBP antenna complexes of cyanobacteria (Gillbro et al., 1985; Suter and Holzwarth, 1987; Holzwarth, 1991; Mullineaux and Holzwarth, 1991; Debreczeny at al., 1995a, 1995b; Sharkov et al., 1996). The calculations that lead to the results shown on the left side of Figure 6, are presented in (Schmitt, 2011). To avoid redundant repetition, we refrain from presenting the exact calculated equations here. Applications, the calculation, and the comparison between experimental results and the theoretical model are shown in chapters 2.6 and 2.7.

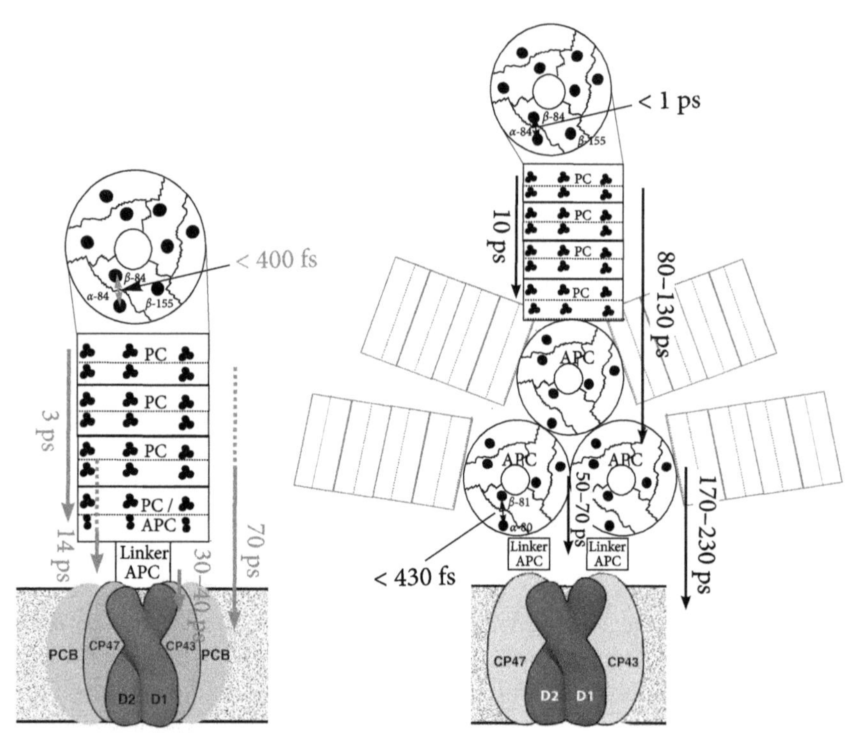

Figure 6. Compartment models describing the kinetics of excitation energy transfer (EET) processes in *A.marina* (left side) and in the cyanobacterium *Synechococcus* 6301 (right side) as published in (Theiss et al., 2011). Image reproduced with permission.

1.5 Top down and Bottom up Signaling

The approach outlined in chapter 1.4 is principally a bottom up description of the system since it explains the behavior of a macroscopic observable as fluorescence light in dependency on the population of substates which could be single molecules or compartments. Assuming that single elements and their probability can be transferred into another state and/or that the interaction between those compounds is the only input that defines the systems properties is typically bottom up because the single constituents and their dynamics are the only elements that define the overall dynamics.

The system constraints are important factors that define the dynamical properties. Constraints can have impact on the microscopic level with or without time delay, however, constraints can also be macroscopic, for instance when the biochemical reactions are restricted to the size of the reaction volume. This could be a single molecule, cell organelle or, for other cases, such as the spread of ROS waves, the whole organism (for ROS waves see chapters 3.2 and 4.6). Importantly such constraints are not necessarily extrinsic system limitations, like the finite capacity or size of the reaction vessel or ecosystem, but can also be understood as the intrinsic interaction between system components that "emerge" due to pattern formation or self-organization in the system. Macroscopic processes that occur in form of self-organized patterns have the potential to "slave" the microscopic dynamics (Haken, 1981).

Such constraints might themselves be time dependent and underlie dynamical changes. One example is the birth of "predators" in an ecosystem due to evolutionary processes. In the moment such a new predator arises, it starts to limit the population of its prey. Therefore, from the point of view of the prey, the emergence of the new predator is a new constraint. In chapter 4.6 we will especially see how prey-predator models can be used to describe the appearance of different properties of ROS, especially ROS waves.

However such obvious interactions of macroscopic species on the microscopic level are not necessarily the only way in which top down messaging occurs. Intrinsic effects can also contribute to a change of dynamics on the macroscopic level due to the influence of a macroscopic phenomenon. This is well known in nonlinear dynamics of nonequilibrium systems, where it is stated that the macroscopic processes might "slave" the microscopic processes. This concept became known as the "slaving principle" or the generalized "order parameter concept". Haken described such phenomena as the enslavement of microscopic effects by macroscopic dynamics.

The bottom up dynamics follows the constitution of the overall system by its individual compounds which are, for example, given by the formation of a radiation field from individual wave elements that are emitted by individual atoms:

"Below laser threshold, light consists of individual wave tracks which are emitted from the individual atoms independently of each other. [...] In order to make contact with other processes of self-organization let us interpret the processes in a lamp or in a laser by means of Bohr's model of the atom. A lamp produces its light in such a way that the excited electrons of the atoms make their transitions from the outer orbit to the inner orbit entirely independently of each other" (Haken, 1981). However, above the laser threshold the system develops a new property:

"Above laser threshold the coherent field grows more and more and it can slave the degrees of freedom of the dipole moments and of the inversion. Within synergetics it has turned out that is a quite typical equation describing effects of self-organization. ...This is probably the simplest example of a principle which has turned out to be of fundamental importance in synergetics and which is called the slaving principle" (Haken, 1981).

Without a real scientific formulation of the underlying principles, everyone should agree that the development of consciousness might be an overwhelming "control parameter" for the slavery of microscopic dynamics. The question that remains is not whether there is any top down messaging but how these forms of top down messaging are conducted. Order parameters such as blood pressure or body temperature can be manipulated by free will and serve as prominent examples of when the activation of genes, and therefore the absolutely basic dynamics in the biological system, is activated top down by order parameters introduced by the free will. While that is not the case in general there might be a link how an organism can influence the activation of certain genes by influencing its state (for example if it consumes a large amount of alcohol which is also a consequence of the free will).

There are simple approaches to show how, for example, protein kinases are activated or deactivated at a certain critical temperature, which might even work as a switch for the activation of genes by phosphorylation on the basic molecular level. These can be activated due to external constraints (temperature), induced stress or by free will in the form of the organism "deciding" to raise its body temperature by conducting a sport program. For an example of a mitogen activated protein kinase (MAPK), see for example (Mertenskötter et al., 2013).

2
Photophysics, Photobiology and Photosynthesis

This chapter aims to start with basic principles of photophysics and its relation to both photosynthesis as well as rate equations. Photophysics as the physical background behind effects like absorption, emission and energy transfer, the mathematical description as an idealized projection into a formalism which is, once settled up, pure logic and inerrable and photosynthesis as the biochemical process and its biological implementation form a triangle between reality, the abstraction into concepts and the mathematical modelling which go hand in hand to make us understand photosynthesis. The chapters 2.1 and 2.2 are kept quite mathematical as they introduce some background that is needed to calculate electromagnetic waves or transition probabilities. Readers who are more interested in the biological aspects of this book can proceed with chapter 2.3

2.1 Light Induced State Dynamics

Fünfzig Jahre angestrengten Nachdenkens haben mich der Antwort auf die Frage "Was sind Lichtquanten?" nicht näher gebracht. Heute bilden sich Hinz und Kunz ein, es zu wissen. Aber da täuschen sie sich.

Albert Einstein (1879–1955), in a letter to M. Besso, 1951

In the following, we show some selected theoretical concepts from classical electrodynamics and quantum mechanics used to describe light and the interaction between light and matter. This will be complemented by an approach to thermodynamics based on rate equations. The concept of rate equations was motivated and shortly introduced in chapter 1.3. It is necessary to briefly show the underlying physical concept from which the probabilities of energy transfer, and thus the values of the "rates" in the rate equation models, are calculated. Therefore, the theory of excitation energy transfer will be introduced.

For me it was a key element during my studies to understand that everything in classical electrodynamics can be derived from the basic equations found by James Clerk Maxwell in 1864. In spite of students learning the complicated optical laws describing diffraction and interference like the Fresnel equations for the reflection and transmission of light on surfaces, just to name one example, everything can be derived from the basic Maxwell equations. However, it is surely not pragmatic to do this in every single case and often it is just necessary to know some special laws themselves. However the ability to use such basic equations as a tool for deriving any law in optics and electrodynamics is a key element in understanding the beauty of nature as it is described by equations like those. The first time dependent Maxwell equation in vacuum (eq. 11) tells us that in vacuum there are no direct sources for electric fields (as there exist no charges which form the source of electric monopols in matter, the divergence of the electric field is always zero):

$$\nabla \cdot \overline{E} = 0. \tag{11}$$

The temporal change of the magnetic field induces rotation of the electric field (eq. 12):

$$\nabla \times \overline{E} + \dot{\overline{B}} = 0. \tag{12}$$

There is no source of the magnetic field (i.e. there exist no magnetic monopols, eq. 13)

$$\nabla \cdot \overline{B} = 0. \tag{13}$$

The temporal change of the electric field induces rotation of the magnetic field (i.e. currents are the sources of magnetic rotation. Eq. 14)

$$\nabla \times \overline{B} = \frac{1}{c^2} \dot{\overline{E}}. \tag{14}$$

From these four time dependent Maxwell equations in vacuum describing the dynamics of electric fields $\overline{E}(\overline{r},t)$ and magnetic fields $\overline{B}(\overline{r},t)$ one can for example derive a wave equation e.g. for the electric field with help of eqs. 11, 12, and 14:

$$\nabla\times\overline{B}=\frac{1}{c^2}\dot{\overline{E}}\xrightarrow{\partial_t}\nabla\times\dot{\overline{B}}=\frac{1}{c^2}\ddot{\overline{E}}\xrightarrow{eq.2}$$

$$-\nabla\times\left(\nabla\times\overline{E}\right)=\frac{1}{c^2}\ddot{\overline{E}}\rightarrow-\nabla\left(\nabla\cdot\overline{E}\right)+\Delta\overline{E}=\frac{1}{c^2}\ddot{\overline{E}}\xrightarrow{eq.1.}$$

$$\Delta\overline{E}-\frac{1}{c^2}\ddot{\overline{E}}=0. \qquad (15)$$

The general solution of eq. 15 is an arbitrary linear combination of plane waves which depend on space coordinate $\overline{r}$ and time t:

$$\overline{E}\left(\overline{r},t\right)=\int\tilde{E}\left(\omega\right)e^{i\left(\overline{k}\left(\omega\right)\overline{r}-\omega t\right)}d\omega. \qquad (16)$$

Interestingly Maxwell derived this wave equation and its solution in 1864 and he realized that he had proved mathematically the existence of electromagnetic waves propagating through electromagnetic fields. In eq. 15 there is the explicite velocity of these waves "c" he could calculate from known parameters as about 310.000 km/s. Maxwell wrote in his publication: "This velocity is so nearly that of light, that it seems we have strong reason to conclude that light itself (including radiant heat, and other radiations if any) is an electromagnetic disturbance in the form of waves propagated through the electromagnetic field according to electromagnetic laws." I found statements like this astonishing as Maxwell surely somewhat better understood the nature of light than his colleagues might have understood during this time. The approach he chose was pure mathematics but the output of his calculations drew new pictures he was able to interpret.

In eq. 16 $\overline{k}\left(\omega\right)$ denotes the wave vector which depends on the frequency ω and $\tilde{E}\left(\omega\right)$ is the frequency dependent electric field amplitude.

Equation 16 solves 15 with the dispersion relation of the vacuum

$$c^2=\frac{\omega^2}{k^2}. \qquad (17)$$

Eq. 17 introduces the most important property of light quanta, i.e. the constant group velocity in vacuum which is the product of wavelength λ and frequency ν as $\left|\overline{k}\right|=\frac{2\pi}{\lambda}$ and $\omega=2\pi\nu$

$$c=\lambda\nu. \qquad (18)$$

The coordinate system can be shifted arbitrarily and the observation can be fixed to one spatial point. If the point $x=y=z=0$ is chosen then the solution 16 simplifies to

$$\bar{E}(t) = \int \tilde{E}(\omega) e^{i\omega t} d\omega \tag{19}$$

which is the Fourier transformed of $\tilde{E}(\omega)$. Due to the Fourier Theorem, eq. 19 and therefore especially eq. 16 represent any desired solution for the electric field $\bar{E}(t)$ denoted in eq. 19 if it is periodic in time or vanishes in infinite time. If we assume a simplified solution $\tilde{E}(\omega) = \bar{E}_0 \delta(\omega - \omega_0)$ with the Dirac Delta function $\delta(\omega - \omega_0)$ for a fixed frequency ω_0 then $\bar{E}(t) = \bar{E}_0 e^{i\omega_0 t}$ becomes an infinitely oscillating function with a constant frequency and vice versa.

This already accounts for an uncertainty relation since any fixed frequency needs to be expanded as an infinite plane wave whereas a fully localized function could only be composed by an infinite number of wavelengths according to the Fourier integral representation of the solution of the wave function as given by eq. 16.

In other words: It is just not possible to get an idea of the wavelength, frequency and direction of a wave if we look at a single point only, or in the particle picture: Any absolutely localized particle has an absolutely uncertain momentum. Interestingly the Fourier transformation between frequency and time or wave vector and space immediately introduces an uncertainty principle for the corresponding two canonical variables.

A pure delta-distribution in the frequency domain does not exist physically. In general in the frequency domain as well as in the time domain we find wave packages with a certain distribution for the time the wave function is present or the frequency interval that it is composed from. Both are connected via a Fourier relation as given by eq. 16 and therefore show indirect proportional to one another as it is stated by Heisenberg's uncertainty principle. For example, a laser pulse can be described by a temporal Gaussian and a spectral Gaussian. If the pulse should be short in time it is necessary to add a large number of frequencies to achieve destructive interference in all areas except the short pulse duration. The pulse becomes white. Vice versa, it is not possible to make short monochromatic pulses. General consequences of Heisenberg's uncertainty principle directly follow, as pointed out here, from the Fourier transformation that results from the mathematical structure of a general solution to our wave equation 15.

If we continue our classical view (which, as pointed out, results in wave packages and an uncertainty relation – the basic principle of the quantization of light fields) we tend to focus on the intensity of the laser pulse in a generalized point of view:

The time dependent energy density $u(t)$ (i.e. the energy per space unit) in the electric field denotes to

$$u(t)=\frac{1}{2}\varepsilon_0\left|\overline{E}(t)\right|^2=\frac{1}{2}\varepsilon_0\overline{E}_0{}^2e^{-\frac{2t^2}{\sigma^2}}. \tag{20}$$

Eq. 21 introduces the spatial photon density $n(t)$ of the photons in the pulse with the expected frequency ω_0 (assuming a fixed frequency as a simplification of an ultrasharp continuous wave laser) equation 20 must be equal to

$$u(t)=n(t)\cdot\hbar\cdot\omega_0=\frac{1}{2}\varepsilon_0\left|\overline{E}(t)\right|^2=\frac{1}{2}\varepsilon_0\overline{E}_0{}^2e^{-\frac{2t^2}{\sigma^2}}. \tag{21}$$

The photon density $n(t)$ is a probability measure for the transfer of electrons by optical excitation from one state into another. We find that $n(t)$ for the incoming photons in time is proportional to the square of the electric field as calculated from the Maxwell equations 11–14:

$$n(t)=\frac{\varepsilon_0\overline{E}_0{}^2}{2\hbar\cdot\omega_0}e^{-\frac{2t^2}{\sigma^2}}. \tag{22}$$

From equation 21 we can calculate the time dependent intensity distribution of the laser pulse employing the continuity equation:

$$I=c\cdot u(t)=\frac{1}{2}c\varepsilon_0\overline{E}_0{}^2e^{-\frac{2t^2}{\sigma^2}} \tag{23}$$

and with help of 22 we find the photon current (photons per area and time) in analogy to equation 23

$$j_{phot}=c\cdot n(t)=\frac{c\varepsilon_0\overline{E}_0{}^2}{2\hbar\omega_0}e^{-\frac{2t^2}{\sigma^2}} \tag{24}$$

which fullfills the relation $I=j_{phot}\cdot\hbar\omega_0$.

Our description focuses on the time dependent evolution of an average electric field, intensity and photon density propagating with the laser pulse. It delivers the relation between a number of photons per square and time unit j_{phot} as it is necessary to activate discrete events that can be described by rate equations like absorption and emission of photons, or the photoeffect and the electric field amplitude $\overline{E}_0$ which is the basic entity for the formulation of electromagnetic phenomena in form of the Maxwell equations 11–14.

With equations 22–24 one can formulate the photon stream that causes the state changes per time and area unit directly from a given laser intensity and derive this formulation from the Maxwell equations.

2.1.1 Light Induced Transition Probabilities and Rate Equations

One can assume rate constants for the absorption and emission of light quanta and describe ground state and excited state of molecules in a very generalized way. The absorption of photons transfers the system from the electronic ground state N_0 to an electronically excited state N_1. We can generally assume to have an infinite number of such states. Recalling eq. 4 and eq. 5 we remember that the general master equation that denotes the dynamical formulation of the temporal change of states N_i might be time dependent $N_i(t)$ and is therefore described by a differential equation that denotes the dynamical change in time $\dot{N}_i(t)$:

$$\frac{dN_i}{dt} = \sum_k T_{ik} N_k(t) \text{ as eq. 4} \tag{25}$$

with the transfer matrix $T_{ik} = \overline{\overline{T}}$. If we separate the supplying processes and the emptying processes we get:

$$\dot{N}_i = \sum_{k \neq i} T_{ik} N_k(t) + T_{ii} N_i(t). \tag{26}$$

This means that the description of our transfer matrix $T_{ik} = \overline{\overline{T}}$ contains all processes that deliver $N_i(t)$ as off-diagonal elements and the process of the decaying $N_i(t)$ as elements on the matrix diagonal. This description is very suitable to calculate time dependent spectra that result from fluorescence decay.

The full master equation denotes to

$$\dot{N}_i = \sum_{k \neq i} T_{ik} N_k(t) - \sum_{k \neq i} T_{ki} N_i(t) \text{ as stated above in eq. 5.} \tag{27}$$

where the diagonal of the transfer matrix is filled with zeros. A generalized description of the motivation of the master equation was given in chapter 1.3 and can be reviewed in (Haken, 1990) and (Schmitt, 2011).

2.1.2 Absorption and Emission of Light

The classical description of light absorption is done by the complex dielectricity describing the phenomena of absorption and refraction with the help of the Maxwell equations. In contrast, a quantum mechanical approach formulates the interaction between the electric field and the dipole strength of atoms and molecules. The spectra for light absorption and light emission can only be understood in a full quantum mechanical picture. Here we use rate equations to model the dynamics of an electron that moves from one state to another. The coupling of photons transferring the system between an electronic ground state N_0 and an electronically excited state N_1 is shown in Figure 7.

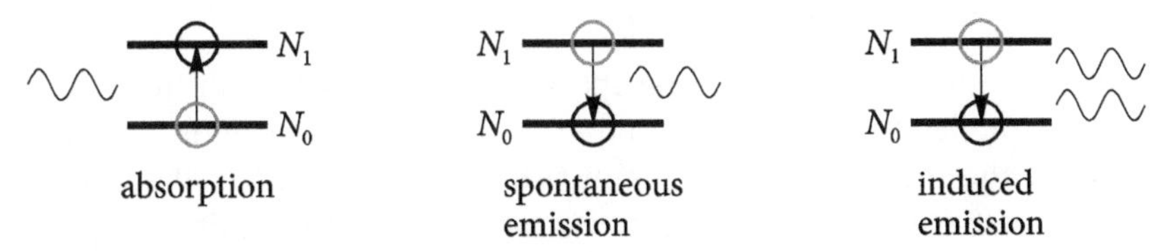

Figure 7. Absorption (left panel) spontaneous emission (middle panel) and induced emission (right panel) of light due to electronic transitions between the electronic ground state N_0 and the excited state N_1.

In full accordance with the concept of rate equations such systems are described by linear differential equations containing the ground state N_0, the excited state N_1, probability coefficients which are called the "Einstein coefficients" and the frequency dependent energy density $u(\omega,t) = n(t)\cdot\hbar\cdot\omega$ where $n(t)$ describes the frequency dependent photon density according to eq. 21 for the monochromatic beam.

If the laser exclusively excites the first excited state (N_1) then the change of the ground state population denotes to:

$$\begin{aligned}\dot{N}_0 &= T_{01}N_1(t) - T_{10}N_0 \\ &= \left(B_{01}(\omega)\cdot u(\omega,t) + A_{01}\right)\cdot N_1(t) - B_{10}(\omega)\cdot u(\omega,t)\cdot N_0(t)\end{aligned} \tag{28}$$

in full accordance with the general master equation formulation presented in eq. 27. If we assume a continuous laser beam the time dependency of the energy density vanishes.

In eq. 28, the transfer matrix elements are $T_{10} = B_{10}(\omega)\cdot u(\omega)$ for the absorption process and $T_{01} = A_{01} + B_{01}(\omega)\cdot u(\omega)$ for the spontaneous emission $\left(A_{01}\right)$ and the induced emission $\left(B_{01}(\omega)\cdot n(\omega)\right)$. A_{01} and

B_{01}, B_{10} are called the Einstein coefficients for spontaneous emission, induced emission and absorption, respectively.

The probability for absorption and induced emission is strongly dependent on the wavelength distribution of the photon flux with the maximum probability for $\omega = \omega_0 = \frac{E_1(N_1) - E_0(N_0)}{\hbar} := \frac{E_1}{\hbar}$, when the ground state energy is defined as zero: $\left(E_0(N_0) := 0\right)$.

To analyse the relation between the Einstein coefficients and the population densities a general system of the two states N_i and N_k is investigated in the equilibrium, i.e. $\dot{N}_i = \dot{N}_k = 0$. Then eq. 28 claims (here the indices i and k stand exemplarily for the two states N_i and N_k):

$$B_{ik}(\omega) \cdot u(\omega) \cdot N_k(t) + A_{ik} \cdot N_k(t) = B_{ki}(\omega) \cdot u(\omega) \cdot N_i(t). \tag{29}$$

We can look at eq. 29 for the special case of a system of quantisized niveaus coupled to a thermal radiation field (i.e. the absorption, the emission and the induced emission is just due to the thermal radiation that couples to the system). In that case we can express the equilibrium population of N_k / N_i according to eq. 10 (see chapter 1.3): $\frac{N_k}{N_i} = \frac{g_k}{g_i} e^{\frac{(E_i - E_k)}{k_B T}}$ and eq. 29 writes as:

$$B_{ik}(\omega) \cdot u(\omega) \cdot \frac{g_k}{g_i} e^{\frac{(E_i - E_k)}{k_B T}} + A_{ik} \cdot \frac{g_k}{g_i} e^{\frac{(E_i - E_k)}{k_B T}} = B_{ki}(\omega) \cdot u(\omega). \tag{30}$$

Eq. 30 can be solved for the energy density of the radiation field $u(\omega)$ to

$$u(\omega) = \frac{A_{ik} \cdot \frac{g_k}{g_i} e^{\frac{(E_i - E_k)}{k_B T}}}{B_{ki}(\omega) - B_{ik}(\omega) \cdot \frac{g_k}{g_i} e^{\frac{(E_i - E_k)}{k_B T}}} = \frac{\frac{A_{ik}}{B_{ik}(\omega)}}{\frac{B_{ki}(\omega)}{B_{ik}(\omega)} \frac{g_i}{g_k} e^{\frac{(E_k - E_i)}{k_B T}} - 1}$$
$$= \frac{\frac{A_{ik}}{B_{ik}}}{\frac{B_{ki}}{B_{ik}} \frac{g_i}{g_k} e^{\frac{\hbar\omega}{k_B T}} - 1} \tag{31}$$

which uses the fact that the energy of the radiation $\hbar\omega$ must be equal to the energetic difference $\hbar\omega = E_k - E_i$. A more accurate derivation would use the spectral line shape functions for $B_{ik}(\omega)$ and $B_{ki}(\omega)$ in eq. 31.

Now the Einstein coefficients in eq. 31 can be identified for special cases. For example in the thermal radiation field one expects that the energy density distribution is in accordance with Max Planck's derivation of the black body radiation field:

$$u(\omega)=\frac{2\hbar\omega^3}{\pi^2c^3}\frac{1}{e^{\frac{\hbar\omega}{k_BT}}-1}. \quad (32)$$

If eq. 32 is compared with eq. 31 it is found: $B_{ki}g_i = B_{ik}g_k$, i.e. the probability per time unit for the absorption equals the induced emission with respect to the degeneration of the states and $\frac{A_{ik}}{B_{ik}}=\frac{2\hbar\omega^3}{\pi^2c^3}$ regarding the relation between spontaneous emission A_{ik} and induced emission B_{ik}, i.e. the induced emission becomes more probable for longer wavelengths with the third power of the photon wavelength.

Most experiments presented in this study were performed with very low radiation intensities and therefore the induced emission is not relevant. All calculations were evaluated taking into account only the absorption and the spontaneous emission. Induced emission is the dominant process for all laser effects. For a monograph regarding lasers see (Eichler and Eichler, 1991).

2.1.3 Relaxation Processes and Fluorescence Dynamics

Excited states can relax along different channels (see Figure 8). These relaxation channels are often visualized by a so-called "Jablonski diagram" which contains the energetic scheme of the relevant molecule's states. The absorption process occurs instantaneously, i.e. it is faster than the resolution limit of the applied optical setup. In the literature there is still discussion about the duration of an absorption process, assumed times for absorption range from the time light needs to cross a molecule (i.e. about 10^{-17} s = 0.01 fs) up to the time that the electron needs to cross the molecule (i.e. about 10^{-14} s = 10 fs). The (one electron) excited states of molecules typically exhibit spin multiplicity 1 (singlet states) or 3 (triplet states). We distinguish this by the notation S_i for the ith excited singlet state and T_i for the ith excited triplet state.

According to the Franck Condon principle the absorption process does not excite the vibronic ground state of an electronically excited singlet state $S_i\nu_0$ but a higher vibronic level $S_i\nu_j$. Due to the shift of the energy parabola between ground and exited state along the reaction coordinate, the overlap between the wave functions of the vibronic ground states ($j = 0$) of the electronic ground and first excited state ($S_0\nu_0$ and $S_1\nu_0$, respectively) is smaller than the overlap between $S_0\nu_0$ and $S_1\nu_k$ ($k > 0$) (see inset of Figure 8).

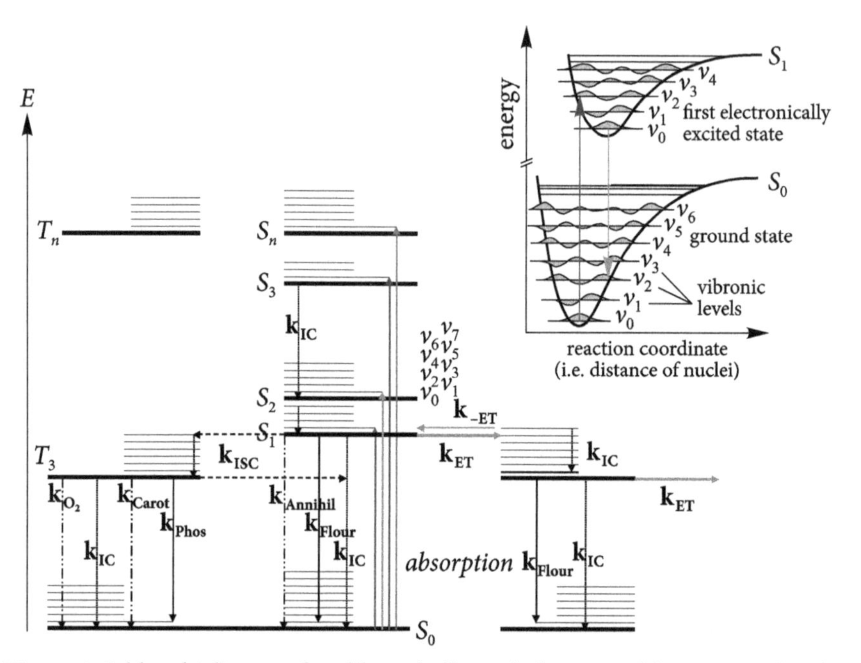

Figure 8. Jablonski diagram for chlorophyll coupled to a neighbouring molecule. The shown energetic states of chlorophyll are the singlet states S_0, S_1, S_2, S_3 and S_n and exemplarily the triplet states T_3 and T_n with vibronic niveaus that are denoted as ν_j. The indicated transitions are explained in the text. The inset is showing the parabola for the ground state and the first excited state with the most probable transitions for absorption and fluorescence according to the Franck-Condon principle.

This energetic picture can be explained by the fact that the molecular configuration of the nuclei is different between the electron in the ground state and in the first excited state. Therefore the nuclei start to move if one electron is excited and the lattice takes up a phonon.

The relaxation by light emission (luminescence) cannot emit the same amount of energy as was absorbed before. Therefore, the fluorescence must be red shifted with respect to the absorption. This red shift of the fluorescence light is called a "Stokes shift".

With typical rate constants $k_{IC} \sim \frac{1}{\tau_{IC}} \sim 10^{12}\ s^{-1}$ the electrons relax from $S_1 v_j$ into the state $S_1 v_0$ due to internal conversion (phonon emission). The subsequent radiative relaxation to the electronic ground state terminates again in a higher vibronic level $S_0 v_k$ due to the Franck-Condon principle.

The electronic relaxation from $S_i > S_1$ into S_1 occurs with $k_{IC} > 10^{11} s^{-1}$ and therefore much faster than the intrinsic fluorescence decay which is about $k_F \approx 6.7 \cdot 10^7 s^{-1}$. From S_1 to S_0 the vibrational relaxation k_{IC} is improbable for chlorophyll and therefore fluorescence can be observed especially from S_1 with wavelengths $\lambda > 680\,nm$ for Chl *a*. The improbable k_{IC} from S_1 to S_0 is correlated with the rigid structure of the porphyrin ring system. Therefore the vibrational modes are energetically separated and the wave functions overlap only slightly.

The highest occupied molecule orbital of the ground state of chlorophyll contains two electrons. Due to the internal spin-orbit (LS)-coupling it should be improbable that a spin flip occurs during the transition from S_0 to S_1 when the chlorophyll molecule is excited. But there exist intramolecular aberrations of the potential which lead to high probabilities for the spin flip which is called "Inter-System Crossing" (ISC) and has a probability of $k_{ISC} \approx 1{,}5 \cdot 10^8$ in chlorophyll. The spin flip results in an electronically excited triplet state, which is most probable T_3 or T_4 (Renger et al., 2009). The rate constants for k_F, k_{IC} and k_{ISC} are in detail investigated for Chl *a* and Chl *b*. Due to the Pauli principle spontaneous emission from the triplet states to the singlet ground state can occur only after a further spin flip. Therefore at least the lowest triplet state T_1 is long-lived with typical decay rates of $k_{Ph} \approx 1000$.

The fast relaxation from excited singlet states is called "fluorescence" whereas the long-lived emission from the triplet states is called "phosphorescence".

For photosynthetic organisms, the existence of long-lived triplet states formed from excited Chl singlet states ${}^1Chl^*$ is a disadvantage because Chl triplet states ${}^3Chl^*$ tend to interact with environmental oxygen which is found in the triplet ground state ${}^3O_2$ forming highly reactive singlet oxygen ${}^1O_2^*$ which afterwards quickly converts to other ROS (Schmitt et al., 2014a):

$$
\begin{aligned}
&{}^1Chl^* \xrightarrow{ISC} {}^3Chl^* \\
&{}^3Chl^* + {}^3O_2 \rightarrow {}^1Chl + {}^1O_2^*
\end{aligned}
$$

chemical equation 2

The singlet oxygen oxidizes neighbouring molecules and cell structures. The rate constant for the formation of singlet oxygen via triplet-triplet interaction is denoted with k_{O_2} in Figure 8.

There exist several other relaxation channels for excited singlet states. The S_1 can for instance interact with other singlet states of coupled pigments and lead to a strong fluorescence quenching via singlet-singlet annihilation ($k_{Annihil}$). This process is a nonlinear relaxation that depends quadratically on the concentration of excited singlet states $\langle S_1 \rangle$ in the sample.

Efficient photosynthesis occurs via the excitation energy transfer (EET) from the antenna pigments to the photochemically active reaction center (see Figure 5). To achieve quantum efficiencies for charge separation in the RC after light absorption up to 99% big EET rate constants in the order of $k_{ET} \approx 10^{10}$ are necessary. This process of EET and the subsequent electron transfer (ET) determines the dynamics of the excited states and therefore the fluorescence dynamics.

In spite of the concurrence of the efficient EET, the photosynthetic active organisms exhibit a small amount of fluorescence of about 1% of the absorbed light energy. Triplet states of the Chl molecules are efficiently quenched by the reaction with carotenoid triplets (k_{Carot} in Figure 8). Carotenoids are very flexible molecules that dissipate their excitation energy fast via internal conversion. It is well known that carotenoids also act as quenchers for excited Chl singlet states (see Belyaeva et al., 2008, 2011, 2014, 2015, and references therein).

The relaxation of an excited singlet state population probability $\langle S_1 \rangle$ occurs with the sum of all decay rates according to the formulation of eq. 33 (see Figure 8):

$$\frac{d\langle S_1 \rangle}{dt} = -\left(k_{Fluor} + k_{Annihil} + k_{ET} + k_{IC} + k_{ISC}\right)\langle S_1 \rangle. \tag{33}$$

Eq. 33 can be integrated and the solution (i.e. the excited state dynamics) is an exponential decay:

$$\langle S_1 \rangle(t) = \langle S_1 \rangle(0)\, e^{-(k_{Fluor} + k_{Annihil} + k_{ET} + k_{IC} + k_{ISC})t} := \langle S_1 \rangle(0)\, e^{-\frac{t}{\overline{\tau}}} \tag{34}$$

with the average apparent fluorescence lifetime

$$\overline{\tau} = \frac{1}{k_{Flour} + k_{Annihil} + k_{ET} + k_{IC} + k_{ISC}}. \tag{35}$$

The fluorescence dynamics depends on all possible relaxation channels. The observable fluorescence $F(t)$ is proportional to the actual population probability of the excited state $\langle S_1 \rangle$:

$$F(t) = k_{Flour} \langle S_1 \rangle (t) = k_{Flour} \langle S_1 \rangle (0)\, e^{-\frac{t}{\bar{\tau}}} := F(0) e^{-\frac{t}{\bar{\tau}}}. \tag{36}$$

The quantum efficiency of the fluorescence Φ_F is defined as the relation between the emitted fluorescence photons per time unit and the overall number of relaxing excited states in the same time interval:

$$\Phi_F := \frac{k_{Flour}}{k_{Flour} + k_{Annihil} + k_{ET} + k_{IC} + k_{ISC}} = \frac{\bar{\tau}}{\tau_0} \tag{37}$$

with the intrinsic fluorescence lifetime $\tau_0 = \frac{1}{k_{Flour}} \approx 15\ \text{ns}$ for chlorophyll.

As mentioned above the radiative relaxation of excited Chl molecules ${}^1Chl^* := \langle S_1 \rangle$ happens with $k_{Flour} = (15\ \text{ns})^{-1} = 6.7 \cdot 10^7\ \text{s}^{-1}$ for Chl *a* and somewhat slower fluorescence relaxation $k_{Flour} = (23\ \text{ns})^{-1} = 4.3 \cdot 10^7\ \text{s}^{-1}$ for Chl *b* (see Renger et al., 2009) and $k_{ISC} = (6.7\ \text{ns})^{-1} = 1.5 \cdot 10^8\ \text{s}^{-1}$ for both species at room temperature. In Chl *b* an additional thermally activated ISC channel might exist that enhances k_{ISC} at higher temperatures (Renger et al., 2009). Therefore the apparent fluorescence lifetime of Chl *a* calculates to $(k_{Flour} + k_{ISC})^{-1} = 4.6$ ns while a slightly faster relaxation is expected for Chl *b* at room temperature, but a slightly slower relaxation in comparison to Chl *a* at 10 K (see Schmitt et al., 2008).

The resulting fluorescence quantum yield calculates to $\Phi_F^{Chla} = \frac{\bar{\tau}}{\tau_0} \approx \frac{k_{Flour}}{k_{Flour} + k_{ISC}} = \frac{4.6}{15} \approx 31\%$ for Chl *a* at room temperature.

2.1.4 Decay Associated Spectra (DAS)

Figure 9b, published in (Theiss et al., 2011), shows typical fluorescence decay curves of whole cells of *A.marina* collected after excitation with 632 nm at room temperature performed with the technique of time and wavelength correlated single photon counting (TWCSPC) (Schmitt, 2011). The number of the registered photons at each wavelength and

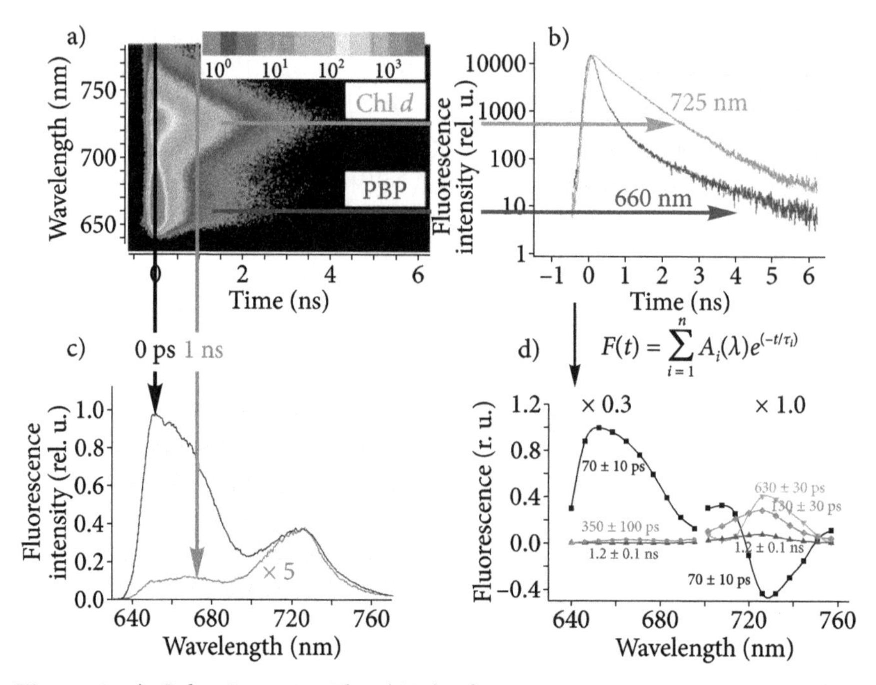

Figure 9. a) Color Intensity Plot (CIP) of a measurement on *A.marina* after excitation at 632 nm at 298 K. b) Fluorescence decay curves at 660 nm and 725 nm. c) Time resolved Fluorescence spectra at 0 ps and 1 ns (at 1 ns multiplied with a factor 5). d) Decay associated spectra (DAS) of a global fit in the range 640 nm – 690 nm (multiplied with a factor 0.3) and a global fit in the range 700 nm – 760 nm (see text). The figure is published in (Theiss et al., 2011 and Schmitt, 2011). Image reproduced with permission.

each time channel was stored in a 2-dim, 256 x 1024 data matrix. In Figure 9a, this data matrix is shown as a color intensity plot (CIP). A CIP is a plot of the fluorescence intensity (pictured by color) as a function of wavelength (y-axis) and time (x-axis). Therefore the CIP contains the time- and wavelength-resolved fluorescence emission data and provides information on the steady state fluorescence spectra and the lifetimes and dynamics of different emitter states. A vertical plot at a constant time t_0 results in the time-resolved emission spectrum $F(t_0, \lambda)$ (Figure 9c) while a horizontal intersection delivers the fluorescence decay at a constant emission wavelength λ_0 (Figure 9b). Time-resolved spectra provide information on the fluorescence of certain fluorophores at distinct times when the fluorescence emission of other pigments or scattered light have already decayed or after energy transfer when the main emission shifted in time from the donor pigment to the acceptor pigment.

The spectral resolution of the spectrometer system used to acquire the data shown in Figure 9 which is described in (Schmitt, 2011) is limited to about 2 nm (spectrometer entrance slit < 0.5 mm) due to the distance of the delay-line meanders and electrical crosstalk between the meander lines.

Figure 9b shows the decay curves of *A.marina* at 660 nm and 725 nm. It is seen that the fluorescence at 660 nm decays much faster than the emission at 725 nm. A closer look at the emission maximum on top of Figure 9b reveals a very small temporal shift between the 660 nm and the 725 nm decay curves. The 725 nm decay is slightly shifted to later times in comparison to the 660 nm decay. A data fit shows that the small temporal shift, in the following called "fluorescence rise kinetics", has a similar time constant as the fluorescence decay at 660 nm. Both curves are convoluted with the instrumental response function (IRF) which leads to the small visible difference.

In Figure 9c the main emission is observed at 645–660 nm (PBP emission) immediately after excitation (0 ps) while after one nanosecond (1 ns) the strongest fluorescence band occurs at 725 nm (Chl *d* emission). For better illustration the spectrum after 1 ns is multiplied by a factor of 5. At longer times the Chl *d* emission exceeds the PBP emission.

Detailed analyses of the fluorescence decay curves were performed by iterative reconvolution of a polyexponential decay model with the IRF using a global lifetime analysis minimizing the quadratic error sum χ^2 employing a Levenberg–Marquardt algorithm (for details see Theiss et al., 2011; Schmitt, 2011). The IRF was measured using distilled water as scattering medium. The multiexponential fits of all decay curves measured in one time- and wavelength resolved fluorescence spectrum were performed as global fits with common values of lifetimes τ_j (linked parameters) for all decay curves and wavelength-dependent pre-exponential factors $A_j(\lambda)$ (non-linked parameters, see Figure 9d). The result of such a fit analysis is usually plotted as a graph of $A_j(\lambda)$ for all wavelength independent lifetimes τ_j. This plot represents the so called "decay associated spectra" (DAS) thus revealing the energetic position of individual decay components (for details also see Schmitt et al., 2008).

2.2 Rate Equations and Excited State Dynamics in Coupled Systems

If there is any physical potential between two states, then these states are coupled. For our present interest, only the general electromagnetic coupling is relevant. This means that two states are coupled if the force field of one of the two states is seen by the other state. In such cases, the dynamics of one state modulate the dynamics of the other.

Coupled 2-level states can undergo several transitions, e.g. by spontaneous emission of light or phonons or by Inter-System Crossing (ISC). These transitions are discussed later and evaluated with regards to the experimental data. Schematic illustrations of excitation energy transfer transitions, including electron delocalisation and electron transfer, are shown in Figure 10, 4th row.

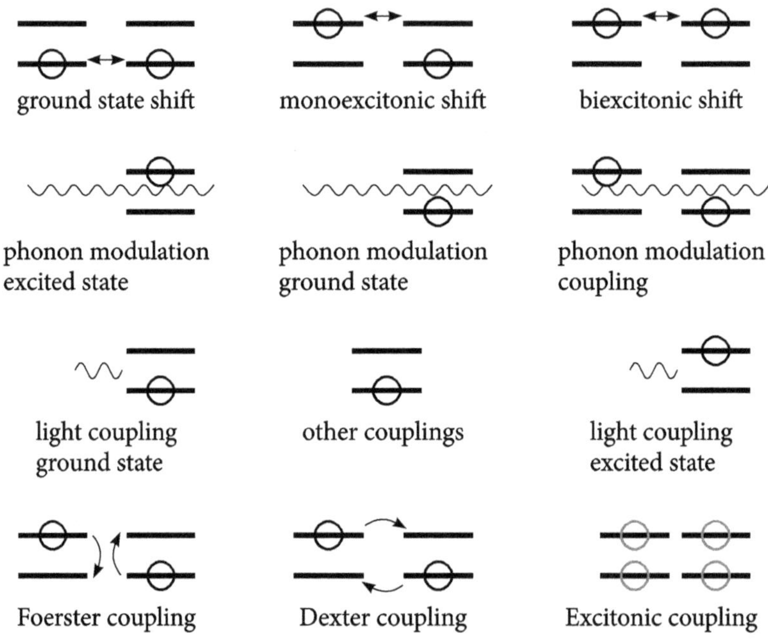

Figure 10. Interactions and EET/ET processes of coupled 2-Niveau systems bound to a protein matrix.

The electron-phonon interaction shown in Figure 10 shifts lifetimes (i.e. the transition probabilities) and the energy of the states. In photosynthetic systems, the formalism of rate equations can be incorporated by treating different molecules as localised states. Generally this will lead to rather complex equation systems of up to a thousand coupled differential equations. A simplified model respects the given resolution of the measurement setup by employing the concept of "compartments" as suggested e.g. by (Häder, 1999).

According to this compartmentation model (see chapter 1.4, Figure 5) a complex system can be treated as a form of "compartment" if the energetic equilibration between the coupled molecules that are treated as a compartment is fast in comparison to the achievable resolution of the measurement setup.

The advantage of the compartmentation model is a reduction of the number of states. For example one can treat the LHC, the core antenna complexes and the reaction center (see Figure 18 and Figure 18) as compartments of thylakoid membrane fragments and therefore as single states instead of treating each single molecule separately. The whole light-harvesting complex, including the core antenna and the Chl molecules in the reaction center, might even be treated as one single compartment.

The states and their couplings are visualized in the scheme as shown in Figure 11. Here five arbitrary states are coupled via EET exhibiting one and the same rate constant k for all transitions to neighbouring states. Additionally, there is a relaxation to the ground state; this is $(k_Q + k_F)$ when the excitation energy interacts with a quenched state as shown in Figure 11. The probability for relaxation to the ground state of all states without quencher is k_F.

In the scheme shown in Figure 11 the state N_4 is bound to a quencher (e.g. a reaction center) and therefore exhibits reduced lifetime. The asterix (*) indicates that initially (time $t = 0$) the state N_1 is excited while all other molecules are found in the ground state.

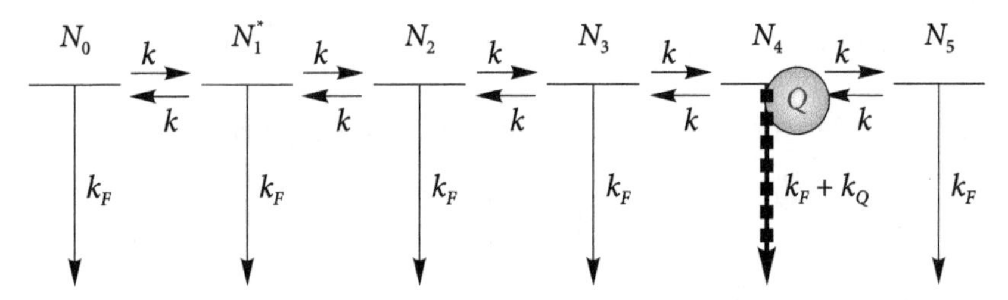

Figure 11. Scheme of five coupled states that can transfer energy to neighbouring molecules. States 1, 2, 3, and 5 have equal relaxation probabilities from the excited to the ground state (rate constant k_F), while state 4 is connected with a quencher and therefore exhibits a higher relaxation probability $k_F + k_Q$.

From the schematic model presented in Figure 11 one can formulate the rate equations for the time derivative of each state. For molecule N_2 this equation exemplarily denotes to

$$\frac{dN_2}{dt} = -2kN_2 - k_F N_2 + kN_3 + kN_1.$$

The final set of five coupled, linear, differential equations of the first order can be solved to obtain the excited state dynamics of all five molecules/states.

The fluorescence intensity is proportional to the time-dependent population density of the fluorescing states. This is calculated from the solution of such a set of coupled linear differential equations of the first order (for mathematical details, see Schmitt, 2011).

This set of equations is solved by a sum of exponential functions with different time constants. In a nondegenerated set with n pairwise different eigenvalues of the transfer (imaging) matrix $\bar{\bar{T}}$ describing the change $\dot{\bar{N}}$ of all n state population densities, the set of differential equations can be written according to eq. 38:

$$\dot{\bar{N}} = \bar{\bar{T}}\bar{N}. \tag{38}$$

If the entries of $\bar{\bar{T}}$ are not known, we at least have some information about the symmetry of $\bar{\bar{T}}$ according to eq. 8. It is a necessary condition for the equilibrium case that $\frac{T_{ki}}{T_{ik}} = e^{\frac{(E_i - E_k)}{k_B T}}$ according to eq. 9. The derivation of eq. 9 is shown in chapter 1.3.

For the sake of simplicity, the relaxing system might be treated as locally equilibrated.

In that case eq. 8 helps us reduce the nondiagonal elements of $\bar{\bar{T}}$ to ½ of their quantity.

The solution for the ith state population density is given by

$$N_i(t) = \sum_{j=1}^{n} U_{ij} e^{\gamma_j t} \tag{39}$$

where γ_j is denoting the jth eigenvalue and U_{ij} the ith component of the jth eigenvector of $\bar{\bar{T}}$. For a nondegenerated system eq. 39 is proportional to the assumed fluorescence decay:

$$F(t,\lambda) = \sum_{i=1}^{n} A_i(\lambda) e^{(-t/\tau_i)} \tag{40}$$

with $\gamma_j = -(\tau_j)^{-1}$ according to the proportionality between fluorescence intensity and excited state population density. If one or more $\gamma_j = 0$, one can reduce the dimensionality of $\bar{\bar{T}}$ to a smaller size with all eigenvalues being nonzero and nondegenerated. In such cases, eq. 40 is suitable to calculate the expected fluorescence emission in the time domain from the transfer matrix $\overline{\overline{T}}$. Vice versa $\overline{\overline{T}}$ can be reconstructed if the fluorescence kinetics of all molecules is measured.

For that purpose the eigenvector matrix $\overline{\overline{U}}$ and the eigenvalues in the diagonal matrix $\overline{\overline{\Gamma}}$ have to be used to calculate the transfer matrix $\overline{\overline{T}}$:

$$\overline{\overline{T}} = \overline{\overline{U}}\,\overline{\overline{\Gamma}}\,\overline{\overline{U}}^{-1}. \tag{41}$$

Eq. 41 is unique if one knows the amplitudes and time constants of all exponential decay components for each compartment in the sample.

The time-dependent fluorescence of emitters described by a coupled system of differential equations as given by eq. 40 occurs multiexponentially with up to n decay components (i.e. exponential decay time constants) for n coupled subsystems (i.e. n coupled excited states that define the dimensionality of $\overline{N}$ with respect to eq. 38).

Therefore the average decay time shown in eq. 35 cannot be used. In the case of coupled states the average decay time can be defined as a time constant proportional to the overall fluorescence intensity (for further details see Lakowicz, 2006):

$$\overline{\tau} = \frac{\int_0^\infty F(t)dt}{F(0)} = \frac{\int_0^\infty \langle S_1 \rangle (t) dt}{\langle S_1 \rangle (0)}. \tag{42}$$

If one evaluates eq. 42 with the multiexponential decay curves as given by eq. 40 it follows that

$$\overline{\tau}(\lambda) = \frac{\int_0^\infty F(t)dt}{F(0)} = \frac{\left[\sum_{i=1}^{n} \tau_i A_i(\lambda) e^{(-t/\tau_i)} \right]_0^\infty}{\sum_{i=1}^{n} A_i(\lambda)} = \frac{\sum_{i=1}^{n} A_i(\lambda)\tau_i}{\sum_{i=1}^{n} A_i(\lambda)} \tag{43}$$

at a certain wavelength position.

Usually one does not find all the eigenvector components (i.e. when some state populations do not show fluorescence or when not all fluorescence decay components can be resolved). In that case eq. 41 cannot be used to find a unique transfer matrix $\overline{\overline{T}}$. But a reasonable assumption of $\overline{\overline{T}}$ with iterative comparison of the eigensystem (i.e. eigenvectors and eigenvalues) ($\overline{\overline{U}}, \overline{\overline{T}}$) with the fluorescence decay helps find a solution for $\bar{\bar{T}}$ which is consistent with the globally observed fluorescence dynamics in all wavelength sections according to eq. 40. Such an iterative comparison of the measurement data (eigenvectors and eigenvalues) and a reasonable transfer matrix $\bar{\bar{T}}$ can be used to calculate a suggestion for the transfer rates of a system of arbitrary complexity which leads to observable fluorescence decay.

Our computation was highly efficient when employing an algorithm that starts with an arbitrary reasonable transfer matrix $\bar{\bar{T}}_{Start}$. First the eigenvectors and eigenvalues of $\bar{\bar{T}}_{Start}$ are calculated. In the next step, one will use the observed decay times for all eigenvalues $\gamma_j = -(\tau_j)^{-1}$ instead of the eigenvalues of $\bar{\bar{T}}_{Start}$. Therefore the eigenvalues are changed but the eigenvectors are not. Doing the transformation according to eq. 41 with the changed (experimentally observed) eigenvalues one will find an improved suggestion $\bar{\bar{T}}$ for the transfer matrix $\bar{\bar{T}}_{Start}$.

There are some symmetry constraints for the transfer matrix $\bar{\bar{T}}$. For example, the rate constant for the energy transfer from pigment one to pigment two has to be the same as the energy reception at pigment two according to the energy transfer. Additionally, we have already mentioned the Boltzmann equilibrium which gives rise to eq. 8 as a symmetry argument for the transposed entries. Therefore, this symmetry condition might also be used as constraint.

According to the suggested algorithm, all violated constraints are corrected in the transfer matrix $\bar{\bar{T}}$. After that one can again calculate the eigenvectors and eigenvalues of $\bar{\bar{T}}$. The observed decay times for all eigenvalues are again set to $\gamma_j = -(\tau_j)^{-1}$. Then the algorithm starts from the beginning. Doing the transformation according to eq. 41 with the changed (experimentally observed) eigenvalues leads to a further improvement of $\bar{\bar{T}}$.

The described algorithm for calculating the transfer rates in *A.marina* converged after five iterations and therefore was faster than other algorithms found in the literature. There are several other correlations between experiment and calculation which could be implemented into the algorithm. For example, one could also use the observed amplitudes of the different fluorescence components to correct the eigenvectors of $\bar{\bar{T}}_{Start}$ instead of the eigenvalues.

2.2.1 Simulation of Decay-Associated Spectra

According to the formalism presented in chapters 1.3, 1.4, and 2.1, the decay-associated spectra can be simulated for each pigment or compartment in a coupled system. First we start with a simulation of the DAS of the coupled chain shown in Figure 11 but without the quencher molecule bound to the state N_4. The DAS of the individual pigments exhibit multiphasic relaxation dynamics at each molecule of the coupled chain:

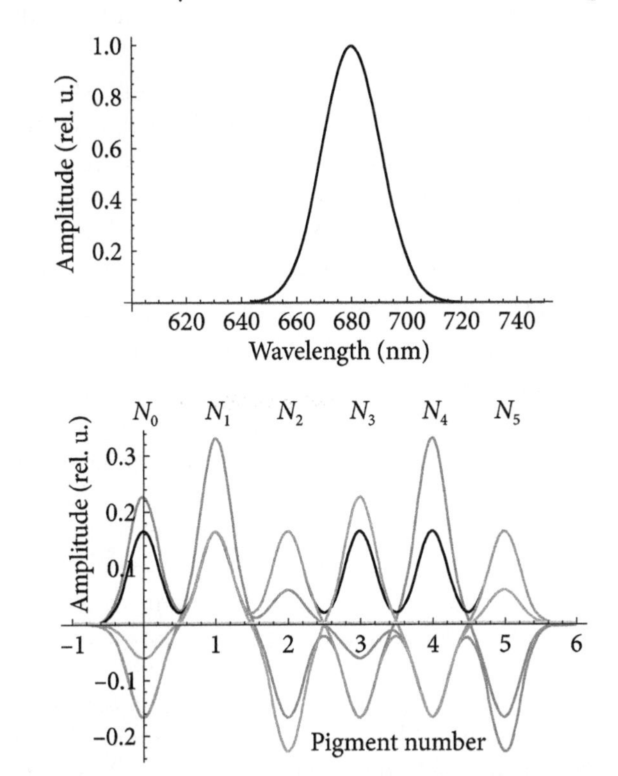

Figure 12. Simulated decay-associated spectra for the system shown in Figure 11 (without the quencher at N_4) with a Gaussian lineshape function in the wavelength domain (upper panel) and in the "pigment" domain (lower panel). The black curve in the upper panel denotes a single exponential decay component with a time constant of 4 ns. In the lower panel we see a complex DAS pattern distributed along all coupled states. The time constants are 2.7 ps (cyan), 3.3 ps (red), 5 ps (green), 37 ps (magenta) and 4 ns (black).

In the simulation shown in Figure 12, it is assumed that k_F = 4 ns and k = 10 ps. The simulation is performed without the quencher located at the molecule N_4. The initial excitation is set to pigment N_1 as indicated in Figure 11. In the spectral domain there is only one monoexponential decay with a typical 4 ns decay time because all molecules are isoenergetic (Figure 12, upper panel). In the spatial domain the DAS is rather complex and each pigment exhibits an individual time dependent population and therefore fluorescence emission. The structure of the spatially resolved DAS pattern, especially the symmetry of the DAS shown in Figure 12, lower panel, is briefly discussed in the following.

The interpretation of the DAS in Figure 12 is in full accordance with the findings that the dominant decay of each pigment exhibits a time constant of 4 ns (decay) and that all other time constants describe the EET inside of the coupled structure. The excited pigment N_1 transfers the excitation energy fast to the neighbouring pigments N_0 and N_2 (red curve, 3.3 ps). These exhibit the characteristic rise kinetics with 3.3 ps time constant (red curve). The red curve shows a mirror symmetry along the axis between N_2 and N_3 (which is not clearly visible as the components with other colors overlay the negative amplitude at pigment N_0 and N_3). Therefore the rise kinetics (negative amplitude) is also found at N_3 and N_5 whereas N_4 contains a high amplitude of the 3.3 ps decay kinetics (positive amplitude) similar to the initially excited pigment N_1.

From the DAS shown on the right side of Figure 12, it can be seen that the DAS components with time constants of 4 ns (black) and 3.3 ps (red) appear mirror symmetric to the center of the chain while the time constants with values of 2.7 ps (cyan), 5 ps (green) and 37 ps (magenta) are anti-symmetric according to the mirror plane between N_2 and N_3. In addition, the evaluation delivers a 10 ps component which has amplitude zero in the whole range of states and which therefore is not visible in the fluorescence.

It has to be strictly pointed out that the complexity that is found in the DAS of Figure 12 cannot yet be resolved experimentally. Even if it would be possible to resolve single molecules in the spatial domain (e.g. the states denoted as N_i in Figure 11; this might be possible with time resolved STED microscopy or with tip enhanced single molecule fluorescence spectroscopy) one cannot expect to find a DAS as depicted in Figure 12, lower panel, because the time constants of 2.7 ps, 3.3 ps and 5 ps are hardly distinguishable with the technique of TSCSPC employing a laser system with 70 ps FWHM of the IRF. Here fluorescence up conversion would be a technique that delivers the time resolution, but applied high resolution techniques like STED or tip enhanced AFM would influence the excited state lifetime of the single molecules and therefore also do not represent the natural state of the system. However the expectation for any given resolution can be calculated from the DAS shown in Figure 12 by mathematic convolution of the temporal and spatial sensor elements with the DAS shown in Figure 12, lower panel.

Most probably the experimenter will find a clearly resolved 4 ns decay kinetic (black curve) and the pronounced 37 ps component (magneta curve) with high positive amplitude at the states N_0 and N_1 and negative amplitude at states N_4 and N_5. All other (fast) rise and decay components will appear as one additional very fast component. The amplitude of this fast component is comparable to the sum of the amplitudes of all fast components (2.7 ps (cyan), 3.3 ps (red) and 5 ps (green)) shown in Figure 12, lower panel.

This situation of a DAS that could be resolved experimentally (still assuming single molecule resolution in the spatial domain) is presented in Figure 13.

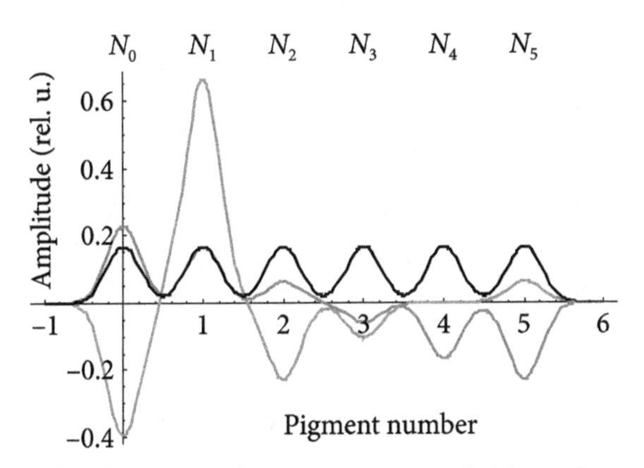

Figure 13. DAS for the system shown in Figure 11 (without the quencher position at N_4) with a Gaussian lineshape function in the "pigment" domain. The fast components shown in Figure 12 (cyan, red, green curve) are summed to one component here (shown in green). In addition the 37 ps (magenta) and 4 ns (black) component are shown.

One expects that the positive decay kinetics of all pigments that are not directly excited is mainly dominated by the 4 ns fluorescence decay (black curve in Figure 13). The initially excited pigment N_1 decays with very fast (3–5 ps, green curve) decay kinetics which appears as fluorescence rise at the neighbouring molecules (negative amplitude). The 37 ps component represents the overall energy transfer along the coupled chain of molecules with positive amplitude at pigment N_0 and N_1 which exhibit mainly donor character, and negative amplitude at pigment N_4

and N_5 which exhibit an acceptor character for the overall energy equilibration process. This overall time spread in the ensemble measurement is restricted by the geometry and symmetry of the pigment chain as shown in Figure 11 rather than by certain transfer "steps" inside the system. The interaction time between the single subunits is much faster than this averaged overall energy transfer (magneta curve) and the amplitude distribution reflects the average pathway of the energy caused by the geometrical constitution of the sample.

It is interesting to see how positive and negative amplitudes of the corresponding fluorescence components are distributed along the pigment chain if the initial excitation changes. One might investigate a situation where N_1 and N_4 are initially excited instead of only N_1 as presented in Figure 14, upper panel.

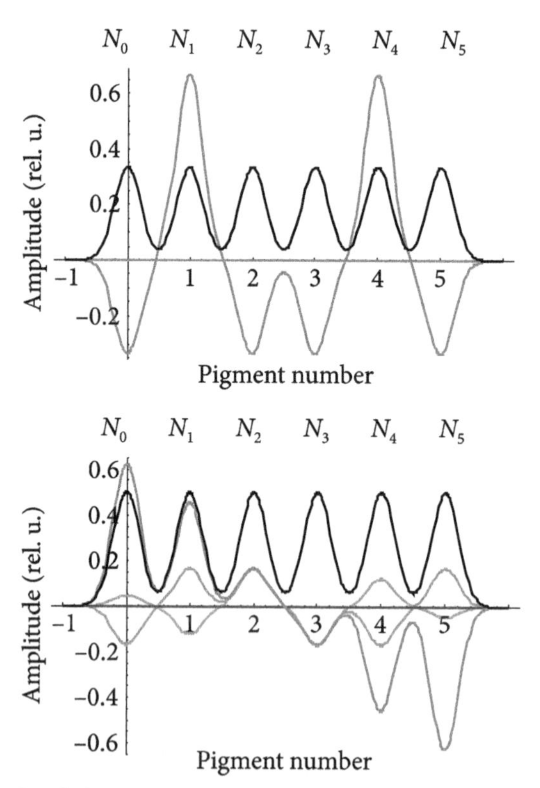

Figure 14. Simulated decay-associated spectra for the system shown in Figure 11 (without the quencher position at N_4) with a Gaussian lineshape function according to Figure 12 but with initial population $N_0(t = 0) = N_2(0) = N_3(0) = N_5(0) = 0$, $N_1(0) = 1$, $N_4(0) = 1$ (upper panel, full symmetric case) and $N_0(t = 0) = N_1(0) = N_2(0) = 1$, $N_3(0) = N_4(0) = N_5(0) = 0$ (lower panel, full antisymmetric case).

In the upper panel of Figure 14, the DAS pattern of a full mirror symmetric initial condition is simulated with $N_0(t = 0) = N_2(0) = N_3(0) = N_5(0) = 0$, $N_1(0) = 1$, $N_4(0) = 1$. In the lower panel, the situation of a primary excitation of N_0, N_1 and N_2 is shown, i.e. a fully antisymmetric condition $N_0(t = 0) = N_1(0) = N_2(0) = 1$, $N_3(0) = N_4(0) = N_5(0) = 0$.

All fast decay components, except the dissipative decay component for the relaxing system after equilibration (which is independent from the initial condition, 4 ns component, black curve), represent the symmetry of the initial condition according to a suggested mirror plane in the middle of the structure.

Functional PS II preparations exhibit a markedly reduced excited state lifetime due to the presence of photochemical quenchers, i.e. the reaction center of PS II where charge separation occurs.

This situation is simulated in Figure 15 for the coupled pigment chain shown in Figure 11 assuming the initial excitation of pigment N_1 and the quencher located at pigment N_4.

The simulation was performed with the rate constants $k_F = (4\ \text{ns})^{-1}$, $k_Q = (2\ \text{ps})^{-1}$ and $k = (10\ \text{ps})^{-1}$.

As seen in the upper panel of Figure 15, the decay is no longer monoexponential in the wavelength domain after introduction of the quencher. The main decay component exhibits a lifetime of 88 ps which is composed by the quencher efficiency k_Q, the EET rate constant k and the distance between the initially excited state and the quencher. The model simulates a "random walk" of the initially excited electron until it is quenched to the ground state (for a description of such random walk models in photosynthesis research see Bergmann, 1999).

On account of the coupling of several pigments, the amplitude distribution of the decay components that are found in the fluorescence kinetics becomes more complex when the quencher is present (see Figure 15, lower panel).

As described, the solution of the rate equation systems delivers the time-resolved fluorescence spectra and DAS. Therefore one can determine k_{FRET} from the experimental data. In chapter 2.2.3 we will see how the Förster Energy Transfer rate constant can be calculated independently from fluorescence spectra. It can also be found with the help of rate eq. 38 and the reverse transformation denoted in eq. 41. The Förster Resonance Energy Transfer can be evaluated independently using the time-integrated spectra of the donor and the acceptor, or it also can be fitted onto the time-resolved spectra comparing the solution of the rate equations with the measurement results.

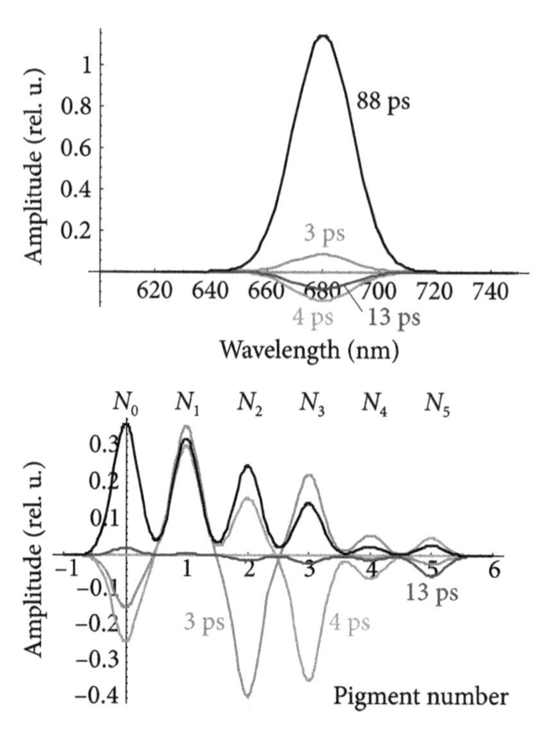

Figure 15. Simulated decay-associated spectra for the system shown in Figure 11 (including the quencher position at N_4) with a Gaussian lineshape function in the wavelength domain (upper panel) and in the "pigment" domain (lower panel). The black curve denotes the main relaxation of the coupled system, occurring as a strictly positive decay component with a time constant of 88 ps. In addition we find fast time constants with 3 ps (red), 4 ps (green) and 13 ps (blue).

In many cases there is a lack of information for the direct calculation of the distance (R_{12}) dependent rate constant $k_{FRET}\left(R_{12}\right)$ due to the fact that there are missing parameters like the pigment dipole orientation factor κ (see chapter 2.2.3). For a full understanding of the advantages of the Förster Resonance Energy Transfer and its limitations we will shortly outline the basic concept of FRET.

2.2.2 Excited States in Coupled Pigments

If two pigments are coupled via an excitation energy or electron transfer, the rate constant k_{ET} can be approximated according to Fermi's Golden

Rule in first order perturbation theory (Messiah, 1991a,b; Cohen-Tanouji et al., 1999 a,b; Häder, 1999; Schwabl, 2007, 2008):

$$k_{ET} = \frac{2\pi}{\hbar} |V_{12}|^2 \rho_{eff}(E). \tag{44}$$

The rate constant k_{ET} which is a probability per time unit (for a transition) can be calculated as the product of the square of the absolute value of the transition matrix element $|V_{12}|^2$ and the effective density of states $\rho_{eff}(E)$ of the initial and final states. In photosynthetic complexes the density of states is determined by the vibrational spectrum of the interacting species and the states of their microenvironment (Renger and Schlodder, 2005). The transition matrix element $V_{12} = \langle \Psi_2 | H_{WW} | \Psi_1 \rangle$ is the scalar product of the initial state $|\Psi_1\rangle$ and the final state $|\Psi_2\rangle$ under the influence of the interacting potential. This potential is represented by the interaction operator H_{WW}, e.g. a Coulomb potential $H_{WW} = V_{Coulomb}$ as denoted in the appendix in eq. 84. If we investigate the situation where the excited state of one pigment in a coupled trimer is transferred to the coupling partner one can formulate the product states

$$\begin{aligned} |\Psi_1\rangle &= \frac{1}{\sqrt{2}} \left| \Phi_1(1)^e \Phi_2(2)^g - \Phi_1(2)^e \Phi_2(1)^g \right\rangle, \\ |\Psi_2\rangle &= \frac{1}{\sqrt{2}} \left| \Phi_1(1)^g \Phi_2(2)^e - \Phi_1(2)^g \Phi_2(1)^e \right\rangle, \end{aligned} \tag{45}$$

where $\left| \Phi_i(j)^{e,g} \right\rangle$ denotes the electronic wave function of the jth electron in the excited (e) or ground state (g) located at the ith pigment.

Evaluating Fermi's Golden Rule (eq. 44) with the product states as denoted in eq. 45 delivers matrix elements of the form

$$\left\langle \Phi_2(2)^e \Phi_1(1)^g \middle| \Phi_1(1)^e \Phi_2(2)^g \right\rangle, \left\langle \Phi_2(1)^e \Phi_1(2)^g \middle| \Phi_1(2)^e \Phi_2(1)^g \right\rangle \tag{46}$$

denoted Coulomb terms and

$$\left\langle \Phi_2(1)^e \Phi_1(2)^g \middle| \Phi_1(1)^e \Phi_2(2)^g \right\rangle, \left\langle \Phi_2(2)^e \Phi_1(1)^g \middle| \Phi_1(2)^e \Phi_2(1)^g \right\rangle \tag{47}$$

denoted exchange terms.

While the Coulomb terms also deliver significant contributions in the weak coupling regime, the exchange terms require a direct overlap of the electronic wavefunctions between the ground state wave functions and the excited state wave functions of both pigments (for details see Renger, 1992).

This direct overlap is only significant if the distance of the interacting fluoropheres is very low. In this case, the electronic wave function is no longer localised at one pigment only but is delocalised among all coupled pigments.

The dominating mechanism for the ET and EET in the short distance regime is called "Dexter transfer", which denotes a direct electron exchange due to the delocalisation of the electronic probability of presence. For the Dexter transfer the selection rules differ from the selection rules of weakly coupled pigments, including triplet-triplet transfer as proposed for strongly coupled Chl and carotenoid molecules.

In agreement with the exponential decay of the electronic wave functions, Dexter transfer occurs at very small distances typically near to the van der Waals-distance in the order of 0.1–1 nm (Häder, 1999). Concomitantly with the amplitudes of the electronical wave functions, the transfer rate constant decreases exponentially with the distance r of the coupled pigments:

$$k_{ET} \sim e^{-\alpha r} \tag{48}$$

where α denotes the so called "Dexter coefficient".

Due to the short distance of the pigments, the excitonic coupling leads to a significant shift of the energetic levels and therefore to a change in the optical spectra. The excitonic coupling leads to excitonic splitting. For excitonically coupled states, the relaxation from the higher excitonic level to the lower excitonic level is typically very fast (<10 ps) and cannot be resolved with time- and wavelength-correlated single photon counting.

In many strong coupled photosynthetic pigments the Dexter transfer is the dominating EET mechanism, i.e. between Chlorophyll molecules in the LHC and in PBS of cyanobacteria and probably also in the PBP antenna of *A.marina* (Theiss et al., 2011). Sometimes it is not fully clear whether the suitable approximation for the dominating EET mechanism is Dexter or Förster. For example in the PBP antenna the transport between different trimers has to bridge distances of about 2 nm which is within both validity areas. It has to be kept in mind that the energy transfer mechanism is neither "Förster" nor "Dexter", but nothing more than an effect resulting from electromagnetic coupling. Just the approximation conducted is in the Förster or Dexter regime.

At distances above 2 nm the electromagnetic potential of the involved chromophores can be estimated with the model of point-dipole approximation (Renger and Schlodder, 2005). In this case a significant probability for energy transfer can be observed up to a distance of 10 nm

(Lakowicz, 2006). The mechanism of EET using the point-dipole approximation for V_{12} in eq. 44 is called "Förster Resonance Energy Transfer" (FRET).

2.2.3 Förster Resonance Energy Transfer (FRET)

The formulation of the point dipole approximation for the resulting energy transfer rate constant was done by T. Förster in 1948 (Förster, 1948). The distance regime of the coupled molecules where this approximation holds (1–10 nm) needs to allow the approximation of the dipole of one pigment seen from the other pigment as point source (i.e. the distance has to be clearly bigger than the dipole length). Förster already suggested that the FRET mechanism could be the dominating process for energy transfer between coupled molecules in photosynthetic organisms. For Chl *a* he calculated the distance R_0 (Förster Radius) to be 8 nm, when the probability for the energy transfer is the same as the sum of all other relaxation processes (Förster, 1948):

$$k_{ET}\left(R_0\right) = k_{FRET}\left(R_0\right) = \frac{1}{\overline{\tau}_D}. \tag{49}$$

At the distance R_0 50% of the donor pigment's energy is transferred to the acceptor pigment (i.e. the probability of the transfer is 50%). In eq. 49 $\overline{\tau}_D$ denotes the apparent average fluorescence lifetime of the excited donor state in absence of the acceptor (i.e. the average excited state lifetime denoted as $\overline{\tau}$ in eq. 35, but without the energy transfer rate constant k_{ET} (absence of the acceptor)). If the fluorescence decay does not occur monoexponentially, eq. 49 has to be treated carefully and $\overline{\tau}_D$ should be calculated from the experimental data according to eq. 43 only after a multiexponential fit of the donor system's fluorescence decay (without the acceptor, see Lakowicz, 2006). For further fluorescence spectroscopic findings of Förster see (Förster, 1982).

Measurements of the depolarisation of fluorescence light in dependency of the fluorophore concentration were done in 1924, mainly by two groups: Gaviola and Pringsheim, and Weigert and Käppler. From these measurements, Perrin in 1925 deduced that a direct electromagnetic coupling between the dye molecules must occur. He realized that the transfer of energy between the pigments is not simply a pure emission and reabsorption effect of real photons that leads to an energy transfer of excited states. In Förster's own words, this energy transfer mechanism "is principally different from such a mechanism which would consist of the trivial one of reabsorption of fluorescence light, because emission and electronic transfers would [in such case] not occur in parallel but subsequently" (Förster, 1948).

Förster used Fermi's Golden Rule, as given in eq. 44, to calculate $k_{ET} = k_{FRET}$. He assumed the dipole interaction as V_{12}. As seen in eq. 45, the wave functions of the two pigments can be expressed as a product state of the wave function of pigment one $\left|\Phi_1^{e,g}\right\rangle$ and pigment two $\left|\Phi_2^{e,g}\right\rangle$ in excited (e) or ground (g) state (Renger and Schlodder, 2005):

$$V_{12} = \left\langle \Phi_1^{\,g}\Phi_2^{\,e} \middle| V_{Coulomb} \middle| \Phi_1^{\,e}\Phi_2^{\,g} \right\rangle. \tag{50}$$

For a distance typically $R_{12} > 2$ nm between the center points of the coupled transition dipole moments of the two coupled pigments, the dipole potential is well described by the point-dipole approximation (see Figure 16):

$$V_{12} = \frac{1}{4\pi\varepsilon_0\varepsilon_r}\left(\frac{\overline{\mu}_1\overline{\mu}_2}{\left(R_{12}\right)^3} - 3\frac{\left(\overline{\mu}_1\overline{R}_{12}\right)\left(\overline{\mu}_2\overline{R}_{12}\right)}{\left(R_{12}\right)^5}\right). \tag{51}$$

In eq. 51 $\overline{\mu}_i$ denotes the dipole moment of the i. pigment and $\overline{R}_{12}$ denotes the vector that points from the center of $\overline{\mu}_1$ to the center of $\overline{\mu}_2$ (see Figure 16). The dipole strength, which is proportional to $\left|\overline{\mu}\right|^2$, can be calculated from the absorption spectrum of the acceptor pigment and from the time-resolved emission of the donor pigment. Due to the vector products in eq. 51 the coupling strength V_{12} strongly depends on the orientation of the interacting dipoles (for the nomenclature see Figure 16):

$$V_{12} = \frac{\left|\mu_1\right|\left|\mu_2\right|}{4\pi\varepsilon_0\varepsilon_r\left(R_{12}\right)^3}\left(\cos\Theta_{12} - 3\cos\Theta_1\cos\Theta_2\right). \tag{52}$$

Using the real part n of the complex refraction index of the medium between both molecules (surrounding medium) (Renger and Schlodder, 2005) and the dipole orientation factor $\kappa := \left(\cos\Theta_{12} - 3\cos\Theta_1\cos\Theta_2\right)$, the energy transfer rate for Förster transfer $k_{ET} = k_{FRET}$ as given in eq. 44 using the potential in eq. 52 can be written as

$$k_{FRET}\left(R_{12}\right) = \frac{2\pi}{\hbar}\frac{\left|\mu_1\right|^2\left|\mu_2\right|^2\kappa^2}{\left(4\pi\varepsilon_0\right)^2 n^4\left(R_{12}\right)^6}\rho_{eff}\left(E\right). \tag{53}$$

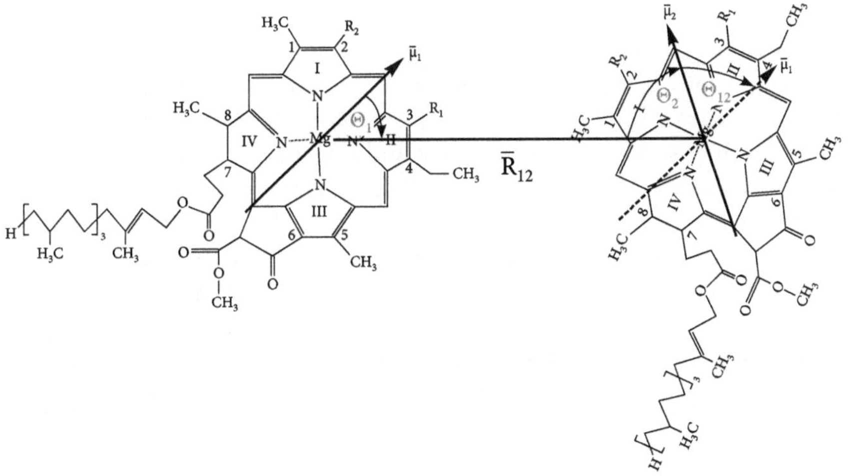

Figure 16. Scheme of the angels Θ_1 between $\bar{\mu}_1$ and the center-center vector $\bar{R}_{12}$, Θ_2 between $\bar{\mu}_2$ and $\bar{R}_{12}$, Θ_{12} between $\bar{\mu}_1$ and $\bar{\mu}_2$. $\bar{\mu}_1$ and $\bar{\mu}_2$ are the two Q_y dipole moments of two coupled chlorophyll molecules.

Eq. 44, i.e. Fermi's Golden Rule, is a formula to analyse the transition probability of any coupled quantum states if the coupling is weak enough. Eq. 53 is a formula to calculate the probability for energy transfer according to the assumption of a point-dipole-approximation of these states (Förster coupling).

The point dipole approximation is good if the distance of the molecules is much bigger than the extension of the dipole moments (~1 nm). Typical distances of coupled pigments in photosynthetic pigment protein complexes can be in the order of some nm. Better approximations take into account higher orders of the multipole extension of the Coulomb potential or describe the dipole extension in the form of the so-called "extended dipole" approximation. This assumes monopole moments from partial charges found along the dipole extension in the molecule (Renger and Schlodder, 2005). When we discuss such approximation, we should keep in mind that Fermi's Golden Rule, as given by eq. 44, is itself an approximation of first order perturbation theory for weak couplings.

Additionally, there are further approximations used in the Förster formula. One prominent simplification is the assumption of a dipole embedded in a homogeneous and isotropic refractive medium which can be described by a simple number of dielectricity. If we look at the two chlorophyll molecules exemplarily shown in Figure 16 and a typical protein environment of the chromophores as, for example, shown for the chlorophylls in the LHC (see Figure 20) it becomes clear that there is no homogeneous and isotropic medium surrounding the chromophores.

The partially classical description of the Förster transfer rate shown in eq. 53 deviates strongly when the wave functions of the electrons located in the excited states of pigments one and two start to overlap directly as mentioned above (i.e. in these short distances the full quantum mechanical Dexter transfer has to be evaluated). This also becomes clear from a quantum mechanical point of view in which the definition of several clearly defined distances and angels, as necessary for the formulation of eq. 53, fails.

As mentioned above, the dipole strength of $\bar{\mu}_1$ and $\bar{\mu}_2$ can be derived from the optical spectra. Förster showed that the term $|\mu_1|^2 |\mu_2|^2 \rho_{eff}(E)$ is proportional to the expression $\frac{\int_0^\infty F_D(\lambda)\varepsilon(\lambda)\lambda^4 d\lambda}{\bar{\tau}_D \cdot \int_0^\infty F_D(\lambda) d\lambda}$, i.e. the energetic resonance between the fluorescence spectrum $F_D(\lambda)$ of the donor pigment and the spectral extinction $\varepsilon(\lambda)$ of the acceptor divided by the (average) lifetime of the donor pigment in the absence of the acceptor $\bar{\tau}_D$ (see Lakowicz, 2006; Förster, 1948, 1982). The dipole strength of the acceptor pigment $|\bar{\mu}_2|^2$ corresponds to the molar extinction coefficient $\varepsilon(\lambda)$ while the dipole strength of the donor $|\bar{\mu}_1|^2$ is expressed by the isolated donor pigment's average excited state lifetime $\bar{\tau}_D$.

The transition probability per time unit k_{FRET} as derived by Förster finally denotes to

$$k_{FRET}(r) = A \frac{\kappa^2}{(R_{12})^6 \tau_0 n^4} \frac{\int_0^\infty F_D(\lambda)\varepsilon(\lambda)\lambda^4 d\lambda}{\int_0^\infty F_D(\lambda) d\lambda} \tag{54}$$

with the normalised overlap integral between the fluorescence of the donor and the extinction of the acceptor $\frac{\int_0^\infty F_D(\lambda)\varepsilon(\lambda)\lambda^4 d\lambda}{\int_0^\infty F_D(\lambda) d\lambda} := J(\lambda)$, the constant $A = \frac{9000(\ln 10)}{128\pi^5 N_A} \text{cm}^3 = 8{,}8 \cdot 10^{-25} \text{ cm}^3\text{mol}$ (dimension of the wavelength $[\lambda] = \text{cm}$) and the extinction coefficient $[\varepsilon(\lambda)] = (\text{M} \cdot \text{cm})^{-1}$. In that case, the overlap integral has the dimension $[J(\lambda)] = \frac{\text{cm}^3}{\text{M}}$. A clear description how to evaluate the Förster formula, as given by eq. 54, numerically is found in (Lakowicz, 2006). Combining eq. 49 and eq. 54 one can rewrite the Förster radius to

$$R_0 = \left(A\Phi_F \frac{\kappa^2}{n^4} J(\lambda) \right)^{\frac{1}{6}} \tag{55}$$

(measured in cm) with $\Phi_F = \frac{\overline{\tau}_D}{\tau_0}$ as found in eq. 37 for the donor molecule. The rate constant for the Förster transfer can then be expressed as

$$k_{FRET}(R_{12}) = \frac{1}{\overline{\tau}_D} \left(\frac{R_0}{R_{12}} \right)^6. \tag{56}$$

In many cases there is a lack of information for the direct calculation of the distance (R_{12}) dependent rate constant $k_{FRET}(R_{12})$ from eq. 56. For example, the value of the orientation factor κ^2 is not known for many pigment-protein complexes. For molecules in solution is $\kappa^2 = \frac{2}{3}$ but in photosynthetic complexes κ^2 can be of much higher values. Certain pathways along the PBP antenna are strongly suppressed due to very small values of κ^2 near to zero, which causes the EET to select pathways along the PBS structures of cyanobacteria (Suter and Holzwarth, 1987). It is found that pairwise coupled pigments with $\kappa^2 \approx 1$ exist in all photosynthetic complexes and therefore FRET helps to explain the structural organisation of pigment-protein complexes.

In the expressions of the Förster formulae given in eq. 54 to eq. 56 there might occur definition problems if the fluorescence decay is multiphasic. For coupled systems of different states of equal or comparable energy (isoenergetic states) the eq. 35 to eq. 37 gathered from integration of the uncoupled system are not valid and the quantum yield for the fluorescence decay cannot be simply calculated by $\Phi_F := \frac{k_{Flour}}{k_{Flour} + k_{Annihil} + k_{ET} + k_{IC} + k_{ISC}} = \frac{\overline{\tau}}{\tau_0}$ as proposed by eq. 37 because

$k_{Flour} + k_{Annihil} + k_{ET} + k_{IC} + k_{ISC} \neq \frac{1}{\tau}$. Eq. 35 is only strictly valid for a monoexponential decay, i.e. if there is not a backwards transfer k_{-ET} in the coupled system (see Figure 8). That means that the inverse average lifetime of a fluorophore is not well described by the sum of all rate constants (see Figure 8) as there is a high probability for a backward transfer from the acceptor molecule to the donor if the donor and the acceptor pigment are isoenergetic or of comparable energy.

If we look at Figure 8 the primary quantum yield for energy transfer from an excited state $\Phi_{ET}(R_{12}) = \Phi_{FRET}(R_{12})$ can be defined according to

$$\Phi_{FRET}(R_{12}) := \frac{k_{ET}(R_{12})}{\sum_i k_i} = \frac{k_{FRET}(R_{12})}{(\overline{\tau}_D)^{-1} + k_{FRET}(R_{12})} \tag{57}$$

focusing on Förster Resonance Energy Transfer $k_{ET} = k_{FRET}$ as dominating EET mechanism (see eq. 54). It is assumed that all other rate constants that appear in the time development of the excited state are concurring mechanisms to the EET via FRET. This is not the case if rate constants exist which populate the state (instead of depleting it, i.e. if the energy is transferred from the acceptor to the donor pigment). In such cases, eq. 57 does not describe the final EET from the donor pigment to the acceptor pigment (i.e. the average amount of energy transferred after equilibration of the system); in these cases it only describes the relative probability that the excitation energy is transferred to the neighbouring molecule directly after excitation. Eq. 57 therefore does not describe the final efficiency of Förster Resonance Energy Transfer from the donor to the acceptor after the full equilibration of the excited state, but only the initial probability for EET.

This is of high importance to understand the difference of average times for energy transfer in comparison to rate constants for energy transfer. Looking at a typical photosynthetic system as shown in Figure 11 (imagine this figure without the quencher positioned at state N_4) we find that if the energy is located at pigment N_0 at zero time and if $k_{FRET}(R_{12}) >> (\overline{\tau}_D)^{-1}$ then $\Phi_{FRET}(R_{12}) \approx 1$. But due to the strong coupling of all 6 states shown in Figure 11, 1/6 (16,7%) of the excitation energy will still relax from state N_0 after equilibration. Therefore this amount is not transferred effectively from N_0 to other pigments of the chain shown in Figure 11 if no quencher exists.

In the following it is noted that the effective energy transfer (the amount of energy that is really transported to the acceptor after equilibration of the system) cannot be calculated by eq. 57 since this equation is only denoting the probability for a single transfer step. Therefore we will suggest an alternative calculation.

$k_{FRET}(R_{12})$ as generally defined is a rate constant describing the faith of a single exciton. We want to derive the typical formulas for the quantum yield $\Phi_{ET}(R_{12}) = \Phi_{FRET}(R_{12})$ as they are found in the literature where they are sometimes incorrectly derived from eq. 57 (see Lakowicz, 2006). For this reason we will not call $\Phi_{FRET}(R_{12})$ as denoted by eq. 57 the "efficiency" of FRET (which would be an expression regarding the equilibrated system) but we will call it the "quantum yield" of FRET (which is an expression valid for a nonequilibrium situation).

Combining eq. 56 and eq. 57 delivers

$$\Phi_{FRET}(R_{12}) = \frac{k_{FRET}(R_{12})}{(\overline{\tau}_D)^{-1} + k_{FRET}(R_{12})} = \frac{\frac{1}{\overline{\tau}_D}\left(\frac{R_0}{R_{12}}\right)^6}{\frac{1}{\overline{\tau}_D} + \frac{1}{\overline{\tau}_D}\left(\frac{R_0}{R_{12}}\right)^6} = \frac{R_0^{\,6}}{R_{12}^{\,6} + R_0^{\,6}}. \tag{58}$$

For the simplified monoexponential chromophores with $(\overline{\tau}_D)^{-1} = \sum_i k_i - k_{FRET}(R_{12})$ and $(\overline{\tau}_{DA})^{-1} = (\overline{\tau}_D)^{-1} + k_{FRET}(R_{12}) = \sum_i k_i$ one obtains

$$\Phi_{FRET}(R_{12}) = \frac{k_{FRET}(R_{12})}{(\overline{\tau}_D)^{-1} + k_{FRET}(R_{12})} = \frac{\sum_i k_i - \left(\sum_i k_i - k_{FRET}(R_{12})\right)}{\sum_i k_i} = 1 - \frac{\overline{\tau}_{DA}}{\overline{\tau}_D}. \tag{59}$$

Looking at eq. 58 and eq. 59 as they are found in the literature (see e.g., Lakowicz, 2006) one can immediately see the contradiction if one interprets a system as shown in Figure 11 without the quencher positioned at the state N_4 if the energies of the electronic states of all coupled pigments are equal and the distance between the coupled pigments $R_{12} = 0$. In that case the fluorescence decay lifetime of the coupled ensemble does not change if no additional dissipative channels are occurring (see the simulation of the DAS of the described system in Figure 12, upper panel). One would get $\Phi_{FRET}(R_{12}) = 1 - \frac{\overline{\tau}_{DA}}{\overline{\tau}_D} = 0$ from eq. 59 but $\Phi_{FRET}(R_{12} = 0) = \frac{R_0^{\ 6}}{R_{12}^{\ 6} + R_0^{\ 6}} = 1$ from eq. 58. Eq. 59 can be understood as a FRET efficiency even if the donor and acceptor pigment do not decay monoexponentially if $\overline{\tau}_{DA}$ and $\overline{\tau}_D$ are calculated according to eq. 43, which is an appropriate calculation of the average decay times for multiexponentially decaying systems (see Lakowicz, 2006). These lifetimes $\overline{\tau}_D$ and $\overline{\tau}_{DA}$ cannot simply be gathered from the DAS (see Figure 12, upper panel); one would need molecular resolution to analyze the excited state lifetime of the donor pigment in presence of the acceptor $\overline{\tau}_{DA}$ as given in the lower panel of Figure 12.

Eq. 58 denotes the quantum yield for an initial FRET process but not an overall efficiency. Independently from the coupling strength, an isoenergetic ensemble of pigments is not necessarily quenched (situation without the quencher at position N_4 in Figure 11) but all pigments exhibit the same apparent fluorescence decay time $\overline{\tau}_D$. This case is found for example in the strongly coupled Chlorophyll complex of the solubilised LHCII trimer. In LHCII $\overline{\tau}_D = \overline{\tau}_{DA}$ as $\overline{\tau}_D$ of the monomeric LHC structure or even of single chlorophyll molecules is similar to $\overline{\tau}_{DA}$ of the solubilized LHC trimers (see Lambrev et al., 2011, and references therein).

Eq. 59 becomes valid if the average lifetimes of the donor and the acceptor are not measured from the ensemble but with single molecule resolution. At the moment this is not possible in experiments for structures as shown in Figure 11 because it would require an optical resolution in the order of 1 nm (less than the distance of 2 pigments). The DAS in the wavelength domain of a structure as shown in Figure 11 (without the quencher located at state N_4) exhibit a single decay constant for the whole spectral range as shown in Figure 12, upper panel. If one could perform a superresolution experiment delivering the possibility to get individual decay curves for the chromophores in the coupled structure shown in Figure 11, then it would be possible to construct spatial DAS indicating the amplitude spread of individual decay components along the space coordinate instead of the wavelength coordinate (see Figure 12, lower panel). In that case, a correct calculation of the average fluorescence lifetimes of the donor in absence or presence of the acceptor $\overline{\tau}_D$ and $\overline{\tau}_{DA}$, respectively, as given by eq. 43, would enable the principal possibility to correct formula 59 so that it is also valid for single isoenergetic molecules in a coupled chain as shown in Figure 11.

Eq. 59 combined with eq. 43 directly delivers the well known intensity formulation for $\overline{\tau}_{DA} \neq \overline{\tau}_D$:

$$\Phi_{FRET}(R_{12}) = 1 - \frac{\overline{\tau}_{DA}}{\overline{\tau}_D} = 1 - \frac{\left[\sum_{i=1}^{n} A_i(\lambda)\tau_i\right]_{DA}}{\left[\sum_{i=1}^{n} A_i(\lambda)\tau_i\right]_D} = 1 - \frac{\left[\int_0^\infty F(t)dt\right]_{DA}}{\left[\int_0^\infty F(t)dt\right]_D} = 1 - \frac{I_{DA}}{I_D}. \quad (60)$$

Eq. 60 calculates the spatially resolved DAS as shown in Figure 12, lower panel and is independent from the restrictions mentioned above if the spectra are normalised, i.e. $\left[\sum_{i=1}^{n} A_i(\lambda)\right]_D = \left[\sum_{i=1}^{n} A_i(\lambda)\right]_{DA}$. This is the case if the absorption of the donor and acceptor in the donor-acceptor pair is the same as the absorption of the isolated donor only.

Evaluating $\frac{\overline{\tau}_{DA}}{\overline{\tau}_D}$ in the spatial "pigment domain" for single chromophores (Figure 12, lower panel, with help of eq. 43) one obtains a value of about $\frac{\overline{\tau}_{DA}}{\overline{\tau}_D} \approx 0.16$ for the initially excited donor pigment N_1 in the exemplary molecule chain shown in Figure 12 if we neglect the contributions $\left[\sum_{i \neq 4ns} A_i(\lambda)\tau_i\right]_{DA} << \left[A_i(\lambda)4ns\right]_{DA}$. In fact, $\frac{\overline{\tau}_{DA}}{\overline{\tau}_D}$ becomes exactly 0.16 for $k \to \infty$ as is expected from theory (see Figure 11). The value of 0.16 is also expected for the relation $\frac{I_{DA}}{I_D}$ if one evaluates the intensity of the donor pigment N_1 in comparison to the fluorescence of all five acceptor pigments N_0 and N_2 to N_5. Therefore the real FRET efficiency after equilibration calculates to 84% as is expected from the structure of the system employing eq. 60.

With the use of eq. 43, eq. 59 (right side) and therefore also eq. 60 are correct for the FRET efficiency after equilibration of the system. In such case the efficiency should be denoted with $\eta_{FRET}(R_{12})$. Eq. 58 describes the quantum efficiency $\Phi_{FRET}(R_{12})$ of a single excited state that undergoes a single FRET transition.

The final formula is therefore suggested to denote to

$$\eta_{FRET}(R_{12}) = 1 - \frac{\overline{\tau}_{DA}}{\overline{\tau}_D} = 1 - \frac{I_{DA}}{I_D} \quad (61)$$

where it would be necessary to determine all parameters from single molecules but not to measure them on the ensemble. That means, one necessarily needs to achieve a signal with superresolution from a single molecule which acts as donor in a strongly coupled chain and calculate the $\overline{\tau}_D$ and $\overline{\tau}_{DA}$, respectively, as given by eq. 43, and

$$\Phi_{FRET}\left(R_{12}\right)=\frac{k_{FRET}\left(R_{12}\right)}{\left(\overline{\tau}_D\right)^{-1}+k_{FRET}\left(R_{12}\right)}=\frac{R_0^{\ 6}}{R_{12}^{\ 6}+R_0^{\ 6}} \tag{62}$$

for the real „quantum efficiency" describing the transfer probability of a single quantum state.

However, there are more problems than the ones already mentioned. Eq. 61 contains additional problems. For example, the fluorescence intensity and average lifetime of the donor pigment in presence of the acceptor, I_{DA} and $\overline{\tau}_{DA}$, respectively, might change due to quenching effects without an energy transfer from the donor to the acceptor pigment. In such cases one might get a kind of "donor depletion efficiency" instead of EET efficiency when one employs eq. 61.

To avoid problems that especially affect the evaluation of the donor fluorescence, it is an advantage to focus on the fluorescence rise kinetics of the acceptor pigment (Schmitt, 2011) and to compare the fluorescence dynamics of the acceptor in absence of the donor with the fluorescence dynamics in presence of the donor pigment. It was found that this method is very stable against uncertainties. Unfortunately, however, the rise kinetics at the acceptor pigment is hard to resolve (Schmitt, 2011).

One additional approximation found in eq. 54 is the fact that the calculation of $k_{FRET}\left(R_{12}\right)$ is usually performed using the fluorescence and absorption spectra of isolated pigments. But the optical properties of donor and acceptor states might be slightly changed by the coupling. Therefore it is always necessary to analyse the stability of the obtained solutions for FRET efficiencies against variations of the parameters used for the calculation and compare the experimental results with complementary methods.

2.3 Light-Harvesting, Energy and Charge Transfer and Primary Processes of Photosynthesis

The properties of chlorophyll molecules in photosynthetic complexes are tuned by binding to specific protein environments. This principle developed evolutionarily in the biosphere and has resulted in pigment-protein structures with the highest efficiency with regards to light-harvesting, transfer of electronically excited states, and the transformation of light into electrochemical free energy. Up until today it remains impossible to assemble a nanomachine of a complexity, stability and regulatory control that can compare to the ones present in a single working thylakoid membrane. A full understanding of the functional relevance of molecular interactions in photosynthetic pigment-protein-complexes (PPCs) might therefore be useful in many areas of application. Artificially designed PPCs can work as nanoscaled tunable antenna systems for photovoltaics, e.g. in the form of light-harvesting complexes with high exciton diffusion coefficients that thus rise and/or regulate the light absorption cross-section. Other applications might include switchable PPCs that undergo transitions from an inactive to an active state when illuminated or when in contact with certain environments. Such PPCs could be applicable in the photodynamic diagnostics and therapy of diseases such as skin cancer.

Photosynthesis appears to be the most important process in solar energy exploitation of the biosphere (Renger, 2008b). It can be described as the light-induced chemical reaction of water with carbon dioxide and glucose:

$$12\,H_2O + 6\,CO_2 \xrightarrow{h \cdot \nu} C_6H_{12}O_6 + 6\,O_2 + 6\,H_2O \qquad \textbf{chemical eq. 1}$$

The photosynthetical reaction summarized by chemical eq. 1 is highly endergonic. It is driven by Gibbs free energy from solar radiation on earth as it is being absorbed by green plants. The light energy is absorbed by chromophores that are bound to the photosynthetic complexes of the photosystems PS I and PS II. These chromophores are mainly chlorophyll and carotenoid molecules. In the following, we will focus on describing PS II as a large fraction of ROS, including singlet oxygen, is produced there. The redox chemistry of PS I lies beyond the scope of this book; therefore a detailed description of PS I will not be presented here. Instead, the reader is referred to the existing literature on the light-harvesting complex and RC of PS I (Byrdin, 1999, and references therein).

The electronically excited singlet states formed by light absorption are not completely transformed into Gibbs free energy of the excited Chl molecules. A fraction is emitted as red fluorescence; the dynamics of the fluorescence emission of all samples containing PS I and PS II is mainly determined by the properties (organization and coupling) of the photosynthetic pigment-protein complexes of PS II (see Schatz et al., 1988). Therefore the structure, electrochemistry and function of PS II will be described briefly.

The PS II core complexes from *Thermosynechococcus elongatus*, a thermophilic cyanobacterium, was resolved at 1.9 A° resolution (see Umena et al., 2011), which showed that the dimer PS II unit is composed of two times 19 protein subunits and two times 25 integral lipids. At room temperature each monomer contains 35 Chl *a* molecules, 2 pheophytin (Phe) molecules, 11 carotenoid molecules, 2 plastoquinones (PQ), the Mn_4CaO_5 cluster, bicarbonate, oxygen atoms, calcium atoms, chloride ions, haem and non haem iron, and at least 1,350 water molecules forming a strong hydrogen-bonding network. The PS II core complex contains the reaction center (RC) (D1/D2/cytb559), the inner antenna (CP43 and CP47), and three peripheral (PsbO, PsbP, PsbQ) proteins; this is the minimal unit capable of oxidizing water and reducing plastoquinone molecules (see Figure 17, first published by Mamedov et al., 2015).

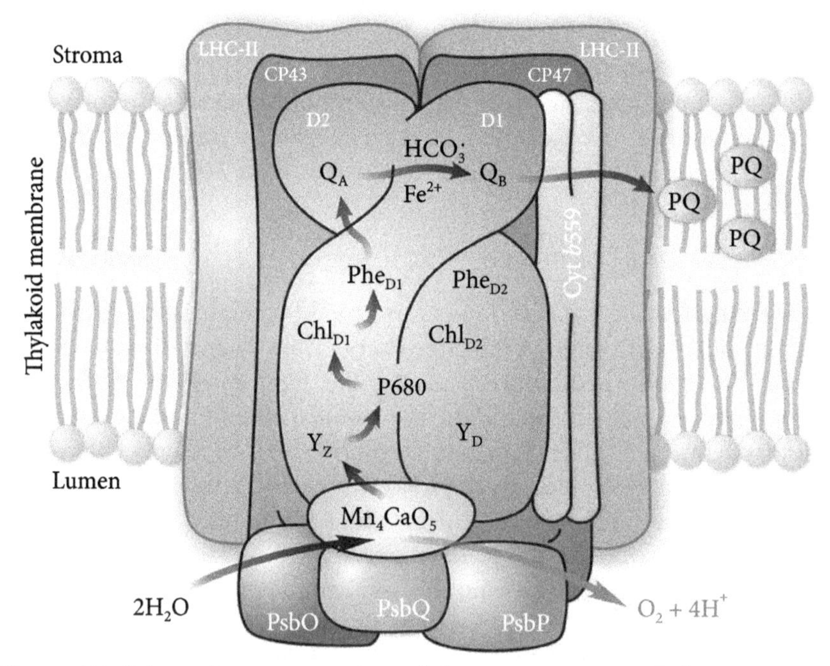

Figure 17. Schematic representation of the core proteins of photosystem II (PS II) in higher plants and green algae illustrating the pathway of the electron flow through PS II by black arrows. Mn_4CaO_5 denotes the inorganic core of the water oxidizing Mn complex; Y_Z the redox active tyrosine residue on the D1 protein and Y_D, a tyrosine, on the D2 protein. PQ the mobile plastoquinone electron carrier CP43 and CP47, the chlorophyll binding core proteins (see text), the light-harvesting complex are denoted LHC; in addition PsbO, PsbP, and PsbQ, the extrinsic proteins of PS II and the redox active cytochrome b559 (Mamedov et al., 2015). Image reproduced with permission.

In Figure 17, the electron transfer is denoted in the form of black arrows that indicate the ET steps that are typically described by rate equations. ET mostly involves the D1 protein, except for Q_A, which is on D2. Figure 17 also shows a bicarbonate (HCO_3^-) ion, which is known to be bound on a non-heme-iron that sits between Q_A and Q_B. HCO_3^- has been shown to play an essential role in electron and proton transport on the electron acceptor side of PS II. Although the light-harvesting systems of the PS II exhibit strong variations between different photosynthetic organisms (e.g. cyanobacteria and higher plants) the architecture of the core complex shown in Figure 17 is very similar among all oxygenic photosynthetic organisms. After light absorption inside the PS II, an excited single state of the chlorophyll localizes inside the RC. The molecular identities of the excited state $^1P680^*$ of the RC, from where electron transfer starts, and the state $P680^{+\bullet}$, where the hole stabilizes, are different. Together with the strongly coupled Chl dimer P_{D1}/P_{D2}, which is called a "special pair", the Chl molecules Chl_{zD1}, Chl_{zD2}, Chl_{D1} and Chl_{D2} and different site energies form an energetic "trap" that is represented by the reaction center in comparison to the excited states in the Chl antenna (for further details see Renger and Renger, 2008, and references therein).

PS II and PS I are functionally connected by the $Cytb_6f$ complex in which plastoquinol PQH_2 formed at PS II is oxidized and the electrons are transferred to PS I via plastocyanin (PC) as mobile carrier (see Figure 4). This process is coupled with proton transport from the stroma to the lumen, thus increasing the proton concentration in the lumen. At PS I, the light driven reaction leads to the reaction of $NADP^+$ to NADPH. The proton gradient provides the driving force for the ATP synthase where ATP is formed from ADP + P. NADPH is used in the dark reactions for CO_2 reduction to produce glucose inside the chloroplast stroma. The spatial separation between lumen and stroma and the oriented arrangement of PS I, PS II and $Cytb_6f$ is the essential symmetry break inside the chloroplast volume, which in its turn enables a directional electron flow that is coupled with the formation of a transmembrane electrochemical potential difference. In this way the absorbed Gibbs free energy of the photons is partially and transiently "stored".

The antenna complexes permit a very efficient adaptation of anoxygenic photosynthetic bacteria (Law and Cogdell, 2007), cyanobacteria (Mimuro et al., 2008) and plants (van Amerongen and Croce, 2008) to different and widely varying illumination conditions. At low light intensities, the few electronically excited singlet states are funnelled with high efficiency to the photochemically active pigment of the RCs, where the photochemical charge separation takes place (Parson, 2008; Setif and Leibl, 2008). An opposite effect is induced under strong illumination; i.e., the radiationless decay of the superfluous excited singlet states is stimulated by opening dissipative channels, and, in addition, (bacterio) chlorophyll triplets formed via intersystem crossing are effectively quenched by carotenoids (Cars) (Belyaeva et al., 2008, 2011, 2014, 2015).

In contrast to the reactions in the antenna, only the excited states in RCs, PS I and PS II are transformed into electrochemical Gibbs free energy via electron transfer. This gives rise to the formation of primary cation-anion radical pairs, which is followed by stabilization steps under the participation of secondary acceptor components (see Parson, 2008; Setif and Leibl, 2008, and references therein). In plants, both of these types of pigment-protein complexes (antennas and photosystems) are highly hydrophobic. Therefore they solve as integral proteins inside the lipophilic, intrinsic part of thylakoid membrane.

Figure 18 represents the protein and pigment protein complexes that form the PS II in a different view than Figure 17. The electron flow/pathways between the cofactors shown in Figure 18 are indicated as black arrows. Next to the Chl-containing core antenna, the LHC-II (Lhcb) complexes are also in contact with protein structures, shown in black in Figure 18, that surround the core complex and contain nearly no pigments. These proteins most likely stabilize the structure of the core complex and they work, like the protein complex S, as linker proteins and seem to be very important for the energy transfer processes. The compounds of the core complex of the PS II (see Figure 17) consists of the reaction center (D1, donor side, yellow and D2, acceptor side, orange in Figure 18) and the Chl-containing core proteins CP 43 (dark green in Figure 18, most probably containing 13 Chl molecules) and CP 47 (red in Figure 18, most probably containing 16 Chl molecules); these form the core antenna. As mentioned above, the core complex exhibits a common architecture among all oxygen photosynthetic organisms (and is also very similar in anoxygenic photosynthetic bacteria) and it has remained nearly unchanged throughout evolution. The main differences are found in the oligomeric macrostructure of the core complex that leads to dimeric or tetrameric supercomplexes as found in several species, e.g. the cyanobacteria *Prochloron didemni* and *A.marina* (Bibby et al., 2003; Chen et al., 2005). In marked contrast, the antenna complexes of cyanobacteria, green sulphur bacteria, purple bacteria and higher plants (just to mention a few examples) are very different in composition and/or architecture.

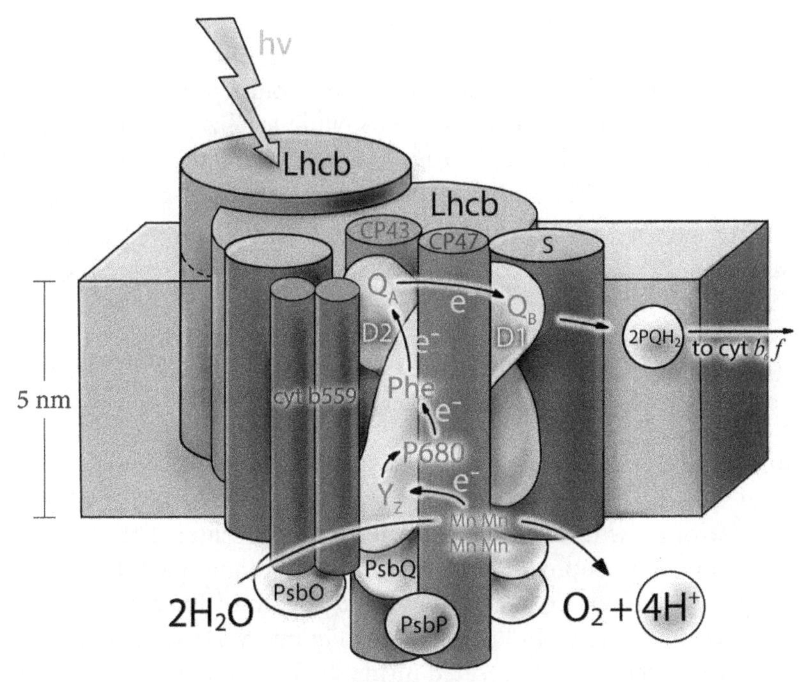

Figure 18. Subunits of the PS II of higher plants, including the LHC-II complexes, as light-harvesting systems and the core complexes (for details see text). Image reproduced with permission.

The substrate water (H_2O) interacts with the water oxidising complex (WOC) that contains four manganese ions (see Figure 18). The interaction between H_2O and the WOC elevates the energetic state of the H_2O molecule. The interaction process activates the release of two electrons per molecule H_2O to an intermediate acceptor Y_Z which fulfills the role of an electron donator for $P680^+$. P680 is the lower excitonic state of the two Chl *a* molecules forming the "special pair" P_{D1}/P_{D2} inside the RC. The oxidation of P680 is possible via interaction with pheophytin (Pheo). The formation of the first radical ion pair $P680^+Pheo^-$ is the primary charge separation. The energy of this radical pair strongly depends on the state of the environment. After formation of the primary radical pair, the environment relaxes and shifts the energy level of $P680^+Pheo^-$ to a lower value (Renger and Holzwarth, 2005).

Due to the strong interaction between $P680^+$ and $Pheo^-$ the recombination rate of $P680^+Pheo^-$ is comparatively high. The charge separation is more stable after the relaxation of the environment and the subsequent "charge stabilization" process, when $Pheo^-$ releases an electron to the plastoquinone Q_A. In Chemical equation 3 charge separation and stabilisation are indicated. Strongly coupled molecules with partial excitonical coupling and EET fast in comparison to the resolution limit of TWCSPC are indicated with a double arrow ($\Leftrightarrow$) while localized electron transitions are indicated with simple arrows for reactions that occur in both directions ($\underset{\leftarrow}{\longrightarrow}$). The index N in Chemical equation 3 indicates that a high number of Chl molecules (N = 100–400) are coupled inside and between the PS II subunits. The electronically excited complexes are marked with an asterisk (*).

$$\left((Chl)_N \Leftrightarrow P680\right)^* \, Pheo\, Q_A \underset{\leftarrow}{\longrightarrow} (Chl)_N \, P680^+ \, Pheo^- \, Q_A \longrightarrow$$
$$\longrightarrow (Chl)_N \, P680^+ \, Pheo Q_A^-$$

chemical eq. 3: most important steps of the charge separation in the reaction center after absorption of light energy in the antenna which interacts strongly with the *P680* inside the reaction center. The electron is released towards pheophytin and (*Pheo*) and plastoquinone (Q_A).

The primary donor P680 and the primary acceptor Pheo are bound to the D1 protein as shown in Figure 17 and Figure 18. D1 is the donor side of the reaction center. Q_A is located inside the D2 protein on the acceptor side. From there the electron is transferred towards the plastoquinone molecule at the Q_B site, and then further on to mobile PQ carriers (see Figure 17 and Figure 18).

2.4 Antenna Complexes in Photosynthetic Systems

The enhancement of the absorption cross section of photosystems is only one function of photosynthetic antenna complexes. They additionally have regulatory functions, such as the dissipation of excess energy in order to avoid the generation of singlet oxygen at high light conditions (see Häder, 1999 for an overview). The structural arrangement of the pigments together with the fine tuning of the Chl molecules by the environment cause a directed energy transfer from the antenna to the photochemically active reaction center (RC). The absorbed light energy is able to bridge distances of 30 nm to the RC with up to 99% quantum efficiency (Häder, 1999). In contrast to the common architecture of the RC, the antenna structures are found to vary in different photosynthetic organisms. The exact pigment compositions of the antennae are not completely determined for a certain species but depend on growth conditions, light intensity and light quality during growth. The latter effect is known as chromatic adaption. Therefore the following subsections present a short outline of some selected antenna systems of anoxygenic bacteria, oxygenic bacteria and higher plants.

It is interesting to consider why the structure of the RC is similar for most photosynthetic organisms while the structures of the antenna complexes vary. There might be several reasons for this discrepancy. It seems most probable that the variations in light spectra and intensity that the antenna complexes are exposed to are the main reason for the difference between structures found in the antennae. The antenna is the protein structure that mainly interacts with sunlight. The RC itself is more accurately understood as the acceptor of excited states of Chl molecules. While all plants that use Chl as light-harvesting pigment in close proximity to the RC are facing the same task to drive water splitting with the free energy of the excited Chl, the task of light-harvesting in different environments is a less clearly defined task and one that is carried out under varying conditions. It seems reasonable that an evolutionary development of the optimized structures forms different structures under different conditions while the same conditions lead to the same structure.

Anoxygenic photosynthetic bacteria contain bacteriochlorophyll (BChl) in the antenna complexes and reaction centers. BChl *a* is found in e.g. the light-harvesting antennae (LH1, LH2) of purple bacteria as presented in chap. 2.4.2 and the Fenna-Matthews-Olson (FMO) complex of green sulphur bacteria (chap. 2.4.3). In contrast to anoxygenic bacteria the oxygenic photosynthetic organisms contain two separated photosystems (PS I and PS II). Both PS have own reaction centers and antenna complexes. The major membrane-extrinsic, light-harvesting antenna of cyanobacteria is phycocyanobilin, which contains protein structures that are mostly associated with PS II but also found to undergo so-called "state transitions" between PS II and PS I. The PBPs organize in different forms: as huge phycobilisomes in most cyanobacteria (chap. 2.4.4), or as minor rod shaped PBP antennae in the cyanobacterium *A.marina* (chap. 2.4.5). Cyanobacteria contain several additional PCB complexes (recently also named "chlorophyll binding proteins" CBP) surrounding the reaction center and core proteins (see chap. 2.3 and Bibby et al., 2003; Chen et al., 2005). Higher plants contain the trimeric major light-harvesting complex II (LHCII), which is located inside the thylakoid membrane (see Figure 18 and chap. 2.4.1) and present in ratios of four or even more trimers per PS II core dimer (Häder, 1999; Lambrev et al., 2011). The LHCII is one of the most prevalent proteins found on earth.

2.4.1 The Light-Harvesting Complex of PS II (LHCII) of Higher Plants

An array of several antenna units is functionally connected with one reaction center complex in the photosynthetic membrane, so that energy transfer between the antenna complexes and from the antenna to the RC occurs. Four or even more (six in Figure 18) trimeric LHCII structures are connected with one reaction center. It is outlined in (Lambrev et al., 2011) that the PSII-LHCII super-complexes can be arranged in large ordered domains in the granal thylakoids. Ordered LHCII domains with a diameter of hundreds of nanometers have been observed (Garab and Mustárdy, 1999). These physical domains contain thousands of pigments, however, it is not clear whether and to what extent the excited singlet states can migrate in these large LHC domains during their lifetime. The size of the functional domain, i.e. the number of pigments connected via EET, depends on the excited state lifetime and on the time of transfer between the pigments. The EET inside the LHCII is found to take place on the fs-ps time scale (Häder, 1999). This dynamic strongly depends on the structural arrangement of the pigments, which is stabilized by the protein matrix that gives rise to a well-defined coupling of the embedded pigments. Structural arrangement and EET are strongly correlated. Even quantum entanglement might influence the ultrafast EET in LHCII (Calhoun et al., 2009).

LHC trimers tend to aggregate strongly depending on the suspension medium and the preparation protocol. Figure 19 shows that huge LHC aggregates can be formed in diluted media. The size of the homogeneous LHCII domains in Figure 19 reaches values of up to 2 μm. The extent of EET between photosynthetic units, termed connectivity, leads to fluorescence rise kinetics of the prompt fluorescence after excitation with short laser pulses. This connectivity has also been proposed to explain the sigmoidal shape of the fluorescence induction transient. Different models have been proposed, for instance the so called "lake model" in which the reaction centres are embedded in a network of interconnected antennae (Garab and Mustárdy, 1999).

According to the concept of connected photosynthetic units, the functional domain size of the LHC antenna is much larger than the number of pigments in one pigment-protein complex. Several attempts have been made to estimate the domain size in thylakoids and isolated LHCII preparations. A common approach is based on studies of the excitation annihilation that results from the interaction between two excited (singlet or triplet) molecules which leads to the dissipation of one excitation

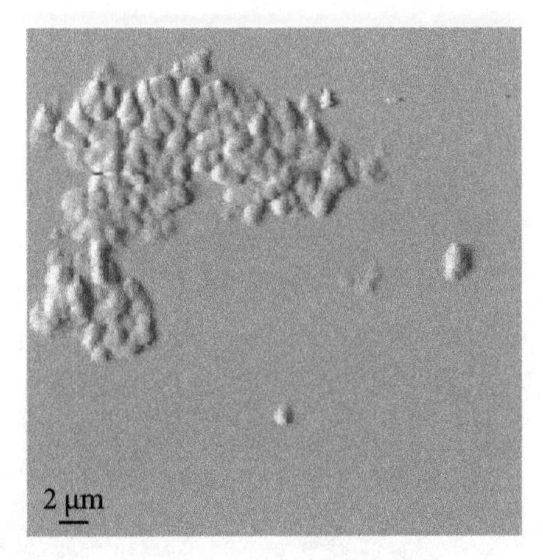

Figure 19. AFM picture of aggregated LHCII trimers prepared by detergent removal. The LHCII samples at original concentration of 10 μg Chl/ml were immobilized and dried on glass plates precoated with poly-L-lysine. The AFM images were taken in close-contact mode. For further information see (Lambrev et al., 2011). Image reproduced with permission.

according to the qualitative equation:

$$S_1 + S_1 \rightarrow S_n + S_0 \xrightarrow{\text{int.}conversion} S_1 + S_0$$
$$S_1 + T_1 \rightarrow T_n + S_0 \xrightarrow{\text{int.}conversion} T_1 + S_0$$

chemical equation 4

where S_i and T_i denote the ith excited singlet and triplet state, respectively. The coexistence of two or more excited electronic states in one domain therefore leads to quenching and a reduction of the average excited-state lifetime. The different dynamics of biexcitons and higher exciton states in comparison to single excitons in coupled quantum configurations primarily contribute to the decrease of the fluorescence quantum yield of LHCII complexes at high excitation intensities. However, the main physical reason for this effect is not singlet-singlet annihilation but Pauli blocking of radiative relaxation channels as described in (Richter et al., 2008).

The molecular structure of the LHCII trimer with a diameter of about 10 nm is shown in Figure 20 (Standfuss et al., 2005). Each monomeric unit contains six Chl *b* and eight Chl *a* molecules. In addition four carotenoid molecules are bound per monomeric subunit, i.e. two luteins, one neoxanthin and one violaxanthin (Liu et al., 2004). The carotenoid molecules have a special regulatory function. Violaxanthin exhibits a slightly higher energetically excited state than the first excited singlet state in Chl *a* (^{1}Chl a^*), while the first excited state in zeaxanthin, which has a larger delocalised π -electron system than violaxanthin, is lower

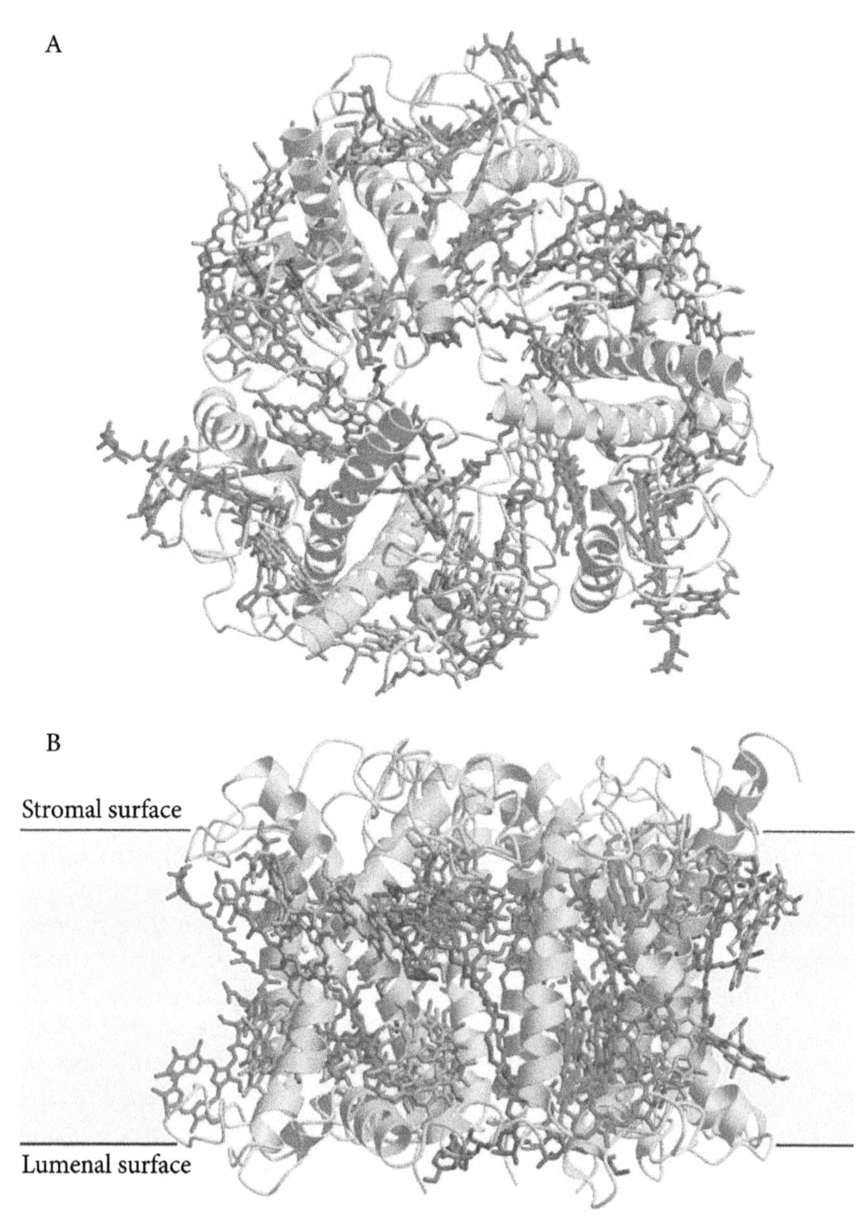

Figure 20. Molecular structure of the trimeric LHC complex according to (Standfuss et al., 2005). Protein structures are shown in grey. The carotenoids are coloured in pink and orange. Chl *b* is hold in light blue and Chl *a* is pictured in green. A) shows the front view of the complex and B) presents the side view of the complex embedded in the thylakoid membrane (see also Figure 18). Image reproduced with permission.

than ^{1}Chl *a**. Violaxanthin can absorb light energy and transfer it to Chl molecules via the mechanism of Dexter transfer (Häder, 1999), whereas the energy flows from Chl to zeaxanthin. This process in the plant LHCII is regulated by the so-called "Xanthophyll cycle" that deepoxilates violaxanthin to zeaxanthin at high light conditions (Häder, 1999) by violaxanthin-deepoxidase while this process is reversed by the zeaxanthin-epoxidase (Jahns et al., 2001; Morosinotto et al., 2002).

Solubilised LHC trimers in buffer containing Beta-DM were measured with a commercial absorption spectrometer (lambda 19). Figure 21 shows the absorption spectrum of solubilized LHCII; it exhibits bands of Chl *a* (Soret band (Bx) with peak at 437 nm, Qy with peak at 675 nm) and Chl *b* (soret band (Bx) with peak at 474 nm, Qy band with peak at 650 nm). Minor absorption around 400 nm and 600 nm is respectively assigned to the By and Qx transitions of the chlorophylls. The strong absorption between 480 nm and 500 nm originates from carotenoid molecules. Below 400 nm, the absorption is dominated by the contribution of the proteins with the typical maximum of tryptophan that peaks at 267 nm. Below 250 nm, the steep increase in the absorption results from the sum of all organic molecules in the complex.

The structure of the Chl molecules is shown in Figure 22. It is characterized by the closed porphyrin ring system complexing the Mg^{2+} ion with a binding motif of 50% covalent and 50% coordinative binding. The phytol chain is bound to ring IV as indicated in Figure 22. The difference between Chl *a*, Chl *b* and Chl *d* is given by different side groups found at the binding sites of R_1 and R_2 at ring I and II, respectively (see inset in Figure 22).

Chl *a* and Chl *b* are found in the LHCII of higher plants, while PS II complexes and PS I contain only Chl *a* molecules.

2.4.2 The LH1 and LH2 of Purple Bacteria

This short outline of the LH1 and LH2 of purple bacteria is meant to give an impression of a typical complex where the energy pathway is mainly formed by pigment-pigment interaction, i.e. excitonic coupling and FRET. In the LH complexes of purple bacteria, the excitonic coupling of pigments results in fast equilibration of electronic excitation along ring-shaped structures in the organized pigments. The absorption of these pigment-rings exhibits maxima at 800 nm and 850 nm and therefore they are denoted as B800 and B850. Figure 23 shows the pigment arrangement. For the nomenclature see also Figure 26. The bacteriochlorophyll

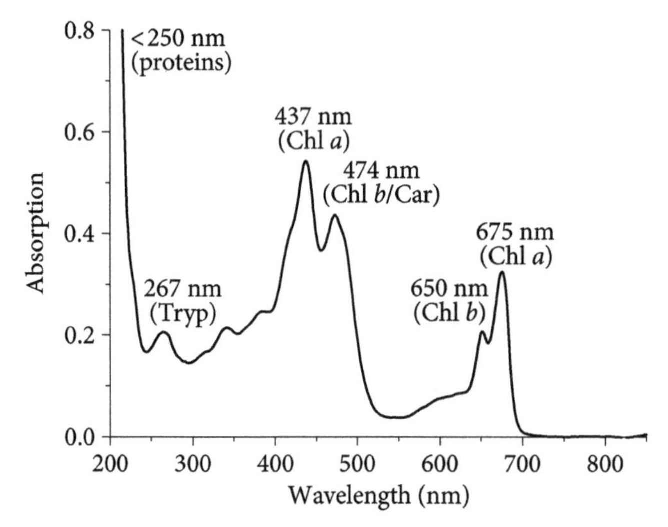

Figure 21. Absorption spectrum of LHC trimers in micelles formed in buffer with Beta-DM. The main absorption bands are directly denoted in the graph and assigned to Chl *a*, Chl *b* or carotenoids and proteins with the absorption band of tryptophan (Tryp) at 267 nm.

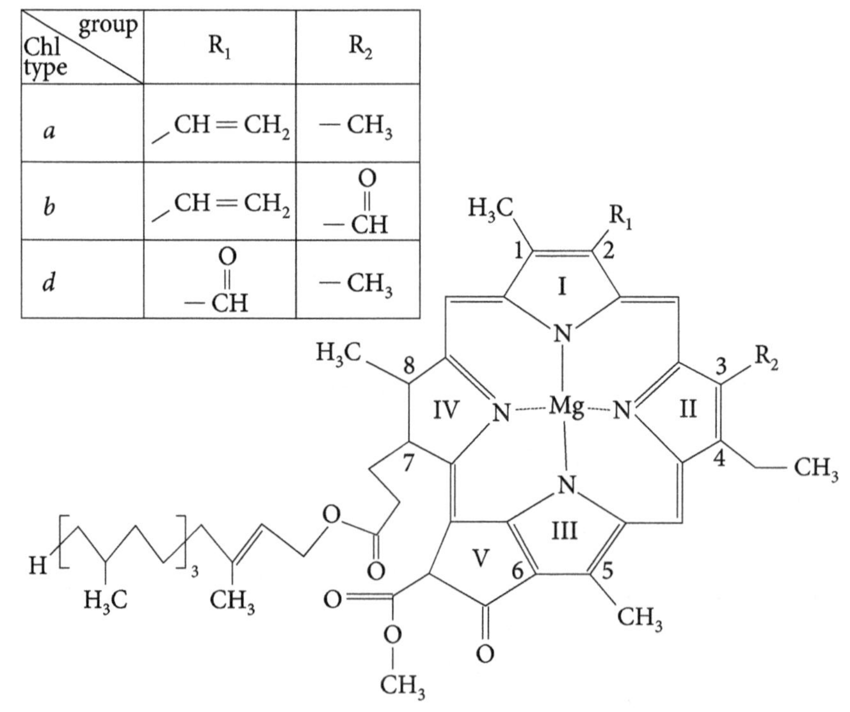

Chl type \ group	R_1	R_2
a	$CH{=}CH_2$	$-CH_3$
b	$CH{=}CH_2$	$-CH{=}O$
d	$-CH{=}O$	$-CH_3$

Figure 22. Molecular structure of the Chls: Chl *a*, Chl *b* and Chl *d* (see inset).

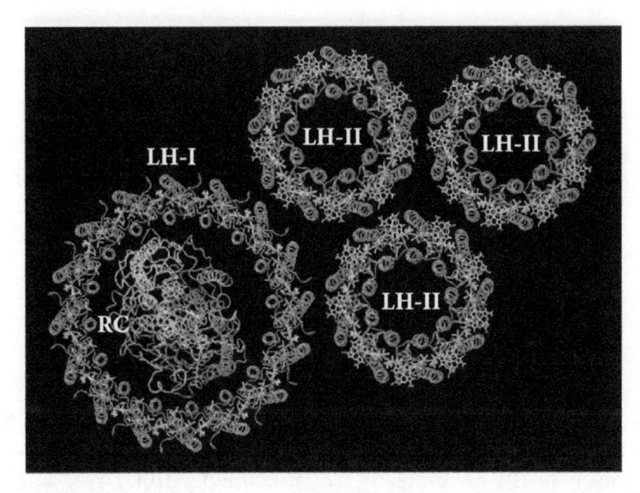

Figure 23. Ring shaped BChl arrangement of the LH complexes found in purple bacteria according to (Schulten, 1999). The LH2 complexes contain an inner ring structure, which is slightly smaller, and an outer ring structure, which is slightly bigger. Similar organisation is found in the bigger LH1 structure surrounding the RC. For more details see text and Figure 24. Image reproduced with permission.

(BChl) molecules B800 in the bigger ring structure of LH2 are less densely packed than those of the smaller structure B850. On account of the shorter distance between neighbouring dipole moments and the steric orientation of the BChl molecules, the intermolecular interaction is much stronger in B850 than in B800. This leads to a stronger coupling and larger excitonic splitting in B850 in comparison with B800. The lowest excitonic state in B850 exhibits a markedly red-shifted absorption maximum at 850 nm (Ketelaars et al., 2001; Ritz et al., 2001). Therefore electronic excitation energy migrates from the outer to the inner ring according to FRET (Freer et al., 1996).

Figure 23 shows the BChl molecules in green and cyan. The proteins are shown in blue and purple (LH1 and LH2) and red and yellow for the L and M subunits of the RC. The BChl organisation in the LH2 is shown in Figure 24. One can see that the BChl molecules in the B850 are more densely packed and orientated in a tilted "sandwich configuration" while the distance of the BChls in B800 is larger and all molecules are lying in the *x*-*y*-plane. AFM pictures of the ring shaped LH2 structures of *Rubrivivax gelatinosus* are shown in Figure 25 (Scheuring et al., 2001). The LH2 shows a typical 9-fold rotational symmetry.

The structural arrangement of the BChls in LH2 (see Figure 24) has strong impact on the pathway of energy transfer. Excitons formed in the

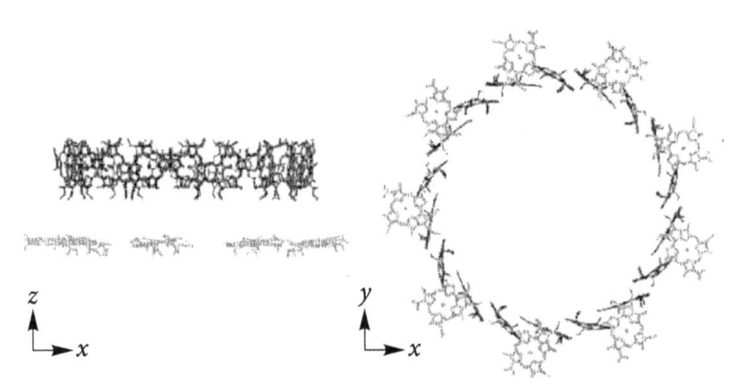

Figure 24. Organisation of the 27 BChl molecules found in the LH2 antenna as published in (Ketelaars et al., 2001). The left side (view in *x-z* plane) shows the organisation of the 18 molecules of the smaller inner ring (B850) placed in *z*-direction above the 9 BChls of the bigger outer ring (B800) while the right side shows the view from the top (as in Figure 23) along *z*-direction on the *x-y*-plane. Image reproduced with permission.

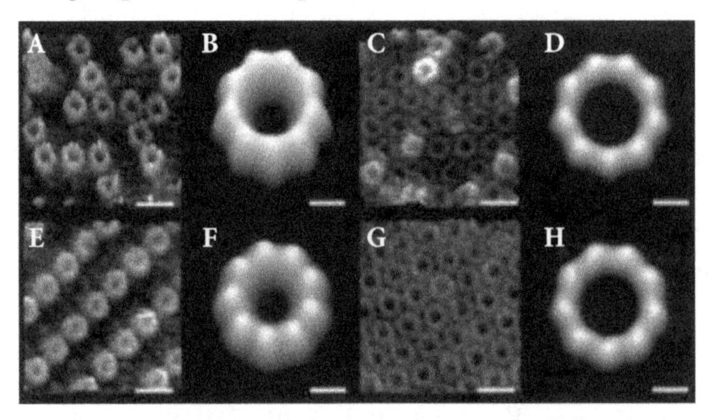

Figure 25. AFM pictures of the LH2 of *Rubrivivax gelatinosus* as published in (Scheuring et al., 2001) (The pictures A, C, E and G show the original AFM data while B, D, F and H represent averaged pictures of A, C, E and G, respectively). Scale bars represent 10 nm in the raw data and 2 nm in the averages. Image reproduced with permission.

B800 are efficiently transferred to the B850 for energetic reasons. Detailed quantum chemical calculations performed by Renger and his coworkers revealed that the excitons equilibrate with a time constant of 500 fs along the B800 BChl while the equilibration in the B850 structure occurs with a faster time constant of 100 fs (Renger and Kühn, 2007). The energy transfer from the B800 to B850 is nearly irreversible and takes 1.2 ps. With a time constant of 3–5 ps the excited states localize in the LH1 and, after 35 ps, in the RC (see Figure 26).

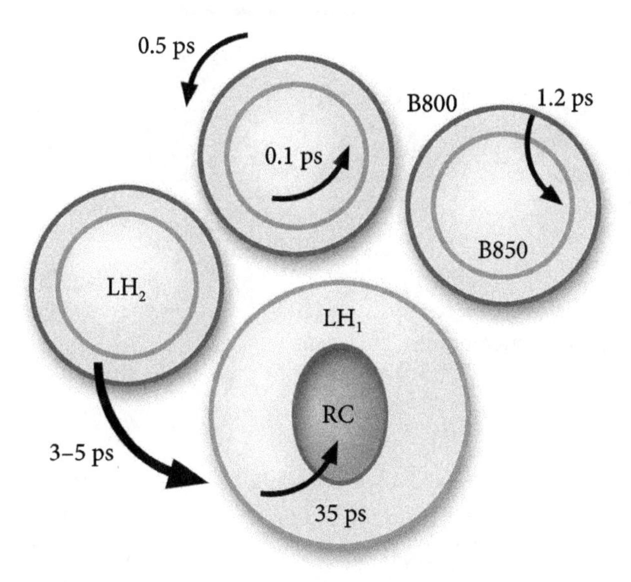

Figure 26. Time constants for the excited state transfer intra and inter the LH2 and LH1 complexes according to (Renger and Kühn, 2007). Image reproduced with permission.

2.4.3 The Fenna–Matthews–Olson (FMO) Complex of Green Sulfur Bacteria

The FMO complex of green sulphur bacteria is another fascinating complex; here the fine-tuning of pigments by surrounding protein plays a more important role for the EET than in B800 and B850. The different site energies of the excitonically coupled BChl in the FMO lead to a stable pathway from the outer antenna complex to the RC. The FMO complex as shown in Figure 27 is described in more detail in e.g. (Adolphs and Renger, 2006; Müh et al, 2007; Wen et al., 2011).

The local protein surrounding of the BChl molecules in the FMO complex (see Figure 27C) modulates their energetic states. In Figure 27C only seven BChl molecules are shown while findings gathered from mass spectroscopic analysis unraveled the existence of an eight BChl (Tronrud et al., 2009; Wen et al., 2011). Fine-tuning the electronic states of the BChls leads to a localization of the lowest energetic states near the chlorophylls number 3 and 4, which are in close contact to the RC (see Figure 27C). Excited electronic states populated at pigments 1 and 6 are rapidly transferred within 5 ps to the pigments number 3 and 4 (Adolphs and Renger, 2006). The pathway of this EET is determined by pigment-protein coupling rather than pigment-pigment coupling. The directed

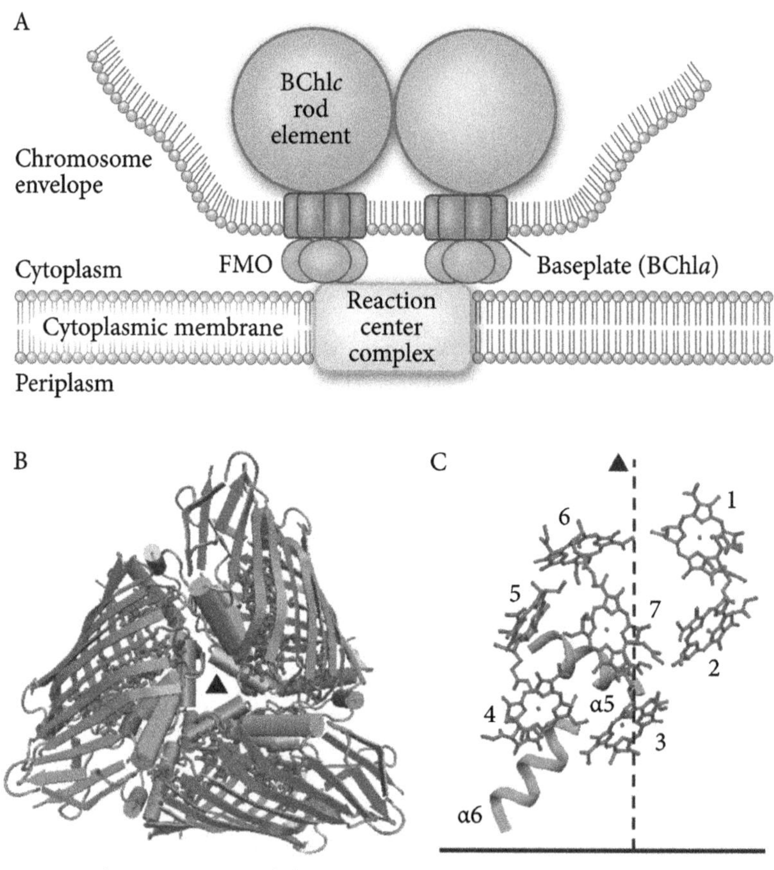

Figure 27. Schematic view of the FMO complex found in green sulfur bacteria (A). The organisation of the trimeric protein structure (green, blue) containing eight BChl molecules (red) per subunit is shown in panel (B). In (Müh et al., 2007) it was assumed that the FMO complex contains only seven BChl per subunit. New findings disclosed the existence of an eighth BChl [Tronrud et al., 2009, Wen et al., 2011]. The detailed structural arrangement of the chlorophylls is presented in panel (C). This figure was published in (Müh et al., 2007). Image reproduced with permission.

EET that is guided along a certain pathway is necessarily correlated with the existence of a pigment array within an asymmetric protein structure.

2.4.4 Phycobilisomes in Cyanobacteria

The antenna complexes of photosynthetic organisms enhance and regulate the effective absorption cross-section by connecting a huge number of pigments to the RC and by regulating the connectivity and excited state lifetime. The modulation by protein environment and the coupling of pigments determines the probability distribution of EET pathways. Additionally, the high number of nondegenerated states of all pigments in these antenna complexes enables light absorption in a wider spectral range than single pigment molecules. In photosynthetic organisms different molecules are bound to peripheral antenna complexes which close the absorption gap of the chlorophyll molecules in the green spectral range and therefore raise the efficiency of the antenna. Such pigments are carotenoids in LHCII (see chap. 2.4.1) and phycobiliprotein (PBP) complexes containing phycoerythrin (PE), phycocyanin (PC) and allophycocyanin (APC) in cyanobacteria. The pigments are open porphyrin ring systems exhibiting strong absorption in the green and yellow spectral range, thus being responsible for the blue color of cyanobacteria.

Figure 28, left side, shows the structure of C-PC of typical cyanobacteria like *Synechocystis* and *Synechococcus*. Each monomeric protein subunit contains three phycocyanobilin (PCB) chromophores (9 pigments in one trimer). In Figure 28, right side, the APC trimer is shown. This trimeric protein binds 6 pigments (two chromophores per monomer).

The PC monomers consist of the α-subunit containing the β-84 chromophore and the β-subunit containing two chromophores, the β-84 PCB and the α-155 PCB (see Figure 28, left side). The APC monomers contain one PCB molecule in the α-subunit (α-80) and one pigment in the β-subunit (β-81) (see Figure 28, right side).

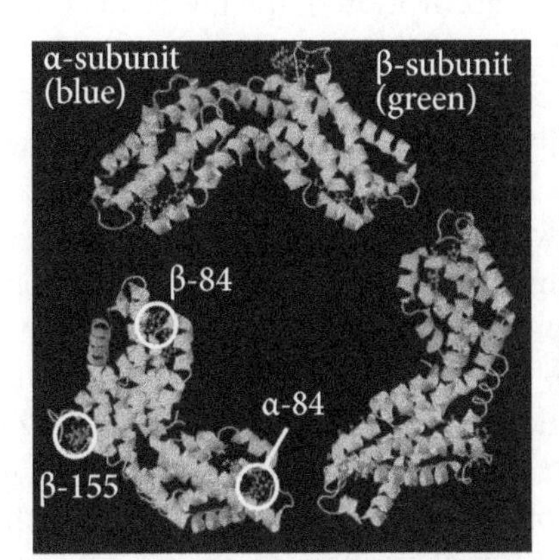

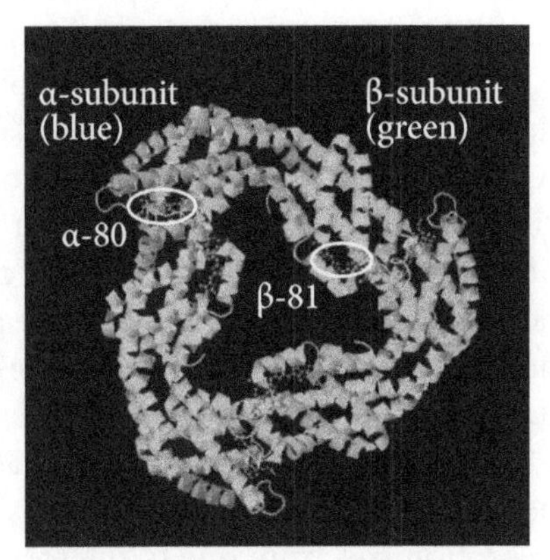

Figure 28. Three monomeric phycocyanobilin containing PCs (left side) and the tetrameric structure of APC (right side). The PC structure was generated with a protein database using the data published in (Nield et al., 2003) and graphically improved with Corel Draw®. The APC structure was generated from the data published in (Murray et al., 2007) improved with Corel Draw®.

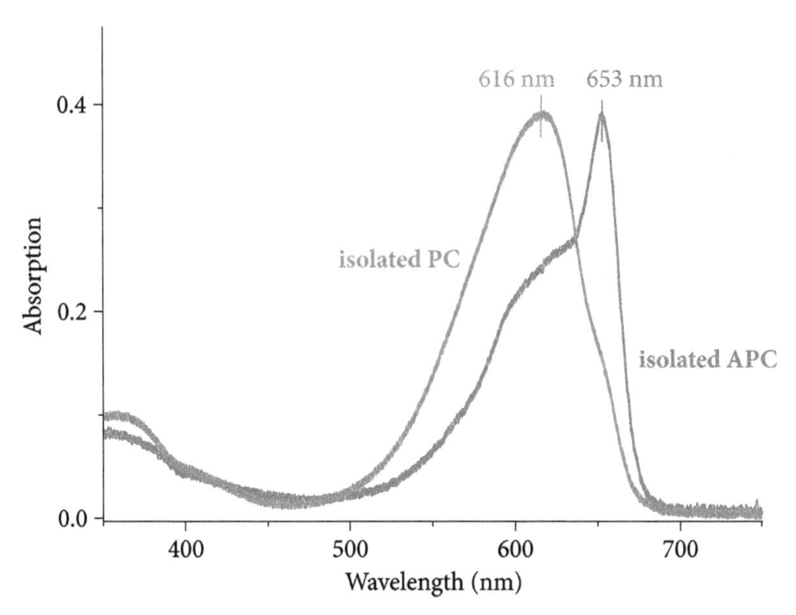

Figure 29. Absorption spectra of isolated PC trimers (green line) and isolated APC trimers (red line) in buffered aqueous solution (data redrawn from Theiss et al., 2011). Image reproduced with permission.

The absorption spectra of PC and APC trimers exhibit characteristic bands that result from strong excitonic coupling of the pigments. The small shoulder in the PC absorption spectrum (green curve in Figure 29) at 650 nm most probably represents the absorption of the lowest excitonic state. This state is strongly visible in the APC spectrum shown in Figure 29 (red curve) at 653 nm. In APC trimers (see Figure 28, right side) there exists a well defined excitonically coupled dimer containing the two PCB chromophores α-80 and β-81 of different monomers while in the PC trimers all three pigments are rather weakly excitonically coupled between different monomers with the α-84 chromophore in the α-subunit of one monomer and the β-84 PC and the β-155 PC in the β-subunit of another monomer (see Figure 30, right side). Sauer and Scheer (1988) calculated that the strongest coupling with a value of 56 cm^{-1} exists between α-84 and β-84 of different subunits. The exact structural constitution of the pigments in the PC and APC trimers is of high relevance for the shape of the absorption spectrum due to the formation of strongly absorbing excitonic states. The coupling makes is impossible to extract the optical properties of an isolated pigment from the absorption spectrum of the pigment-protein complex (e.g. a PC monomer). This is shown in (Debreczeny et al., 1993). To analyse the properties of individual compounds, the sample has to be decomposed into the individual constituents. This leads to a loss of information on the coupling of the pigments, however, since this coupling is an essential physical property determining the functionality of the structure, it is not possible to analyse the function of the whole complex by studying the decomposed compartments. The functionality is directly connected with the whole structure. This is in some way related to the fact that a whole structure cannot be explained by its (isolated) parts, as was mentioned at the outset of this book (Heisenberg, 1986).

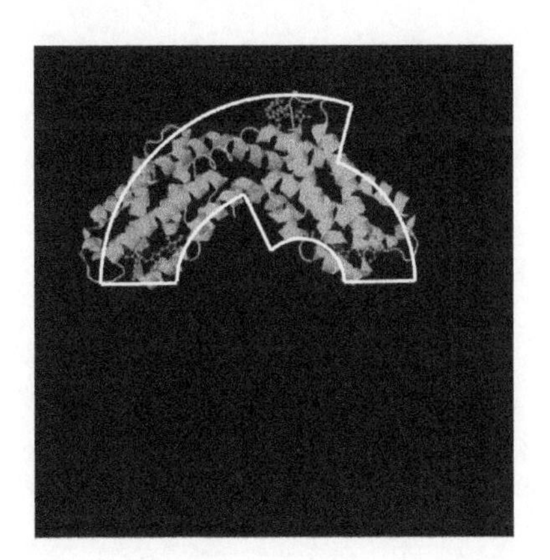

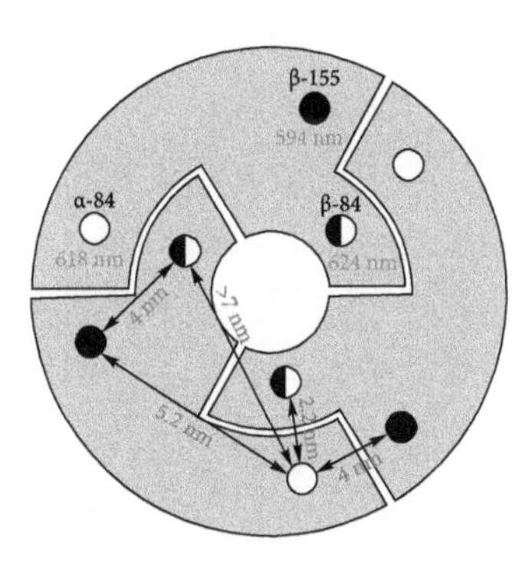

Figure 30. Structure of monomeric PC (left side) generated with protein data and graphically improved with Corel Draw®. The protein monomer consists of two subunits: The β-subunit shown in green binding the two chromophores β-84 and β-155 and the α-subunit (shown in blue) which binds only one chromophore, the α-84 PCB. The absorption maxima and distances between the different PCB molecules in the trimeric PC calculated by (Sauer and Scheer, 1988) and (Suter and Holzwarth, 1987; Holzwarth, 1991) are shown on the right side. Image reproduced with permission.

The effect of excitonic coupling of PCB molecules in different protein subunits can be analysed if the absorption of PC monomers is compared with PC trimers as shown by Debreczeny et al. (1995a, 1995b). An accurate comparison revealed that the small shoulder at 650 nm in the spectrum of isolated PC trimers (see Figure 29) is the result of a coupling between α-84 and β-84 in different monomers. With high resolved fs-absorption spectroscopy Sharkov et al. showed that, in APC trimers, excitonic relaxation in the coupled α-80/ β-81 dimer between different monomers occurs with 440 fs (Sharkov et al., 1992, 1996). Much slower excitation energy transfer processes, with time constants of 140 ps, were observed between the different chromophores in one monomeric subunit. The coupling strength of the different chromophores in PC monomers and trimers was analysed by (Sauer and Scheer, 1988). The absorption wavelengths and calculated distances of the PC molecules inside the trimeric PCB containing protein disc are shown in Figure 30.

The strongest interaction between pigment molecule and protein occurs via a sulfur bridge between the pyrrole ring of the PCB pigment α-84 and a cysteine group of the protein as shown in Figure 31. Hydrogen bonds help to stabilize the structure and in-plane geometry of the chromophore molecules and the protein.

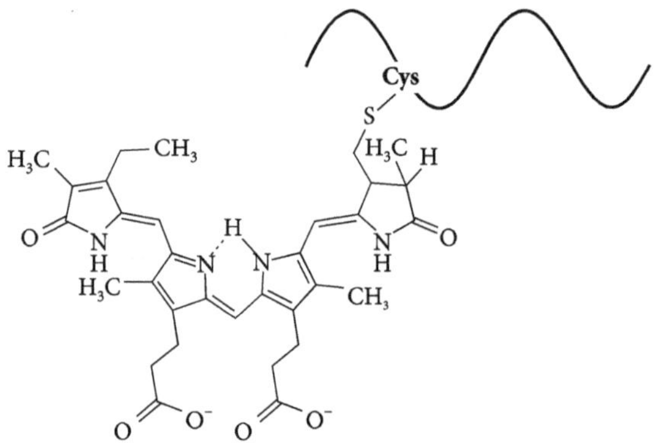

Figure 31. PCB chromophore α-84 bound to the protein matrix via a sulfur bridge between PCB and the cysteine of the protein. (Figure 31 was generated with ChemSketch V®).

In addition to the sulfur bridge shown in Figure 31, there are several strong hydrophobic interactions between pigments and protein structure. Such attractive van-der-Waals forces decay with the sixth power of the distance between pigment and protein. Therefore only the strongest interactions determine the structure.

Figure 32 shows a view on the β-84 PCB chromophore found in the monomeric PC structure that interacts with leucine along the shortest hydrophobic interaction. From this illustration one can get an idea of the flexibility of the protein matrix comparable to an ensemble of coupled springs. This spring like structural arrangement results in a broad distribution of oscillation modes that efficiently dissipates electronic excess energy in the pigments and modulates the electronic properties of the pigments bound to the protein quasistatically and dynamically. The quasistatic modulation leads to a broad Gaussian distribution of chromophore site energies while the dynamic modulation leads to a homogeneous broadening of the chromophore ensemble and even is assumed to be involved in dynamic excited state localisation effects that supports the EET.

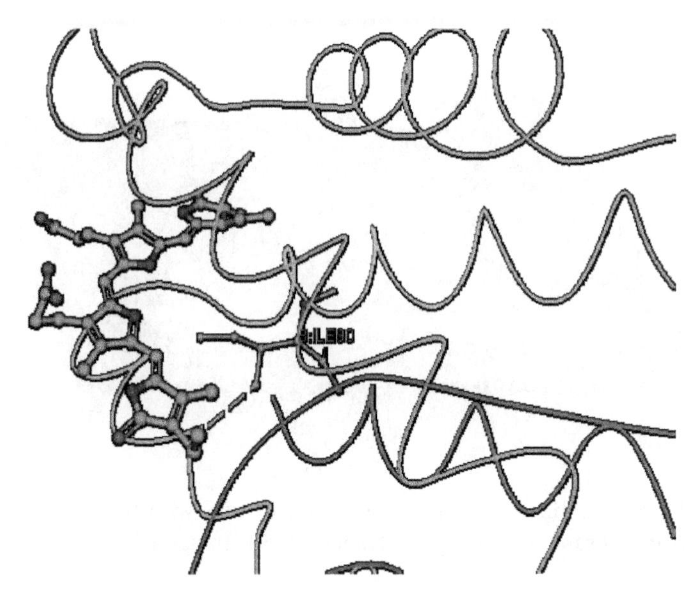

Figure 32. β-84 chromophore in PC bound to the protein β-subunit. The shortest hydrophobic interaction (3.2 °A) between the chromophore and the protein is shown as green dashed line. The figure was generated with protein data base according to the data published in (Nield et al., 2003) and graphically improved with Corel Draw®.

The PC and APC trimers (see Figure 28) tend to aggregate further in the so-called "face-to-face" dimerization which forms hexameric structures (for further detail see e.g. Theiss, 2006; Holzwarth, 1991; Glazer, 1985). These hexamers undergo further aggregation to form so-called PBP antenna rods, which in their turn aggregate to huge (up to 50 nm in diameter) phycobilisomes (PBS) as shown in Figure 33 and Figure 37 (left side). In Figure 37, the APC-containing hexameric structures are present only in the form of single discs, but they are also known to form rod-shaped staples of hexameric discs which "lie" on the outer thylakoid membrane. Some cyanobacterial PBS additionally contain phycoerythrin (PE) as well as PC and APC hexamers in the PBS.

The properties of PBS are well understood and therefore PBS is quite a good system for reference measurements of energy transfer. In the study presented here measurements of PBS from the cyanobacterium *Synechocystis* PCC 6803, containing PC and APC as schematically shown in Figure 37, were performed. It was found that energy migrates in about 200 ps from PC to Chl *a* in *Synechocystis* which is a typical value for PBS of common cyanobacteria in agreement with (Holzwarth, 1991; Trissl, 2003; Glazer, 1985).

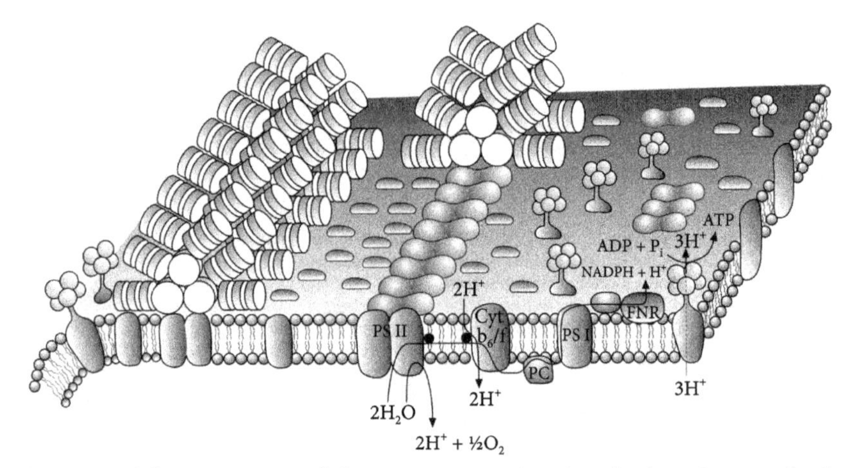

Figure 33. Schematic view of the association of PBS with the PS II inside the thylakoid membrane according to (Häder, 1999). Image reproduced with permission.

It is assumed that the PBS are mainly associated with the PS II core complex of the thylakoid membrane (see Figure 33). Most cyanobacteria contain high amounts of PBS; this makes the PBS chromophores the most important light-harvesting pigments found in cyanobacteria. PBS undergo state transitions from the state indicated in Figure 33, where the PBS are associated with the PS II, and a state where they are associated with PS I. Other studies show that similar behaviour is also approved for the LHCII in higher plants (Minagawa, 2010). Figure 33 indicates the electron pathways after light absorption. Details for the electron transfer chain are found in chap. 1.4.

2.4.5 Antenna Structures and Core Complexes of *A.marina*

As an example of an exotic cyanobacterium that has been analysed and modelled in detail with special attention given to EET processes in its antenna system and subsequent ET we use the cyanobacterium *A.marina*. The *A.marina* is an exception because it contains mainly Chl *d* in the membrane intrinsic antenna and RC complexes. The Qy band of Chl *d* (Qy,(0,0)) is red-shifted to about 699 nm in ethanol solution, as shown in Figure 36, compared to the absorption maximum of Chl *a* at 675 nm (see Figure 21). The absorption maximum of the Chl *d* antenna in *A.marina* was found red-shifted by another 20 nm to an absorption maximum at 719 nm for the Qy(0,0) transition according to our measurements (see Figure 34) while according to (Miyashita et al., 1997) the Chl *d* absorption maximum appears at 714 nm in *A.marina*. Former measurements showed a peak at 717 nm for the Qy band of Chl *d* in living cells of *A.marina* (Schmitt, 2011). This red shift most probably influenced by the surrounding protein matrix.

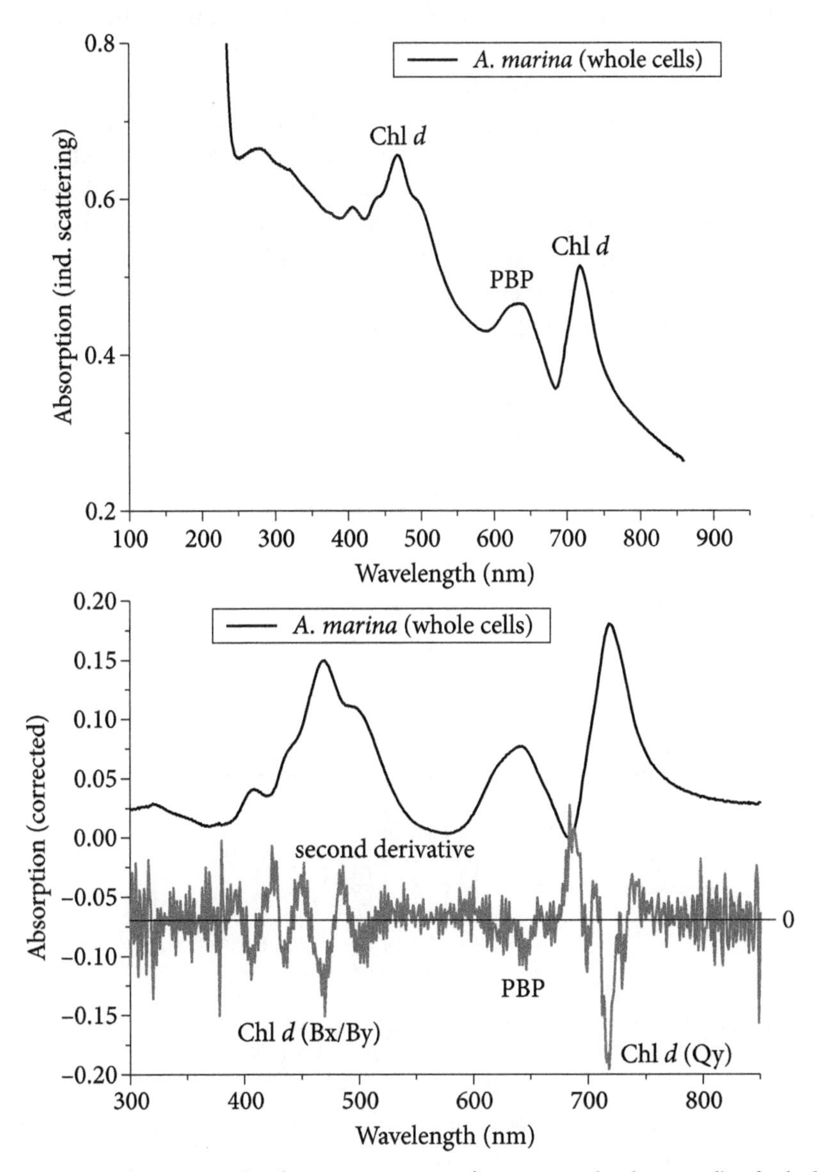

Figure 34. Absorption (without correction of scattering background) of whole cells of *A.marina* (upper panel) and corrected spectrum with second derivative (lower panel).

The spectrum shown in Figure 34, upper panel, is not corrected for scattered light. The whole cells with a diameter of about 1 μm exhibit strong scattering; this leads to an increase of the background level at shorter wavelengths. Figure 34 also shows the absorption of the PBP antenna in *A.marina* between 610 nm and 650 nm (see chap. 2.4.4.) and the Bx and By absorption band of Chl *d* which exhibits highest absorption at 468 nm in living cells (see Figure 34 in comparison to Figure 36 for the calculated Chl *d* spectrum).

Figure 35 shows the energy levels and selected electronic transitions, including vibrational states of Chl *d*, in 40:1 methanol: acetonitrile at 170 K according to (Nieuwenburg, 2003). In (Nieuwenburg, 2003) the dipole strengths and spectral linewidths of the different transitions are published as shown in Figure 35. This information was used to plot the relative contributions of these transitions to the absorption spectrum (see Figure 36, for further details see Schmitt, 2011).

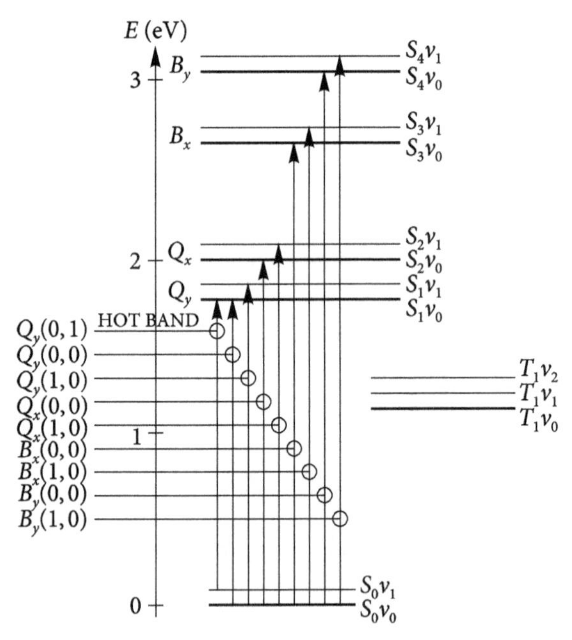

Figure 35. Energetic states and characteristic transitions of Chl *d* in 40:1 methanol: acetonitrile at 170 K according to (Nieuwenburg, 2003).

The broad absorption spectrum of pigments caused by intramolecular vibrational states is of importance for achieving an efficient solar energy exploitation and fulfilling the resonance condition between the fluorescence spectrum of donor pigments and the absorption of acceptor pigments. The latter is necessary for an efficient energy transfer along the antenna pigments according to the FRET-mechanism.

The thylakoid membrane in *A.marina* is ring-shaped and contains densely packed PBP antenna rods. *A.marina* has a uniquely composed light-harvesting system, as schematically shown in Figure 37 (right side). The PBP antenna has a simpler rod shaped structure than that of PBS in typical cyanobacteria (Schmitt, 2011).

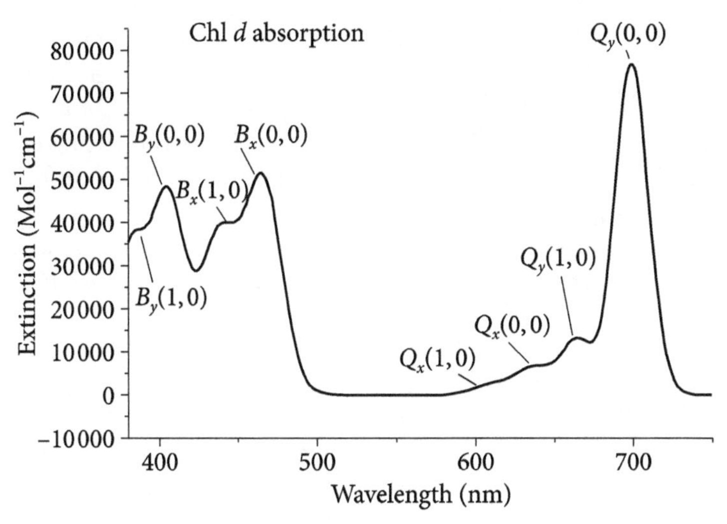

Figure 36. Calculated absorption spectrum of Chl *d* using the transitions shown in Figure 35 and values for the dipole strengths, spectral linewidths and extinction maximum at 699 nm.

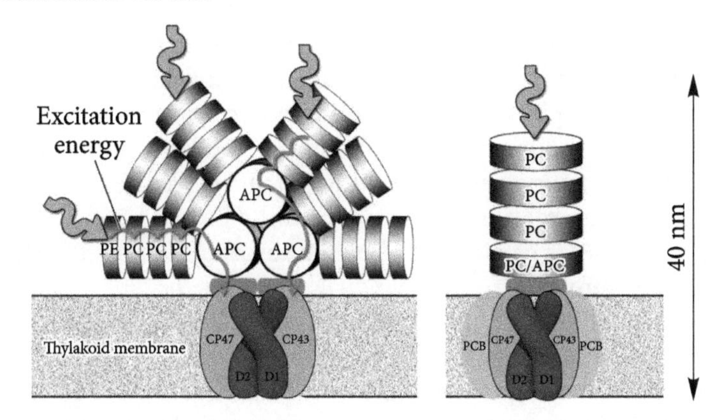

Figure 37. Overall geometry of PBS of typical cyanobacteria (left side) and rod shaped PBP antenna structure of *A.marina* (right side) according to (Marquardt et al., 1997), Both PBP contain PC and APC. The PBS additionally contains PE trimers.

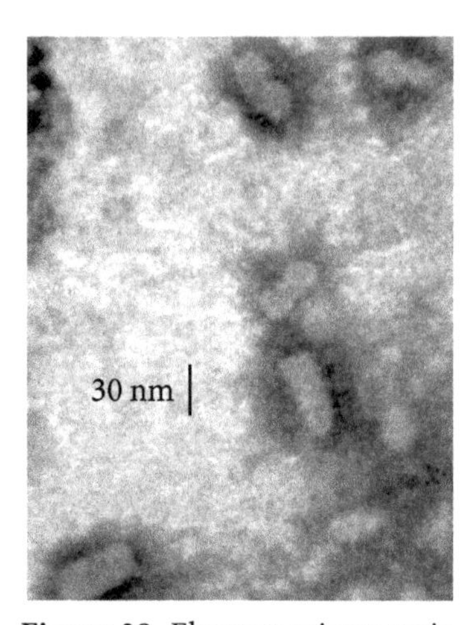

Figure 38. Electron microscopic study of PBP preparations of *A.marina* in buffer containing phosphate after negative staining with $Na_4[W_{12}SiO_{40}]$. The PBP antenna complexes appear transparent (white) due to the process of negative staining while the staining salt leads to a dark green contrast.

This work confirms the PBP structure suggested by (Marquardt et al., 1997), (Figure 37) through transmission electron microscopy of the PBP antenna complexes of *A.marina* as shown in Figure 38. In comparison to the pictures presented in (Marquardt et al., 1997), the PBP complexes appear slightly larger with extensions of up to 40 nm in length and 20 nm in diameter. The contrast of the protein structure in the electron beam is low. Therefore a procedure of negative staining had to be performed to achieve a picture of the PBP structure.

A carbon film on a copper net acts as a sample holder in transmission electron microscopic studies and was incubated into a concentrated PBP solution in phosphate buffer for 24 hours. Then the carbon film was put into a 4% (w/w) solution of Sodium-Silicotungstate ($Na_4[W_{12}SiO_{40}]$) which contains the heavy metal tungsten (W) which has a high scattering cross section for keV electrons. The specimen is left in the $Na_4[W_{12}SiO_{40}]$ solution for about 30 min. After this negative staining procedure, the sample is washed for circa two minutes in distilled water and left to dry for 24 hours. The carbon film can be directly used in the electron microscope.

Typical absorption spectra of PBP antenna complexes of *A.marina* dissolved in aqueous phosphate containing buffer are shown in Figure 39 (black line). For the purpose of comparison, the absorption of isolated PC trimers (green curve) and isolated APC trimers (red curve) are shown in Figure 29. The large absorption gap of Chl *d* between 500 nm and 650 nm (see Figure 36) is partially filled by the absorption of the PBP antenna as shown in Figure 39.

Interestingly, the absorption spectrum of isolated PBP antenna complexes from *A.marina* (black curve in Figure 39) is less broadened than the spectrum of isolated PC trimers (green curve in Figure 39). Since the PC trimers couple to hexamers, one would expect a slight broadening of the PBP absorption compared to isolated PC trimers. In addition, the PBP antenna of *A.marina* contains APC which should contribute to the red edge of the spectrum at 653 nm (see Figure 29 and red line in Figure 39).

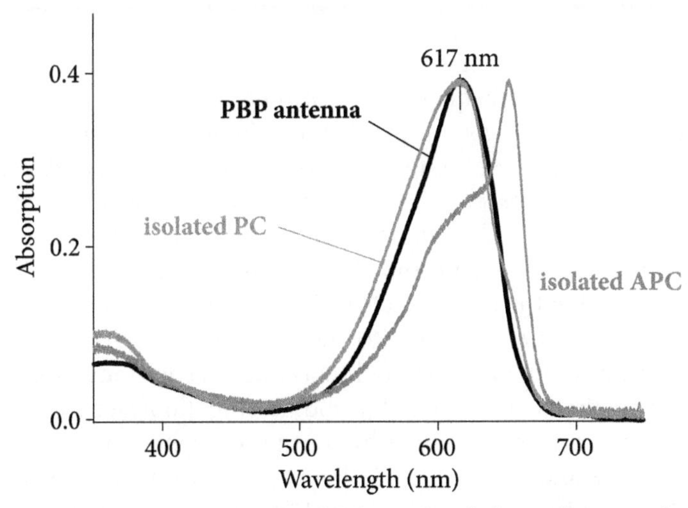

Figure 39. Absorption spectrum of PBP isolated from the cyanobacterium *A.marina* (black curve) in comparison to isolated PC trimers (green line) and isolated APC trimers (red line) as shown in Figure 29 according to (Theiss, 2006). The trimers were diluted in buffered aqueous solution. Image reproduced with permission.

The narrow PBP spectrum (black curve) could not be explained fully. However, it seems that there is some absorption at wavelengths >680 nm. These long wavelength absorption states play an important role for the energy transfer between the PBP antenna and Chl *d* (see chap. 2.6)

The PBPs of *A.marina* have been reported to consist of four hexameric units (Marquardt et al., 1997) (see Figure 37, right side). Each homohexamer covalently binds 18 PCB molecules as chromophores. The PC/APC heterohexamer is found to contain 9 PCB and 6 APCB chromophores (see Figure 28). Isolated PBPs from *A.marina* exhibit a fluorescence maximum at 665 nm (APC) with a shoulder at about 655 nm (PC) at room temperature (Marquardt et al., 1997).

2.5 Fluorescence Emission as a Tool for Monitoring PS II Function

The radiative emission from Chl *a* offers an invaluable tool to study the dynamics and the efficiencies of the primary processes of photosynthesis described here. Important information can be gathered from various techniques of fluorescence measurements depending on the mode of excitation and detection (Belyaeva et al., 2008, 2011, 2014, 2015). Information on excitation energy transfer among antenna pigments, trapping, charge separation and charge stabilization within the reaction centers (according to Chemical equation 3), forming of channels for nonphotochemical quenching, pigment–pigment or pigment-protein coupling can be gathered from analyses of the time decay and wavelength dependence of the emitted fluorescence (Schatz et al., 1988; Roelofs et al., 1992; Renger et al., 1995). Often short (typical ps) light pulses are used and time resolved analysis of prompt fluorescence decay after excitation is often performed in the mode of TWCSPC (Schmitt, 2011).

Another technique is to monitor light-induced transients of fluorescence quantum yield, often referred to as fluorescence induction. A great variety of methods is applied for monitoring fluorescence induction curves: samples are excited either by continuous light at different intensities (Govindjee and Jursinic, 1979; Renger and Schulze, 1985; Bulychev et al., 1987; Neubauer and Schreiber, 1987; Baake and Shloeder, 1992; Strasser et al., 1995, 2004; Bulychev and Vredenberg, 2001; Chemeris et al., 2004; Lazár, 2006) or by light pulses of different duration and intensities (Schreiber et al., 1986; Schreiber and Krieger, 1996; Christen et al., 1999, 2000; Goh et al., 1999; Steffen et al., 2001, 2005a, 2005b).

In general, the prompt fluorescence decay and the transient changes of the fluorescence yield are complicated functions depending on many parameters. Therefore model-based data analyses are required to deconvolute both, the fluorescence decay (Schatz et al., 1988; Roelofs et al., 1992; Renger et al., 1995), fluorescence induction (Shinkarev and Govindjee, 1993; Strasser et al., 2004; Lazár, 2006) and the transient changes of the fluorescence yield (Belyaeva et al., 2008, 2011, 2014, 2015) into a set of parameters for individual reactions. Numerous models have been developed to simulate the experimentally determined fluorescence curves. One essential feature for data analyses is the finding that fluorescence emission of oxygen-evolving photosynthetic organisms is dominated by processes connected with photosystem II (PS II) (see Belyaeva et al., 2008). For a review of different models see (Lazár, 2006).

It has been shown that the prompt fluorescence decay is mostly determined by the kinetics of EET in antenna complexes, charge separation and charge stabilisation as shown in Chemical equation 3 (Schatz et al., 1988; Roelofs et al., 1992).

A much greater variety of models exists for the simulation of transient fluorescence change from ns–s range since these are monitored under very different excitation conditions and depend, in a complicated manner, on very different processes that take place on different time scales. The models typically cover a time range of up to about 10 s. At time periods longer than 500 ms, formation and decay of an electric potential and the pH difference across the thylakoid membrane have to be taken into account for an accurate description of transient fluorescence changes (Van Kooten et al., 1986; Bulychev et al., 1987; Bulychev and Vredenberg, 1999, 2001; Leibl et al., 1989; Dau and Sauer, 1992; Gibasiewicz et al., 2001; Lebedeva et al., 2002; Belyaeva et al., 2003; Vredenberg and Bulychev, 2003; Belyaeva, 2004).

2.6 Excitation Energy Transfer and Electron Transfer Steps in Cyanobacteria Modeled with Rate Equations

The formalism based on rate equations as presented in chapters 1.3 and 1.4 was used in studies during recent years to model time dependent fluorescence data captured in form of time- and wavelength-correlated single photon counting (TWCSPC) or time dependent fluorescence quantum yield measurements covering 9 orders of magnitude in time. The described samples and experiments as partially described in (Schmitt, 2011) include:

1) preparations of the phycobiliprotein (PBP) antenna from of *Synechocystis* sp. PCC6803 (Maksimov et al., 2013; 2014a) and *A.marina* (Theiss et al., 2007a, 2008, 2011; Schmitt et al., 2006; Schmitt, 2011);
2) different solubilized and aggregated LHCII trimers from spinach (Lambrev et al., 2011),
3) the water soluble chlorophyll binding protein (WSCP) that was genetically expressed in *E.coli* and reconstituted with Chl *a*, Chl *b* or mixtures of Chl *a* and Chl *b* (Theiss et al., 2007b; Schmitt et al., 2008; Renger et al., 2009, 2011; Pieper et al., 2011);
4) artificial PPCs formed from eosin as chromophores bound to proteins (Barinov et al., 2009);
5) semiconductor-pigment-protein-hybrid-complexes (Schmitt, 2010; Schmitt et al., 2011, 2012) and
6) whole cells of different cyanobacteria, especially the cyanobacterium *Acaryochloris (A.) marina* (Petrášek et al., 2005; Schlodder et al., 2007; Schmitt et al., 2006, 2013; Theiss et al., 2007a; Theiss et al., 2007a, 2008, 2011);
7) whole cells of the green algae *Chlorella pyrenodoisa* Chick and on the higher plant *Arabidopsis thaliana* (Belyaeva et al., 2008, 2011,

2014, 2015) as well as fluorescence quenching in lichens (Maksimov et al., 2014b);

8) the application of rate equations to describe the image signal gathered by modern microscopic techniques like photoacoustic imaging (Märk et al., 2015a, 2015b).

As an example in this chapter the fluorescence dynamics in the PBP complexes of *A.marina* and of whole cells of *A.marina* are analysed. The next chapter (2.7) focuses on the question how more complex systems like the whole Photosystem II dynamics of higher plants can be described with basically the same rate equation formalism (Belyaeva et al., 2008, 2011, 2014, 2015).

The main focus was not the spatial structure and dynamics of traceable fluorophores but the analysis of time resolved excitation energy transfer (EET) and electron transfer (ET) steps on the temporal scale of picoseconds.

The unusual cyanobacterium *A.marina* was discovered in 1996 and it is up today the only known organism which mainly contains Chl *d* instead of Chl *a* in the membrane intrinsic Chl antenna and reaction centers (Miyashita et al., 1996, 1997; Marquardt et al., 1997). This feature was reason enough for us to investigate EET and ET processes in A-marina with the technique of TWCSPC.

Figure 9 (see chapter 2.1.4), published in (Theiss et al., 2011), shows typical fluorescence decay curves of whole cells of *A.marina* collected after excitation with 632 nm at room temperature performed with TWCSPC (Schmitt, 2011). In Figure 9a, this data matrix is shown as a color intensity plot (CIP). A vertical plot at a constant time t_0 results in the time-resolved emission spectrum $F(t_0, \lambda)$ (Figure 9c) while a horizontal intersection delivers the fluorescence decay at a constant emission wavelength λ_0 (Figure 9b). Figure 9b shows the decay curves of *A.marina* at 660 nm and 725 nm. It is seen that the fluorescence at 660 nm decays much faster than the emission at 725 nm. A closer look at the emission maximum on top of Figure 9b reveals a very small temporal shift between the 660 nm and the 725 nm decay curves. The 725 nm decay is slightly shifted to later times in comparison to the 660 nm decay. A data fit shows that the small temporal shift in the following called “fluorescence rise kinetics” has a similar time constant as the fluorescence decay at 660 nm. Both curves are convoluted with the instrumental response function (IRF) which leads to the small visible difference.

The multiexponential fits of all decay curves measured in one time- and wavelength resolved fluorescence spectrum were performed as global fits with common values of lifetimes τ_j (linked parameters) for all decay curves and wavelength-dependent pre-exponential factors $A_j(\lambda)$ (non-linked parameters, see Figure 9d representing "decay associated spectra" (DAS) thus revealing the energetic position of individual decay components (for details also see Schmitt et al., 2008).

In whole living cells of *A.marina* the average time constant for the excitation energy transfer (EET) between the uniquely structured and rod shaped phycobiliprotein (PBP) antenna and the Chl *d* containing membrane intrinsic antenna complex was found to occur within an overall time constant of 70 ps (Petrášek et al., 2005; Theiss, 2006; Theiss et al., 2011; Schmitt, 2011) (see Figure 9). This value is 3 times faster than in other cyanobacteria containing phycobilisomes (PBS) assembled from several PBP rod structures and Chl *a* containing membrane intrinsic antenna complexes (Glazer, 1985; Holzwarth, 1991; Mullineaux and Holzwarth, 1991; Trissl, 2003; Theiss et al., 2008).

Figure 40 represents DAS characterized by EET from PC and APC containing hexamers of the membrane extrinsic PBP antenna to spectrally red shifted APC molecules which are most probably located in the linker protein representing the so-called terminal emitter (TE). This EET was resolved to occur within 20 (±10) ps in *A.marina*. These values were found in living cells of *A.marina*. Further elucidation of the fast equilibration processes in the PBP antenna of *A.marina* is not easily possible with TWCSPC as can also be shown by numerical simulations of the expected time resolved spectra. Measurements of flash induced transient absorption changes performed in the fs-ps time domain by C. Theiss on isolated PBP complexes showed fast energy equilibration between different phycocyanin (PC) molecules (equilibration time <1 ps) which are bound to a hexameric protein structure in the PBP of *A.marina* and in PBS. The equilibration between the PC and the APC containing hexamers occurs with a time constant of 3–5 ps (Theiss et al., 2008, 2011).

Rate equations can now be used to simulate the exciton diffusion inside the rod-shaped structures of *A.marina* and to elucidate the real relationship between the observed fluorescence dynamics and intrinsic EET steps between single molecules or compartments. These simulations showed that the EET velocity depends critically on entropy effects caused by the small distance of the energetic states of the coupled PC chromophores. This effect is the main reason for slower exciton diffusion inside the PBS containing more than 300 PBP chromophores (Holzwarth, 1991) and additionally Chl *a* molecules, in contrast to the PBP complexes of *A.marina* containing 63 PC and 6 APC chromophores coupled to Chl *d*.

The geometry and small size of the *A.marina* antenna in contact to Chl *d* leads to faster energy transfer in comparison to the structure of the much larger PBS.

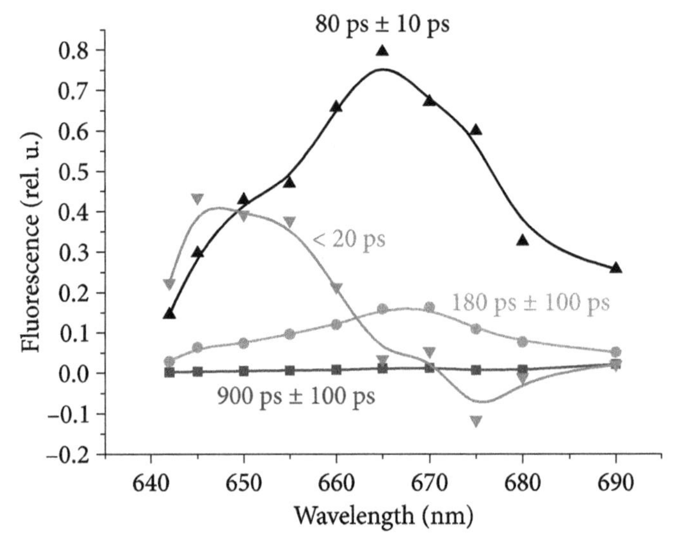

Figure 40. Decay associated spectra (DAS) of a global fit (4 exponential components) of the fluorescence emission at 20 °C of whole cells of *A.marina* in the 640 nm – 690 nm range after excitation at 632 nm. The fluorescence below 645 nm is cut off by a long-pass filter. The graphics is redrawn from the data published in (Schmitt, 2011). Image reproduced with permission.

Entropy effects are the consequence of equation 10 (see chapter 1.3) indicating that the probability for a transfer towards a compartment is proportional to the quotient of the number of empty states that can take up the excitation energy in the acceptor compartment and the number of states for the exciton in the donor compartment. As there are 4 times more states in an antenna structure with 280 pigments as compared to an antenna with 70 pigments EET from such large antenna is delayed by 4 times even if the energetics and transition probabilities for all single transfer steps are the same in both cases which is the case if the coupling of all molecules is of similar strength.

As shown by the DAS depicted in Figure 40, the fluorescence dynamics in the PBP antenna of living cells is characterized by four exponential decay components with lifetimes of <20 ps, 80 ps (±10 ps), 180 ps (±100 ps) and 0.9 ns (±0.1 ns).

The DAS of the <20 ps component has a positive amplitude in the PC and APC emission band (645–660 nm) and a negative amplitude at

675 nm that is red shifted in comparison to the APC emission. This suggests that in living cells of *A.marina* the EET from PC to the most red chromophore of the PBP antenna occurs with a time constant of less than 20 ps. The 80 ps decay component reflects the EET from the PBP antenna to PS II (see below) and the slower components (180 ps, 0.9 ns) are probably due to conformationally distorted and/or functionally decoupled PBP complexes in *A.marina* (Schmitt et al., 2006).

Figure 41a shows a typical DAS gathered from TWCSPC spectroscopy on whole cells of *A.marina*.

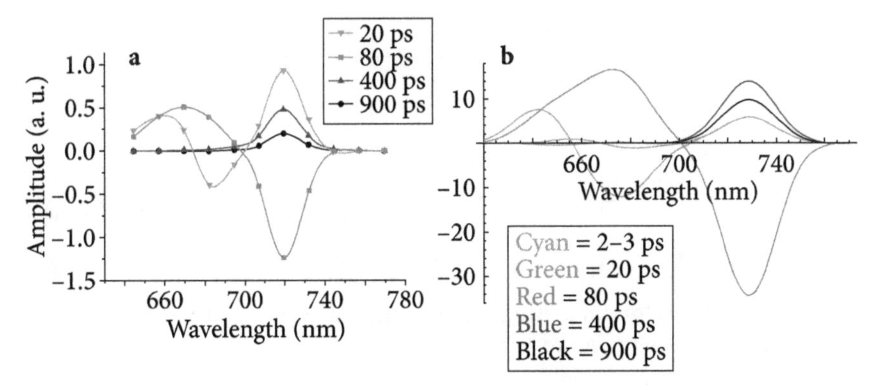

Figure 41. a) Measured DAS of *A.marina* after excitation with 632 nm at room temperature. The simulated DAS is shown in panel b). The graphics was published in (Schmitt et al., 2011; Schmitt, 2011). Image reproduced with permission.

The values of the long decay components in the Chl *d* regime are given by time constants of 400 ps (blue triangles) and 900 ps (black circles) that differ slightly from the statistically evaluated data exhibiting time constants of 630 ps (±30 ps) and 1.2 ns (±0.1 ns) measured in 2005 (see Figure 9). It has to be pointed out that these values differ between different cells and different illumination conditions. The main focus of the simulated exciton dynamics is the fast fluorescence decay with time constants of 20 ps (green triangles in Figure 41a) and 80 ps (red squares Figure 41a). These values are found to be widely invariant between all investigated cells of *A.marina*. Therefore the mentioned deviations in the Chl *d* dynamics which are caused by the different redox states of the Chl containing reaction center are not relevant for the following conclusions.

The simulation shown in Figure 41b is based on the assumption, that the EET between the PC trimers occurs almost exclusively through the β-84 chromophores (see chapter 2.4.4; Mimuro et al., 1986; Holzwarth, 1991). In addition about 3-6 chromophores of the trimeric PC disk should be involved into the EET inside the PC disc which occurs from outside to inside of the PC trimer (Mimuro et al., 1986; Holzwarth, 1991). Therefore the excitation energy is trapped at the β-84 chromophore inside of a trimer first and subsequently funneled to the APC chromophores and the terminal emitter (TE). The pathway of EET is guided by "fluorescing" chromophores in accordance with calculations of Suter and Holzwarth that were done for PBS (Mimuro et al., 1986; Suter and Holzwarth, 1987). The scheme that was used for the simulation shown in Figure 41b is presented in Figure 42.

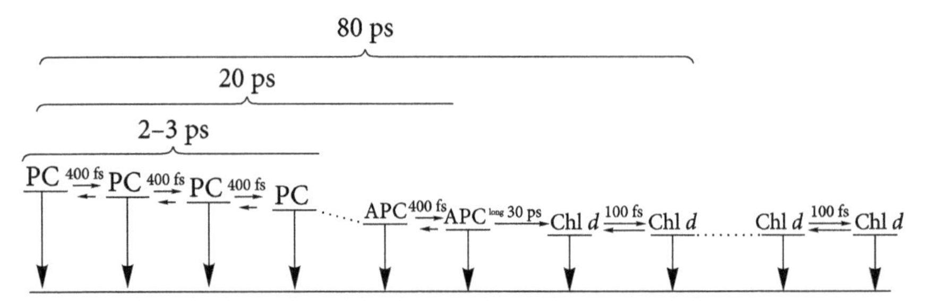

Figure 42. Scheme for simulating the data presented in Figure 41a. The simulated DAS are shown in Figure 41b.

Figure 42 indicates the situation where 22 out of the 69 chromophores found in the PBP antenna of *A.marina* are assumed to be involved in EET along the PBP antenna of *A.marina.* The simulations were performed by a simplified model of the PBP antenna containing 17 PC chromophores with emission wavelengths between 638 nm and 654 nm, 5 APC chromophores with emission wavelengths between 660 nm and 680 nm, 5 Chl *d* pigments with emission wavelengths between 723 nm and 732 nm and two states of the radical pairs.

For the simulation the EET between different PC and APC molecules in the PBP antenna is set to 400 fs as found by Theiss (Theiss, 2006; Theiss et al., 2007a; Theiss et al., 2011). The EET from the "red" APC molecule (TE) to Chl *d* is set to 30 ps and the EET between the Chl *d* molecules in the Chl containing core antenna is set to 100 fs. For the sake of simplicity only four PC, two different APC molecules and four acceptor states of Chl *d* are explicitly shown in this scheme (Figure 42) while the rest of states added to the simulation is omitted and substituted by a series of black dots.

The simulation results presented in Figure 41b show that the experimental data can be explained by assuming equally coupled PC and APC molecules in the PBP antenna as shown in Figure 42. It turned out that all values found in the fluorescence decay kinetics should be understood as "effective" transfer times which cannot necessarily be identified with rate constants for single EET steps. The 2–3 ps kinetics most probably describe the overall relaxation between several PC pigments while the 20 ps component describes the overall relaxation from PC, APC and red APC (TE) in the linker protein. The exact value of the EET transfer step from the TE to Chl *d* depends on the number of chromophores involved in the EET. If only 8 instead of 17 PC molecules are used in the simulation, this value prolongates to 40 ps instead of 30 ps, respectively (data not shown). The TE, which most probably is the long wavelength APC around 680 nm (APC^{long}) found by Theiss et al. gives rise to the negative amplitude (rise kinetics) of the 20 ps component at 680 nm. Variations of the time constants in the model of Figure 42 reveal that the 20 ps component of the experimental data (Figure 41a) can be explained by EET along equally coupled PC/APC molecules. A rate limiting step between PC and APC or APC and APC^{long} can also explain the 20 ps component (data not shown). But in all cases the overall EET time from the PBP to the Chl *d* is found to be 80 ps in the model only if a transfer time of max. 30–40 ps is set for the inverse transition probability between APC^{long} and Chl *d*.

Finally the apparent time constants are not present on the molecular level in the PBP antenna of *A.marina*. They "emerge" from the coupling scheme and can already be explained by a simple rate equation system as it is a synergetic study of the whole complex. This example demonstrates how the underlying microscopic EET is hidden by a macroscopic observable that finally determines the behavior on dynamical time scales that are chosen for observation on a given system.

The decay time constants of the Chl *d* (400 ps and 900 ps at 725 nm) are explained by the electron transfer processes in the PS II (see Schmitt, 2011). The diffusion along the PC molecules leads to additional fast components of 2–3 ps with very small amplitude (cyan curve in Figure 41a) which can not be resolved in our experiment.

The simulation presented in Figure 41b fits the data presented in Figure 41a, except of the large amplitude of the 20 ps component (green curve) dominating in the measurement at 725 nm. This effect probably originates from PS I fluorescence at 725 nm. The simulations are performed for the dynamics of the PS II only and therefore can not reproduce the large fraction of the 20 ps component which mainly results from fast PS I fluorescence.

The simulation exhibits that in fluorescence spectroscopy the measurement data reflects apparent time constants for EET processes. If an ensemble of several energetic states couples to an acceptor then the fluorescence dynamics summarizes the diffusion of excitation in the donor states and the EET process. The 10–20 ps EET component (green curve of Figure 41) is also found in the fluorescence emission of isolated PBP. Therefore the fluorescence spectroscopic data reveals that the EET along the whole PBP antenna takes place within a time of about 20 ps. The overall EET from the PBP to the Chl *d* is found to appear as a 70–80 ps component in the fluorescence data if the inverse transition probability between APC^{long} (the TE) and Chl *d* is about 30 ps. The reason for this prolongated "apparent" decay time is the entropy effects of the degenerated PC states found in the PBP antenna structure.

In the PBP antenna of *A.marina* for example the equilibration between PC and APC molecules can still be resolved with a time constant of 20 (±10) ps. The information of faster equilibration processes simply cannot be found in the fluorescence dynamics observed by TWCSPC. In that case it is intrinsically impossible to extract information on the accurate values of the single energy transfer steps because the dynamics of different coupled states takes place in the same time domain between spectrally strongly overlapping chromophores. The states of the systems can mix in a way that makes it impossible to separate the processes even if the experimental resolution is high enough. This fact mostly limits the available information content. TWCSPC is principally limited by the amount of accessible information as shown with information theoretical approaches regarding single molecule fluorescence microscopy. The desired goal of the experimenter should be to extract all experimentally accessible information to have the general possibility to proof any predicted effect of an applied theory. In general the experimenter is not able to accumulate an infinite amount of information and it would be economic to choose exactly the right amount. Therefore the experimenter has to care that the accumulated information represents the accessible information.

An improvement can only be achieved by a study of the decomposed sample or by the employment of complementary experimental techniques which localise the observed state. But in both cases the experimental results might depend on the special configuration of the sample and the measurement setup and it is necessary to investigate the influence of the sample preparation and experimental manipulation onto the experimental results.

It might be useful to accomplish a kind of methodical statistics of the experimental techniques. Then the experimenter can shift systematic aberrations to statistical uncertainties to a certain amount. For this purpose the experimenter has additionaly to be able to reproduce the sample often enough without any variation.

In most cases the principal resolution is not limited by the setup but by the uncertainty of the sample. Doing high resolved fluorescence nanoscopy we shift the uncertainty from the optical wavelength representing our measurement setup to the observed states which is an ultimate resolution limit.

According to the presented results the excitation energy transfer in the antenna complexes of *A.marina* can be summarized as shown in Figure 43, left side. In this scheme it is assumed that the PBP antenna of *A.marina* resembles the rods of the phycobilisomes in common cyanobacteria with the main exception, that the hexamer closest to the thylakoid membrane is a heterohexamer consisting of one PC- and one APC-trimer as suggested by Marquardt et al. (1997). The EET from PC to Chl *d* is characterized by four kinetic components with lifetimes of <400 fs, 3 ps, 14 ps and 70 ps. The red arrows in Figure 43 exhibit our own evaluation results while literature values are marked black.

Figure 43, right side, shows a scheme with the range of time constants for EET processes in *Synechococcus* 6301 as reported in literature (black arrows) in comparison to our own findings obtained on *Synechocystis* (red arrows, compare Figure 44 and Figure 45). The equilibration along the trimeric PC disks in the PC containing rod appears with a typical rate of about $(10\ ps)^{-1}$ (Gillbro et al., 1985). Based on experimental results from ps-studies (Suter and Holzwarth, 1987) presented a random walk model for the EET in the PC rods of *Synechococcus* 6301 and calculated rates of $(10\ ps)^{-1}$ to $(3.3\ ps)^{-1}$ for single step EET between trimeric PC disks and about $(25\ ps)^{-1}$–$(40\ ps)^{-1}$ for the EET from the innermost PC trimer to the APC core. For the overall EET from a PC rod containing 4 hexamers to the APC core the literature reports rate constants of $(80\ ps)^{-1}$ to $(120\ ps)^{-1}$ (Gillbro et al., 1985; Suter and Holzwarth, 1987; Holzwarth, 1991; Mullineaux and Holzwarth, 1991) (Figure 43, right side).

Our own findings obtained at a slightly different organism, *Synechocystis*, did not resolve the involvement of an APC bound to the linker molecule but was sufficiently simulated with time constants of 70 ps for the EET from PC to APC and 280 ps from APC to the Chl *a* containing membrane integral antenna complex (Figure 44 and Figure 45).

In contrast, in *A.marina* where APC and PC share the same hexamer, the PC to APC EET occurs with a rate constant of only $(3\ ps)^{-1}$ (Figure 43, left panel). This is finally at least 8 times faster than the fastest EET rate reported from the innermost PC trimer to the APC core in phycobilisomes as reported in (Suter and Holzwarth, 1987) and even faster than the single step EET between adjacent trimers in pure PC rods of other cyanobacteria. The 14 ps component is observed in the DAS of the transient absorption spectra as a decay of a bleaching at 640 nm and a rise of a bleaching at 670 nm. We assign this component tentatively to an EET from APC (absorbing at 640 nm) to a low energy APC (bound to a linker protein) that transfers it´s energy to Chl *d* with a time constant of 30–40 ps and therefore facilitates an efficient overall EET from the PBP antenna to Chl *d* which occurs with a rate constant of $(70\ ps)^{-1}$. This is more than three times faster than the EET transfer from the APC core to Chl *a* in *Synechococcus* 6301 which occurs typically with $(170–120\ ps)^{-1}$ (Mullineaux and Holzwarth, 1991).

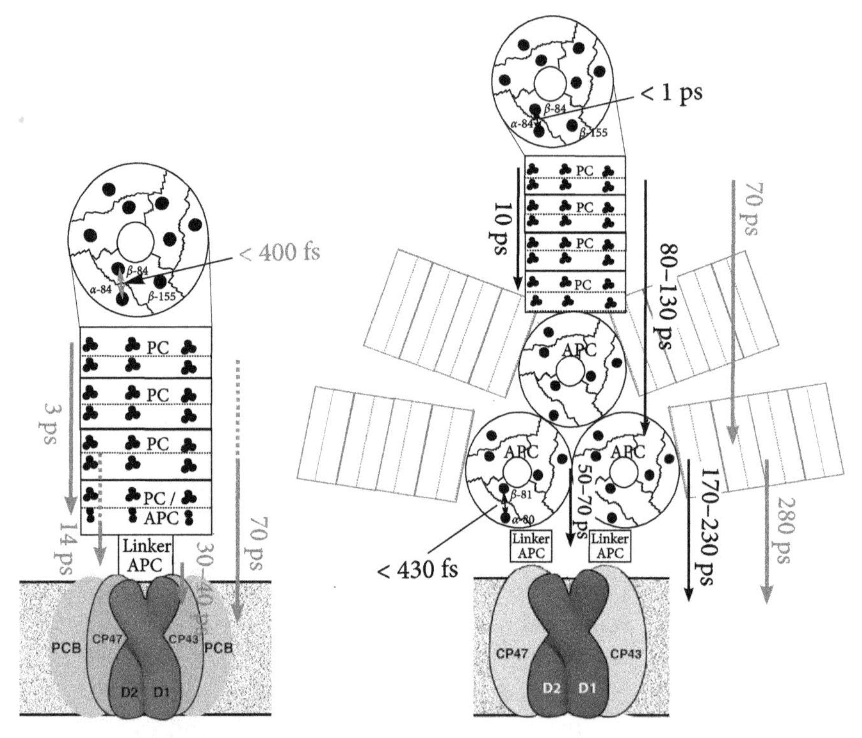

Figure 43. Kinetics of excitation energy transfer (EET) processes in *A.marina* and in the cyanobacterium *Synechococcus* 6301 as published in (Theiss et al., 2011). Left panel: Model Scheme for the excitation energy processes (EET) inside the PBP antenna rod and between the PBP antenna and Chl *d* in *A.marina*. At the top the model scheme of the PC trimer with its bilin chromophores is shown. Right panel: Model Scheme and time constants for the EET inside the phycobilisomes of *Synechococcus* 6301 and from there to the RC giving a résumé of the literature (Gillbro et al., 1985; Suter and Holzwarth, 1987; Holzwarth, 1991; Mullineaux and Holzwarth, 1991; Sharkov et al., 1994; Debreczeny et al., 1995a, 1995b). Image reproduced with permission.

The very fast EET in *A.marina* from PC to APC then to Chl *d* reflects the unique feature of *A.marina* due to its tiny rod-shaped PBP antenna where PC and APC share the same heterohexamer instead of huge phycobilisomes with rods containing PC only and an additional APC core.

As comparison we applied the formalism of rate equations onto the same problem (studying EET and ET transfer processes, in a different cyanobacterium, *Synechocystis* PCC 6803 available in our laboratory which should have similar properties like the above studied *Synechococcus* 6301).

The DAS of whole cells of *Synechocystis* are shown in Figure 44, upper panel. The energy transfer from PC to APC is associated with a characteristic fluorescence time constant of 60 ps (red curve in Figure 44). The fluorescence exhibits an additional 150 ps component (black curve) associated with an EET from the pigments around 640–660 nm (PC and APC) to Chl *a* of PS II (685 nm). The Chl *a* fluorescence decays with time constants of 300 ps (green curve) and 1 ns (blue curve).

Figure 44, upper panel, shows the result of a simulation that is performed according to the scheme given in Figure 45. The scheme comprises the photochemical light reaction in the PS II according to chemical equation 2 with an additional PC and APC containing PBS antenna.

In common Chl *a* dominated cyanobacteria such as *Synechococcus* 6301 (which exhibits a similar phycobilisome antenna like *Synechocystis*) the energy transfer from the PBS to PS II occurs on an at least 3 times slower time scale in comparison to *A.marina*. In these cyanobacteria the fluorescence decay of the PBS is characterized by multiphasic kinetics with lifetimes of typically 120 ps for the energy transfer from the PC-containing rods to the APC-core, 70 ps for the energy transfer from the APC-core to the terminal emitters and 200 ps for the energy transfer from the terminal emitter to Chl *a* of PS II.

Based on experimental results from ps-studies Holzwarth et al. calculated a time constant of 102 ps to 124 ps for the overall EET from a PC rod containing 4 hexamers to the APC core (Suter and Holzwarth, 1987; Holzwarth, 1991). Comparable rate constants found in literature are $(80–120\ ps)^{-1}$ (Gillbro et al., 1985) and $(90–120\ ps)^{-1}$ (Mullineaux and Holzwarth, 1991).

The simulation and the experimental results are in agreement if the EET time constant is set to 70 ps for the transfer from PC to APC which is slightly faster than the values given in literature, as mentioned above (see Figure 45). The time constant for the APC to Chl *a* transfer is simulated with 280 ps which resembles the sum of 70 ps for the energy transfer from the APC-core to the terminal emitters and 200 ps for the energy transfer from the terminal emitter to Chl *a* of PS II (Gillbro et al., 1985; Mimuro et al., 1986; Mullineaux and Holzwarth, 1991).

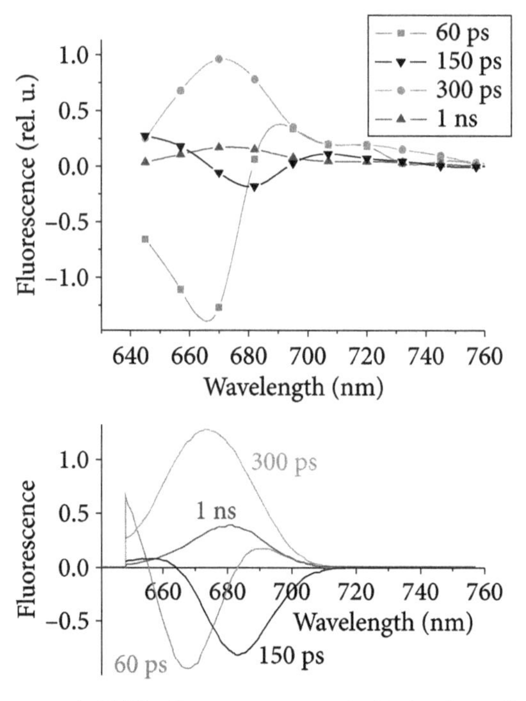

Figure 44. Upper panel: DAS of a measurement obtained on *Synechocystis* at room temperature after excitation with 632 nm laser light. The time constants of the components exhibit values of 60 ps with a minimum in the fluorescence regime of APC (670 nm, red curve), 150 ps with a minimum in the fluorescence regime of Chl *a* (680 nm, black curve) and additional positive time constants in the Chl *a* regime with 300 ps (green curve) and 1 ns (blue curve). Lower panel: Simulation of the DAS according to the scheme given in Figure 45, assuming a temperature of 300 K and a spectral bandwidth of the Gaussian emitter states of 25 nm. An edge filter width a cut off wavelength at 648 nm is simulated comparable to the edge filter used in the measurement data (DAS of the measurement is given in the upper panel).

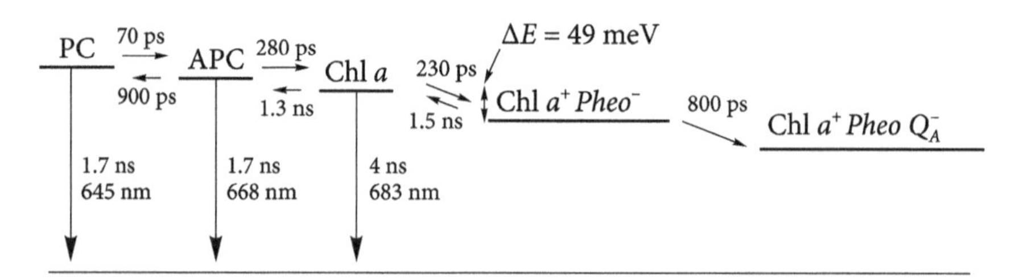

Figure 45. Scheme for the simulation of the DAS shown in Figure 44, lower panel.

In addition a time constant of 230 ps is found for the primary charge separation and ΔG of the relaxed primary radical pair in comparison to the excited Chl state in the antenna system is found to be about 50 meV. The charge stabilisation (see also chemical equation 2) appears to be quite slow with 800 ps. Especially the time constant of the 1 ns component (see Figure 44) depends critically on this ET step. The reason for the high amplitude of the 1 ns component and the concomitant slow 800 ps charge separation may be the existence of partially closed reaction centers during the measurement.

The main difference of the simulated DAS (Figure 44, lower panel) and the measurement (Figure 44, upper panel) is the apparent amplitude deviation of the rise kinetics of the fast 60 ps term (red curve in Figure 44). The determination of the amplitude of the rise kinetics is very difficult and exhibits confidential intervals that are of comparable size to the amplitude itself. This situation is even more critical for the measurements presented in Figure 44, as there are two overlapping rise terms with time constants of 60 ps and 150 ps that are resolved in the DAS. Therefore the agreement between the experiment, Figure 44, upper panel and the simulation Figure 44, lower panel is sufficient within the confidence interval of the experimental results. The time constants determined as given in Figure 45 show that there is an agreement between the results obtained with our setup and the concomitant analysis via rate equations and literature values. Therefore our experiments and the theoretical approach are supported by the successful determination of the EET steps in the PBS of *Synechocystis*.

2.7 Excitation Energy and Electron Transfer in Higher Plants Modelled with Rate Equations

R. Steffen et al. developed a compact fluorescence spectroscopic setup suitable to monitor single (laser) flash induced transient changes of the fluorescence quantum yield (SFITFY) in cell suspensions and whole leaves in the time domain from 100 ns to 10 s (Steffen et al., 2001, 2005a, 2005b) (see Figure 46).

In our former studies we used data gathered by this setup to describe the PS II dynamics in whole cells of *Chlorella pyrenoidosa Chick* according to a reaction pattern based on the formalism of rate equations as described in chap. 1.3 and 1.4 (Belyaeva et al., 2008). Later it was shown that the intensity dependent pattern of SFITFY traces measured in whole leaves of *Arabidopsis thaliana* at four different intensities of the actinic laser flash can be described with one set of model parameters (Belyaeva et al., 2011). These findings are briefly outlined in this chapter and extended to an approach enabling the comparison of single transfer probabilities in different species. In (Belyaeva et al., 2014, 2015) a detailed comparison of the evaluated rate equation scheme of green alga and higher plants which was fit to the measured SFITFY curves of these species allowed for the comparison of single ET transfer steps in the proposed reaction scheme to distinguish the kinetics of single steps between green algae and higher plants. Details regarding the measurement setup (Steffen et al., 2001), the sample condition (Steffen et al., 2005a, 2005b) and the discussion of the metabolic relevance of the gathered results (Steffen et al., 2005a; 2005b) are described in the cited references. The following results are already published in (Belyaeva et al., 2014, 2015) and focus on the applicability of rate equation systems on the synergetic network analysis of ET steps in higher plants.

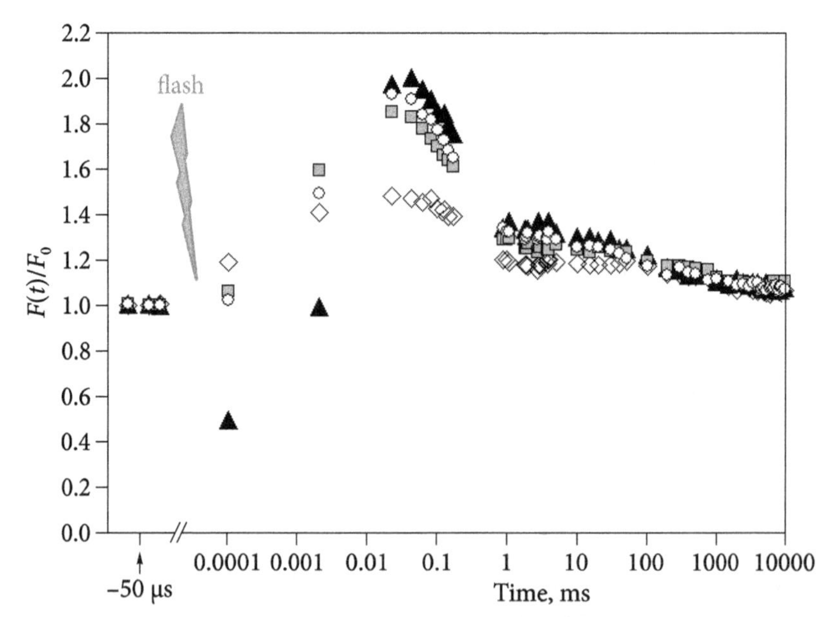

Figure 46. Experimental data of fluorescence yield changes (SFITFY curves) induced in whole leaves of *Arabidopsis thaliana* wild type plants by excitation with a single actinic 10 ns laser flash of different energy: $7.5 \cdot 10^{16}$ photons/ (cm^2·flash) (triangles), $6.2 \cdot 10^{15}$ photons/ (cm^2·flash) (circles), $3.0 \cdot 10^{15}$ photons/ (cm^2·flash) (squares) and $5.4 \cdot 10^{14}$ photons/ (cm^2·flash) (diamonds). The data are redrawn from (Belyaeva et al., 2011). The green arrow symbolizes the excitation by the actinic laser flash. Image reproduced with permission.

The time courses of the photosystem II (PSII) redox states were analyzed with a model scheme as shown in Figure 47. Patterns of Single Flash Induced Transient Fluorescence Yield (SFITFY) measured for leaves (spinach and *Arabidopsis (A.) thaliana*) and the thermophilic alga *Chlorella (C.) pyrenoidosa* Chick (Steffen et al., 2005a; Belyaeva et al., 2008, 2014, 2015) were fitted with this PS II model.

Figure 46 shows experimental SFITFY data of Steffen and coworkers (Steffen et al., 2005a) obtained by excitation of dark adapted whole leaves of *Arabidopsis thaliana* wild type with single actinic laser flashes of different energy. The signal F_0 (−50 μs) before the actinic flash monitors the fluorescence induced by the weak measuring pulses. It reflects the normalized fluorescence quantum yield of dark adapted PS II complexes and is used to normalize the time dependent SFITFY curve monitored by a sequence of weak measuring light pulses in the range from 100 ns to 10 s after the actinic flash (which is symbolized by the green arrow in Figure 46). The photon densities per flash and unit area of the actinic laser flash of the four different SFITFY data sets were: $5.4 \cdot 10^{14}$ (diamonds), $3 \cdot 10^{15}$(squares), $6.2 \cdot 10^{15}$ (circles) and $7.5 \cdot 10^{16}$ (triangles) photons/(cm^2·flash) (0.7%, 4%, 8% and 100% of reference).

The SFITFY patterns exhibit an "instantaneous" change of the fluorescence yield within 100 ns that strongly depends on the energy of the actinic flash (see Figure 46). This instantaneous change is a rise at low energies and turns into a drop below F_0 at high energies of the laser flash. This drop is explained by the population of ^{3}Car states that act as highly efficient quenchers of the fluorescence (Steffen et al., 2005a, 2005b; Belyaeva et al., 2008; 2011). A normalized maximum value $F_{max}(t)/F_0$ of about two is reached about 50 μs after the actinic flash. In the subsequent time domain the normalized fluorescence yield declines to a level slightly above its original value F_0.

Simulations based on a model of the PS II reaction pattern presented in Figure 47 provide information on the time courses of population probabilities of different PSII states. The appropriate application of the formalism of rate equations provides a flexible basis for comparative analyses of time dependent fluorescence signals observed on different photosynthetic samples under various conditions (e.g. presence of herbicides, other stress conditions, excitation with actinic pulses of different intensity and duration). The general formalism is principally suitable to describe any system that can be described by states and transitions between these states.

The kinetic scheme of Figure 47 allows the calculation of the transient PS II redox state populations ranging from the dark adapted state, via excitation energy and electron transfer steps induced by pulse excitation,

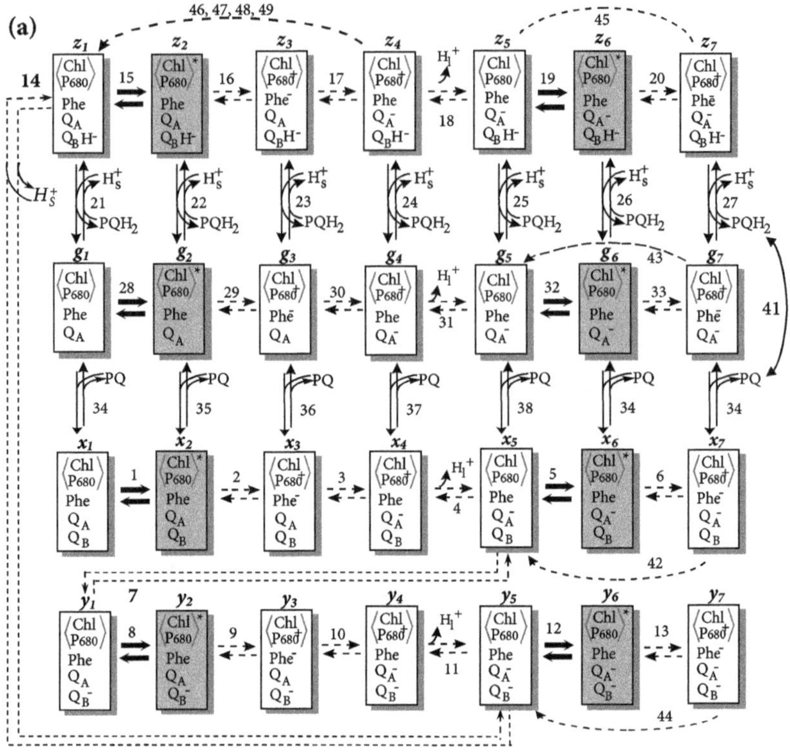

Figure 47a. Kinetic scheme of Photosystem II as presented in (Belyaeva et al., 2008, 2011, 2014, 2015). Each rectangle refers to one of the states. $\left\langle \begin{matrix} \text{Chl} \\ \text{P680} \end{matrix} \right\rangle$ denotes the total PS II chlorophyll including the antenna and the P680 pigments and $\left\langle \begin{matrix} \text{Chl} \\ \text{P680} \end{matrix} \right\rangle^{*}$ is used to determine singlet excited states $^{1}Chl^{*}$ delocalized over all pigments in antenna and RC. Further components are P680 – photochemically active pigment, Phe – the primary electron acceptor pheophytin. Q_A and Q_B – the primary and secondary quinone acceptors. PQ – plastoquinone, PQH_2 – plastoquinol; H_L^+ – protons, which are released into lumen, H_S^+ – protons in stroma. The letters above rectangles $(x_i, y_i, z_i, g_i, i = 1, \ldots, 7)$ correspond to the model variables. Shaded areas symbolize the excited states that are capable of emitting fluorescence quanta. Dashed arrows designate fast steps (characteristic time values less than 1 ms). Bold arrows mark the light induced steps. Numbers at the arrows correspond to the step numbers. Dashed arcs designate two types of irreversible reactions of the processes of non-radiative recombination: $Phe^{-\bullet}$ with $P680^{+\bullet}$ (42–45:= k_{Phe}) and $Q_A^{-\bullet}$ with $P680^{+\bullet}$ (46–49). Image reproduced with permission.

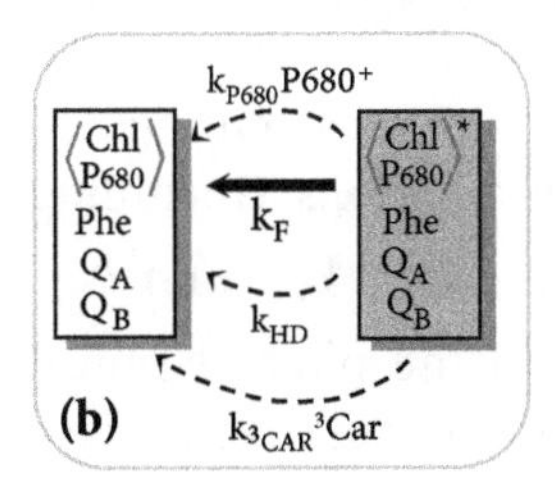

Figure 47b. The decay into ground state occurs via i) radiative fluorescence emission (k_F), ii) nonradiative dissipation of excited chlorophyll singlets by quenching due to cation radical $P680^{+\bullet}$ and/or by the triplet states of carotenoids with rate constants k_{P680+} and $k_{3_{Car}}$, respectively and iii) radiationless dissipation of excitation to heat (k_{HD}). Image reproduced with permission.

followed by final relaxation into the stationary state eventually attained under the measuring light. The shape of the actinic flash was taken into account by assuming that an exponentially decaying rate constant simulates the time dependent excitation of the PS II by the 10 ns actinic flash. The maximum amplitude of this excitation exceeds that of the measuring light by 9 orders of magnitude.

Figure 47 shows the whole scheme of redox states and their transitions as described in detail in (Belyaeva et al., 2011). Each box in Figure 47 represents the redox state of the corresponding component (compare chap. 1.2): antenna and RC chlorophyll – $\left\langle \begin{matrix} \text{Chl} \\ \text{P680} \end{matrix} \right\rangle$, pheophytin – Phe, primary quinone acceptor – Q_A, secondary quinone acceptor – Q_B. The model comprises the processes of light induced charge separation (reaction numbers: 2, 9, 16, 29), charge stabilization by $Q_A^{-\bullet}$ formation (reaction numbers 3, 10, 17, 30), electron transfer from $Q_A^{-\bullet}$ to Q_B (reaction number 7) and from $Q_A^{-\bullet}Q_B^{-\bullet}$ to Q_BH^- (reaction number 14), protonation of Q_BH^- under PQH_2 release (reaction numbers 21-27) and refilling of the empty Q_B-site with oxidized plastoquinones (PQ) (reaction numbers 34–40). For the sake of simplicity we assume that for each electron, transferred from the water oxidizing complex (WOC) via tyrosine Y_Z to the oxidized RC chlorophyll $P680^{+\bullet}$ (reaction numbers 4, 11, 18, 31), one proton is released into the lumen.

The model scheme of Figure 47a comprises 28 redox states of the PS II RC together with two states of the PQ pool. Therefore 30 variables (metabolites $N_i(t)$, $i = 1...30$) and a set of 30 differential equations describe the rate of production and consumption of $N_i(t)$ which is a function of the variables $N_j(t)$ involved into population and deactivation of $N_i(t)$ ($i, j = 1...30$) and of the rate constants, i.e. probabilities for each transition per time unit k_n, k_{-n} ($n = 1...49$) for forward and backward transfer steps, respectively (see chapter 1.3 and 1.4). Some of the reactions in the model scheme (Figure 47a) involve protons in lumen and stroma ($[H_l^+]$ and $[H_S^+]$) as model parameters.

The exact mathematical structure of the set of ordinary differential equations and the method of data variation and fitting is outlined in ref. (Belyaeva et al., 2008, 2011). The dissipative reactions (including inter-system crossing), given by the rate constants k_{P680^+}, $k_{3_{Car}}$ and k_{HD}, quench the fluorescence from PS II in addition to the photochemical quenching via electron transfer to the acceptor quinone molecule Q_A (see Figure 47b). The fluorescence emission is symbolized in Figure 47b by the radiative rate constant k_F. Actinic flash induced formation and decay of non-photochemical quenching states $P680^{+\bullet}$, ^3Car and radiation-less dissipation as heat was the main focus of (Belyaeva et al, 2011).

The numerical fits of SFITFY data presented as colored solid lines on Figure 48 are seen to describe with high precision the SFITFY patterns at the four different values of the actinic flash energy. The energies of the actinic flashes and the corresponding relative values are compiled in columns 1 and 2 of Table 1. The maximal light constant values ($k_{L\text{-}Max}$)

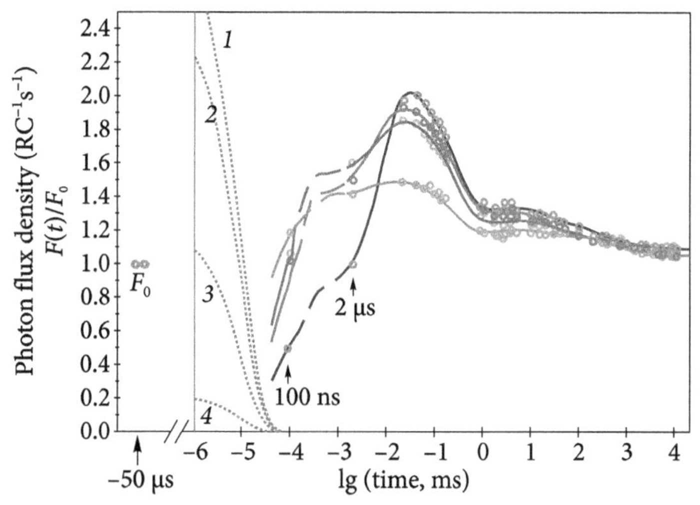

Figure 48. Simulation of the experimental SFITFY data of Figure 46 by the PS II model presented in Figure 47 (a, b) (redrawn from Belyaeva et al., 2011). SFITFY curves in whole leaves of wild type plants of *Arabidopsis thaliana* are shown by symbols at the different laser flash energies: $7.5\cdot10^{16}$ photons/cm^2·flash (dark-blue), $6.2\cdot10^{15}$ photons/cm^2·flash (magenta), $3.0\cdot10^{15}$ photons/cm^2·flash (beige) and $5.4\cdot10^{14}$ photons/cm^2·flash (light-green). The numerical fits are shown accordingly by lines calculated with the rate constant $k_{L\text{-}Max}$ values (see Table 1): $7.2\cdot10^9$ s^{-1} (dark-blue), $6.0\cdot10^8$ s^{-1} (red), $2.9\cdot10^8$ s^{-1} (brown), $5.2\cdot10^7$ s^{-1} (green) and the parameters as shown in Table 2. The dotted magenta lines represent the time courses of $k_L(t)$. The measuring light of low intensity was simulated with $k_{L\text{-}Min} = 0.2$ s^{-1} (see Table 1). Image reproduced with permission.

in column 3 imitate the rate of light quanta exciting the Chl molecules in the PS II model (Figure 47). The measuring light beam of the LED pulses, is not visible in Figure 48 because this value is smaller by several orders of magnitude compared to the activating actinic flash.

Table 1. Values of parameters used for the fitting of SFITFY data (see Figure 48) according to the model of PS II shown in Figure 47 as published in (Natalya, 2011). (PPFD stands for photosynthetic photon flux density; all other variables are described in the text).

curve number	$\lambda = 532$ nm; fwhm = 10 ns		k_n (s^{-1})			a_{Car}	pH_{Stroma}	$\Delta\Psi_0$, mV
	photons/ (cm^2·flash)	%	$k_{L\text{-}Max}$	k_{HD}	k_{Phe}	τ_{3Car} = 5.5 μs		(τ_Ψ, ms)
1	**$7.5 \cdot 10^{16}$**	100	$7.225 \cdot 10^9$	$1.95 \cdot 10^8$	$8 \cdot 10^8$	$2.08 \cdot 10^9$ 1	7.5	20 (800)
2	$6.2 \cdot 10^{15}$	8.27	$5.975 \cdot 10^8$	$1.95 \cdot 10^8$	$7 \cdot 10^8$	$5.6 \cdot 10^8$ 0.27	7.5	15 (500)
3	$3.0 \cdot 10^{15}$	4	$2.89 \cdot 10^8$	$1.95 \cdot 10^8$	$6.4 \cdot 10^8$	$3.5 \cdot 10^8$ 0.17	7.5	14 (500)
4	$5.4 \cdot 10^{14}$	0.72	$5.202 \cdot 10^7$	$1.65 \cdot 10^8$	$3 \cdot 10^8$	$1.2 \cdot 10^8$ 0.06	7.3	14 (400)
Measuring. light PPFD 0.8 μmol photons $m^{-2} \cdot s^{-1}$			0.2					
	1	**2**	**3**	**4**	**5**	**6**	**7**	**8**

Table 2. Values of parameters used for quantitative fits with the PS II model (Figure 47) simulations of SFITFY curves for whole leaves of *Arabidopsis thaliana* plants (see Figure 48).

reaction number, *n*	k_n (s^{-1})	K_{eq}	k_{-n} (s^{-1})	Processes References
2, 9, 16, 29	$3.2 \cdot 10^{11}/125$	40	$6.4 \cdot 10^7$	Charge separation (open RC)
6, 13, 20, 33	$2.56 \cdot 10^9 / 2.25$	40/4	$1.14 \cdot 10^8$	Charge separation (closed RC) (Schatz et al., 1988; Roelofs et al., 1992; Renger et al., 1995)
3, 10, 17, 30	$3 \cdot 10^9$	10^8	30	Charge stabilization on $Q_A^{-\bullet}$ (Schatz et al., 1988, Roelofs et al., 1992; Renger et al., 1995; Renger and Holzwarth, 2005)
4, 11, 18, 31	$(3.3 \div 1.5) \cdot 10^7$	80		Electron donation from tyrosine Z to $P680^{+\bullet}$ (Renger, 2001)
7	5000	16	312.5	ET from $Q_A^{-\bullet}$ to Q_B
14	1750	10	1750	ET from $Q_A^{-\bullet}$ to $Q_B^{-\bullet}$ (Renger and Schulze, 1985; Crofts and Wraight, 1983)

Table 2 shows the results gathered from the fit procedure for the values of PS II electron transfer parameters. The parameter values for processes of charge separation, stabilization and Q_B -site reactions could be kept invariant to actinic flash energy without deviations between experimental data and the numerical fit using values from the literature (see column 5 of Table 2). In marked contrast the parameters of dissipative reactions in the antenna and PS II and the rate constant for charge recombination (a_{Car}, k_{HD}, Figure 47b, k_{Phe}, Figure 47a) had to be changed when varying the relative actinic flash energy (see columns 4, 5, 6 of Table 1).

To simulate the fluorescence decay the values of $\Delta\Psi$ (transmembrane electrical potential difference) and pH_S (pH of the stroma) had to be varied lightly according to the values given in Table 1. The pH on the luminal side was set $pH_L = 6.5$ for all energies of the actinic flash.

The SFITFY curves monitored in the range from 100 ns to 10 s after excitation of dark adapted samples with a single actinic flash of a definite energy can be explained by time courses of the PS II redox states that are illustrated in Figure 49a and Figure 49b for 100% light intensity and 4% light intensity, respectively. In closed RCs with reduced Q_A^- the states $P680^{+\bullet} Phe^{-\bullet} Q_A^{-\bullet}$ are populated as a nonlinear process of two photon absorption during the actinic flash followed by a decrease due to the exponential decay of $k_L(t)$. Figure 49a (100% light intensity) shows full saturation (100%) of the closed RC and reduced overall Q_A^- population ($\sum Q_A^-$). The value of 100% light intensity refers to the value $k_{LMAX} = 7.5 \cdot 10^9\ s^{-1}$. $\sum Q_A^-$ stays near 100% until the maximum of the SFITFY curve is reached at about 50 μs.

Using $k_{LMAX} = 6.2 \cdot 10^8\ s^{-1}$ (corresponding to 8%·light intensity) the level of $\sum Q_A^-$ is still very close to saturation (95%) (data not shown). The calculations with $k_{LMAX} = 3 \cdot 10^8\ s^{-1}$ (4% light intensity) give rise to redox state population kinetics for $\sum Q_A^-$ as shown in Figure 49b. Here the levels of maximal $\sum Q_A^-$ are diminished to about 88%.

The results as presented in Figure 48 and Figure 49 (see summary in Table 2 and ref. Belyaeva et al., 2011) show that a consistent description of the experimental SFITFY data obtained at different excitation intensities can be achieved with invariant rate constants of electron transfer steps for all excitation pulse energies. In marked contrast, an increase of the actinic flash energy by more than two orders of magnitude from $5.4 \cdot 10^{14}$ photons/ (cm^2·flash) to $7.5 \cdot 10^{16}$ photons/ ($cm^2 \cdot$ flash), leads to an increase of the extent of fluorescence quenching due to carotenoid triplet (^3Car) formation with factor 14 and of the recombination reaction between reduced primary pheophytin (Phe⁻) and P680⁺ with factor 3 while the probability for heat dissipation in the antenna complex remains virtually constant.

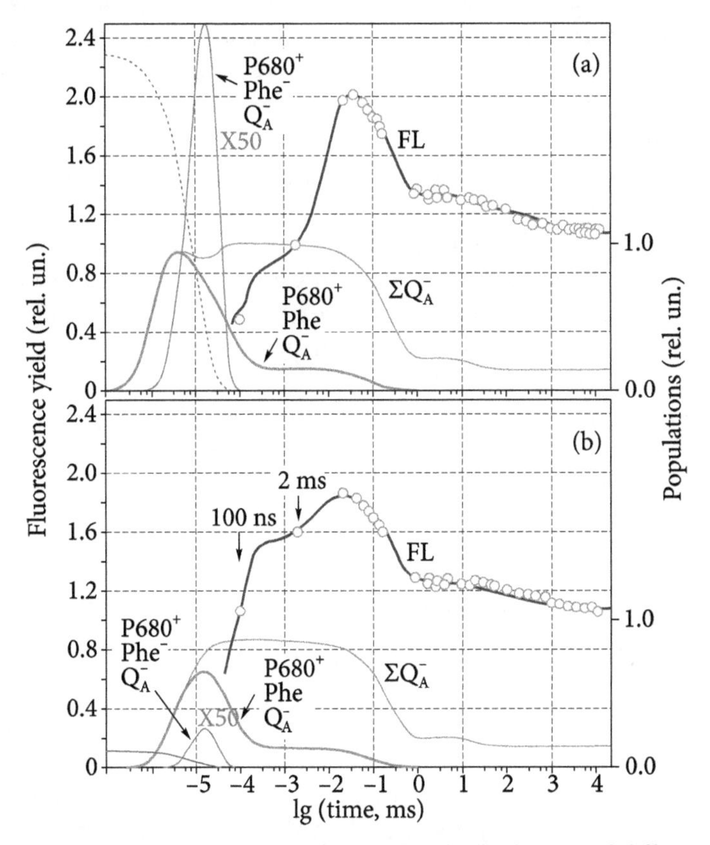

Figure 49. Calculated time course of normalized populations of different redox states in the PS II to simulate the SFITFY data of Figure 46 (circles) in whole leaves of *A.thaliana* wild type plants after illumination with an actinic laser flash (fwhm = 10 ns) at two different energies described by $k_{L\text{-}Max}$ values of $7.225 \cdot 10^9\,s^{-1}$ (panel a) and $2.9 \cdot 10^8\,s^{-1}$ (panel b) (redrawn from ref. Belyaeva et al., 2011). The measuring light is described by $k_{L\text{-}Min} = 0.2\,s^{-1}$. The PS II model parameters used are presented in Tables 1, 2. The time courses of $k_L(t)$ are shown as dotted purple lines for $k_{L\text{-}Max} = 7.225 \cdot 10^9\,s^{-1}$ (panel a) and $k_{L\text{-}Max} = 2.9 \cdot 10^8\,s^{-1}$ (panel b). $\Sigma Q_A^{-\bullet}$ represents the sum of the closed RC states ($x_4 + g_4 + y_4 + z_4 + x_5 + g_5 + y_5 + z_5$) (dark green curve, see nomenclature in the scheme of Figure 47). All states including oxidized Chl *a* in the RC ($P680^{+\bullet}$) are presented in the Fig.: $P680^{+\bullet}PheQ_A^{-\bullet}$ denoting the sum of the closed RC states ($x_4 + g_4 + y_4 + z_4$) (light green curve), $P680^{+\bullet}Phe^{-\bullet}Q_A^{-\bullet}$ – the sum of the closed RC states with reduced pheophytin ($x_7 + g_7 + y_7 + z_7$) multiplied with a factor 50 for better visibility (red curve). Image reproduced with permission.

However, when applying such a large parameter state as given in Figure 47 to fit a bunch of fluorescence curves as shown in Figure 46 and Figure 48 some of the parameters used to reproduce the measurements can be varied in a broad range without a significant change of the simulated SFITFY trace. Parameters describing fast processes (like e.g. k_{HD}) do not influence the shape of the SFITFY decay on a longer timescale (and vice versa). This phenomenon offers the opportunity to analyze a wide range of processes occurring on different time scales simultaneously by precise evaluation of different time domains of the measured SFITFY data or TCSPC traces.

On the other hand, the analysis of processes taking place on similar timescales requires complementary information gathered from independent experiments if there is not a well defined parameter different for these processes (for example wavelength, polarization).

Thus, taking into account these properties, the PS II model offers new opportunities to compare electron transfer and dissipative parameters for different species (e.g. for the green algae and the higher plant) under varying illumination or growth conditions.

2.8 Nonphotochemical Quenching in Plants and Cyanobacteria

The activity of PS II can be efficiently examined by measuring the fluorescence of Chl *a*, most frequently in form of prompt fluorescence or transient fluorescence measurements by recording Chl fluorescence induction curves imaging the transient development of photochemical and nonphotochemical quenchers in dark-adapted samples after excitation by intense light (Genty et al., 1989, Roháček, 2002, Stribet and Govindjee, 2011).

Several other techniques of fluorescence detection are used to analyze PSII activity. Delayed fluorescence (DF) is a suitable indicator to measure the recombination fluorescence response at the reaction centers of PSII that can be monitored as a very weak fluorescence signal with characteristic kinetics (Bigler and Schreiber, 1990; Goltsev et al., 2009). Prompt Chl fluorescence is analyzed to obtain the rate constants for charge separation and charge stabilization directly by evaluating time- and wavelength-resolved fluorescence measurements with model-based kinetic analysis (Schatz et al., 1988; Roelofs et al., 1992; Renger et al., 1995; Kramer and Crofts, 1996; Schmitt, 2011). Generally fluorescence is suitable to quantify photochemical or nonphotochemical quenching of excited states.

During evolution, cyanobacteria and plants have developed various mechanisms of acclimation, in particular regulatory pathways for defense to stress induced by unfavorable environmental factors. One important regulation mechanism is nonphotochemical quenching (NPQ) which already starts in the antenna complexes as a regulatory mechanism for EET and excited state accumulation under high light conditions. In such way plants decrease the rate of ROS generation in their leaves, increase the rate of ROS scavenging and accelerate of the repair of damaged cell structures.

The different protection mechanisms operate in markedly different time domains and light intensities. The fastest response is the annihilation process of multiple excitons which are excited in plant antenna complexes, however, such multiexcitons just occur at extreme light intensities or at rare probability.

Light harvesting systems of plants are highly optimized structures measuring coupled domains as the optimum tradeoff between light-harvesting, excitation energy transfer and energy storage to optimize the EET efficiency to the reaction center (Lambrev et al., 2011).

At low light intensities light-harvesting is optimized, whereas at high-light conditions the excess energy needs to be dissipated by photoprotective mechanisms where NPQ is probably the most direct and most efficient mechanism. NPQ starts in the antenna complexes as a regulatory mechanism for excitation energy transfer and excited state accumulation under high light conditions.

The most important mechanisms of NPQ might be quenching of superfluous excitation energy by carotenoids (Cars) and the induction of NPQ processes due to acidification of the thylakoid lumen by formation of a transmembrane pH difference (Niyogi and Truong 2013; Demmig-Adams and Adams, 1996; Demmig-Adams et al., 2014; Ruban et al., 2012). In this context, the large ΔpH across the thylakoid membrane (with acidic luminal pH) that builds up under extreme light due to the limited capacity of the F_0F_1-ATPase system triggers NPQ mechanisms (Müller et al., 2001; Szabo et al., 2005).

A regulation of excitation energy funnelling to PS I and PS II in oxygen-evolving organisms occurs via a phenomenon designated "state transitions" which comprises reversible phosphorylation/dephosphorylation of light-harvesting complexes II (Iwai et al., 2010).

Interestingly carotenoids seem to be of optimized chemical structure as dissipation molecules for excess energy. In contrast to the closed porphyrin molecule forming the light-harvesting system, chlorophyll, the carotenoids are long open carbon chains that exhibit a much larger rate constant for internal conversion. All carotenoids with more than ten conjugated C=C bonds have an excited singlet S1 state low enough to accept energy from excited Chl. However, the S1 state cannot be populated by one-photon absorption, but it can be reached upon rapid internal conversion from the S2 state or by direct Dexter-transfer of excitation energy from Chl molecules. In higher plants the dominating NPQ mechanism seems to be the light-induced and pH-dependent xanthophyll cycle (Härtel et al., 1996; Demmig-Adams et al., 1996). Additionally carotenoids effectively reduce the population of ^{3}Chl in antenna systems as well as PS II of plants (Carbonera et al., 2012) and act as direct ROS scavengers. The interaction with singlet oxygen ($^1\Delta_gO_2$) does not only lead to NPQ, but also to oxidation of carotenoids by formation of species that can act as signal molecules for stress response (Ramel et al., 2012; Schmitt et al., 2014a).

In the violaxanthine cycle of plants and green or brown algae xanthine deepoxidases associate with thylakoid membranes at low pH. In that configuration violaxanthin deepoxygenase converts violaxanthin via antheraxanthin to zeaxanthin. Distinct cycles developed in other organisms as the diadinoxanthin cycle in diatoms (Müller et al., 2001). Zeaxanthin deactivates excited Chl molecules more efficiently than violaxanthin.

For an analysis of the hierarchy of light induced kinetic steps in the PS II by measurement of single flash induced transient quantum yield and modelling with a PS II reaction scheme respecting different NPQ mechanisms by rate equations see (Belyaeva et al., 2008, 2011, 2014, 2015, and references therein).

Also mobile carotenoid binding proteins are found in plants and cyanobacteria which can be pH activated or directly induced by high light due to photoswitchable complexes. The carotenoid binding PsbS subunit of PSII in higher plants acts as a pH activated excess energy quencher after conformational changes by rearranging the PS II complexes. The semi-crystalline ordering and increased fluidity of protein organization in the membrane leads to NPQ (Bergantino et al., 2003; Müller et al., 2001; Niyogi and Truong, 2013; Schmitt et al., 2014a; Goral et al., 2012).

A prominent example of a mobile protein in cyanobacteria is the photo-switchable orange carotenoid protein (OCP) in cyanobacteria (Wilson A. et al., 2008; Wilson et al., 2010; Boulay et al., 2010; Stadnichuk et al., 2013; Maksimov et al., 2014a, 2015). OCP is a directly photoswitchable protein containing 3'-hydroxy-echinenone as cofactor, but also echinenone and canthaxanthin lead to photoconversion and quenching of phycobilisomes (PBS). During its photocycle OCP undergoes a spectral shift by more than 20 nm from the orange OCP^O to the photoactive red form OCP^R. The structure of the active OCP^R form is still unknown, which is likely due to substantial structural flexibility of the OCP^R state. OCP^R assumes a molten globule-like state with a larger hydrodynamic volume than the inactive OCP^O form and attaches quickly to phycobilisomes after photoactivation (Maksimov et al., 2015; 2016). The fluorescence decay curves of phycobilisomes (PBS) interacting with activated OCP^R are characterized by short decay components with $(170\ ps)^{-1}$ at strongest NPQ by OCP^R. PBS, which are strongly interacting with OCP^R, are lacking excitation energy transfer to the terminal emitter of the PBS antennae indicating that OCP quenches mainly the transfer from allophycocyanin in the PBS (Maksimov et al., 2014a). This fact was interpreted as intermolecular interaction between the OCP^R and its binding site in the PBS core induced by blue light. Detailed spectroscopic studies and investigations of OCP mutants unraveled most probable H-bonds between two residues, Trp-298 and Tyr-203 and an oxygen localized at the beta-ring of 3'-hydroxyechinenone as the most important interaction to stabilize the orange form OCP^O (Kirilovsky and Kerfeld, 2013; Leverenz at al., 2014; Maksimov et al., 2015; 2016).

Conclusively, Cars play a pivotal role (for reviews on the key role of Cars in photosynthesis, see Polívka and Sundström, 2004, Pogson et al., 2005) for NPQ developed under light stress (for a review, see Ruban et al., 2012) thus effectively reducing the population of ^3Chl in antenna systems as well as PS II of plants (Carbonera et al., 2012). Cars, in addition, act as direct ROS scavengers. The interaction between $^1\Delta_gO_2$ and singlet ground state Cars does not only lead to photophysical quenching, but also to oxidation of Cars by formation of species that can act as signal molecules for stress response (Ramel et al., 2012).

Conformational changes of pigment-protein complexes are typically induced under high light conditions leading to the depletion of excited singlet states by internal conversion and interaction with quenching groups in the protein backbone. Recently, such conformational changes were artificially introduced by freezing of PBS of cyanobacteria and it was shown that this can reduce the fluorescence quantum yield of the PBS by 90% (Maksimov et al., 2013).

Light harvesting complexes containing phycobiliproteins are not prone to triplet formation since phycocyanobilins (linear tetrapyrrols) do not undergo inter-system crossing. However triplet states are the prerequisite for the formation of highly reactive singlet oxygen ($^1\Delta_gO_2$)-.

As they do not form triplets and $^1\Delta_gO_2$, the PBS must not necessarily be quenched by carotenoids at high light conditions. The system just prevents EET to the Chl containing core antenna systems. Decoupling mechanisms seem also to occur that have been extensively studied for PBS and the rod-shaped phycobiliprotein antenna of the cyanobacterium *A.marina* including the EET processes on a molecular level (Schmitt et al., 2006; Theiss et al., 2008, 2011; Schmitt, 2011). It was found that the phycobiliprotein antenna of *A.marina* decouples from the PSII under cold stress (Schmitt et al., 2006) therefore reducing the influx of excitation energy into the PS II.

A detailed description of Chl *a* fluorescence as a reporter of the functional state of the PS II is found in (Strasser et al., 2000; Papageorgiou and Govindjee, 2005).

2.9 Hierarchical Architecture of Plants

The dynamics and structural organisation of photosynthetic organisms covers more than 10 orders of magnitude in space and up to 26 orders of magnitude in time, ranging from the pm-dimension of electronic orbitals up to plants that are tens of meters large and from the fs dynamics of light absorption to the life time of trees that can span thousands of years. Our aim is to contribute to our understanding of the complicated hierarchical network of communication that enables the stability of such structures over time.

In the following chapters, we will focus on the role of ROS as a messenger product and substrate in its contribution to communication and feedback in hierarchical networks. The described formalism of rate equations will be an approach to model such behaviour in the form of communication trees or flux diagrams that will make the observed behaviour understandable with reference to only the basic molecular entities of the system. Top down messaging in such systems occurs in the form of regulated gene expression that is originally caused by macroscopic parameters. One example is the regulation of genes by MAPKs that respond to ROS that are produced under high light conditions. In that sense the high light condition changes the protein environment in cells and is therefore a type of top-down communication in the plant. Generally all kind of feedback loops, from macroscopic structures to the microscopic environment, are top down. Such feedback does not necessarily need to target the basic molecular level. However, looking at the regulation of genes, the composition and, consequently, the behaviour of molecules on the microscopic level is partially induced from the macroscopic level.

These phenomena might also be constraints that develop from collective behaviour that are behind the structure parameters or the slaving principle mentioned in the first chapter. Plants turn out to be the optimized system for such consideration on the way to explain living matter as they exhibit features of top down signalling without the need to model consciousness or focus on the emergence of consciousness as plants do not seem to exhibit consciousness in a measurable way.

It is proposed that the ability to react to external stimuli by active mobility strategies is a necessary prerequisite to develop a strategic and existential advantage from consciousness and therefore supports its emergence. This assumption is in agreement with psychoinformatic studies performed in cognitive sciences such as e.g. the proposition of the anticipatory drive (Butz et al., 2003).

Figure 50 presents a schematic overview on the hierarchical organisation of structures that form the photosynthetic organism (e.g. a photosynthetic tree of 10 m) that is a sum of its cells (typically 10–100 μm). The photosynthesis takes place inside the leaves which are formed by an inhomogeneous 3-dim. array of cells on the μm-scale. The single cell is the smallest organized form that is still named "living system". It is a whole organism and all higher individual organisms are formed by a symbiotic biosystem of such cells.

In eukaryotic organisms the cells contain chloroplasts (chloroplasts have very different sizes and geometries, see e.g. (Häder, 1999) where photosynthesis takes place inside the thylakoid membrane. The cells contain the endoplasmatic reticulum that connects ribosomes and the cell organelles, i.e. mitochondria, chloroplasts, the nucleus and the vacuole. Additional minor components like hormones, proteins, free DNA and RNA and a number of different small molecules are not shown in Figure 50. The chloroplasts have their own DNA, lipid droplets, starch, ribosomes and they contain the Grana stacks of the thylakoid membrane. The thylakoid membrane is formed by a lipid bilayer with hydrophilic surface. Inside the thylakoid membrane the lipophilic environment contains the membrane proteins. They are hydrophobic and therefore bound to the membrane (membrane intrinsic proteins).

Figure 51 schematically shows the substructure of the chloroplasts containing the thylakoid membrane where the photosynthetic pigment-protein-complexes are located as membrane intrinsic proteins.

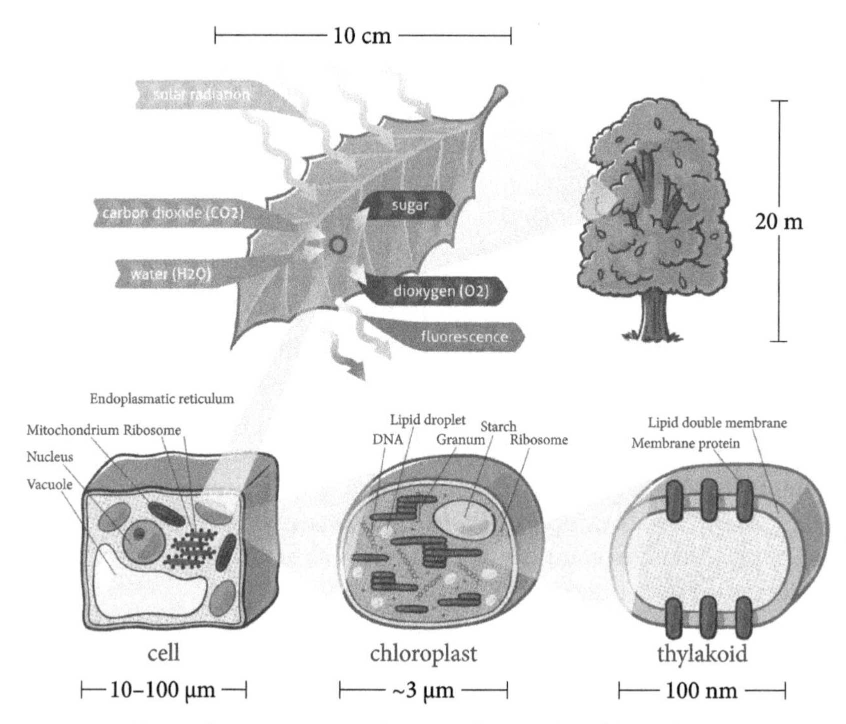

Figure 50. Hierarchic structures of green plants. The chloroplasts contain the Grana stacks of the thylakoid membrane where photosynthesis takes place.

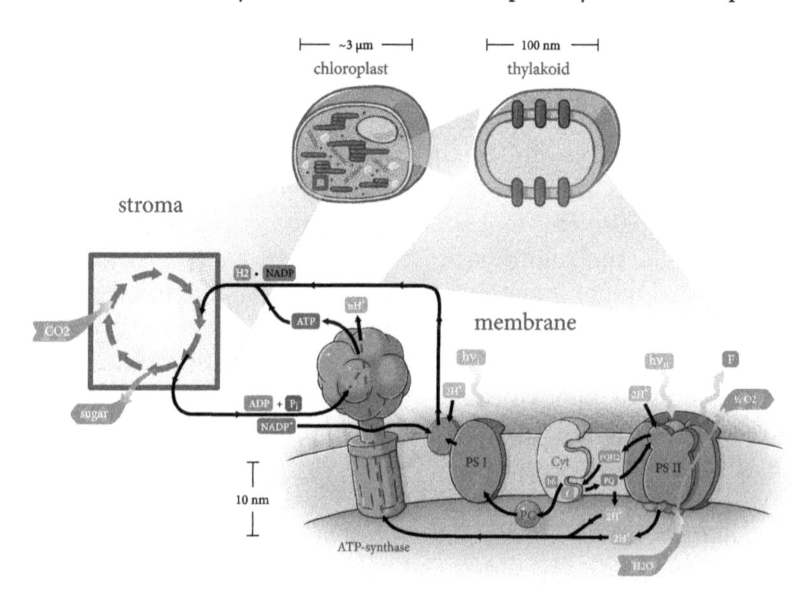

Figure 51. Membrane proteins inside the lipid double membrane of the thylakoids.

The dynamics of the thylakoid membrane is assumed to have impact on the transient changes of the quantum yield of photochemical light transformation. Therefore the thylakoid membrane is typically the highest organised structure taken into account for the description of fluorescence measurements that cover a time scale up to several ms after light excitation like single flash induced transient fluorescence yield measurements in whole cells of *Chlorella pyrenoidosa* Chick or higher plants (Belyaeva et al., 2008, 2011, 2014, 2015).

The photosynthetic reactions leading to CO_2-fixation comprise the light driven reactions which take place inside the thylakoid membrane and the "dark" reactions which take place inside the chloroplast stroma (indicated as a red cycle process in Figure 51).The thylakoid membrane divides the aqueous phases of the chloroplasts into the thylakoid lumen and the chloroplast stroma. Protons released into the thylakoid lumen and protons transferred from the stroma into the lumen form a trans-membrane electrochemical potential difference which drives the adenosine triphosphate (ATP) synthase that phosphorylates adenosine diphosphate to the energy rich compound ATP required as free energy source in the dark reaction for carbon reduction.

The "light" reaction performs the exploitation of solar energy by highly functionalized PPCs. Solar energy represents the unique Gibbs free energy source of the biosphere on earth. The Gibbs free energy is converted into high energetic chemical compounds via the process of photosynthesis. This goal is achieved in a perfect manner by incorporation of suitable chromophores into protein matrices. The PPCs are optimized to energy absorption and transfer producing the high energetic compounds ATP and NADPH.

3
Formation and Functional Role of Reactive Oxygen Species (ROS)

In the following chapters we will focus on the role of ROS as a messenger product and substrate in contributing to communication and feedback in hierarchical networks. The next two chapters provide an overview on recent developments and current knowledge about monitoring, generation and the functional role of reactive oxygen species (ROS) – H_2O_2, $HO_2^{\bullet}$, $HO^{\bullet}$, OH^-, $^1O_2$ and $O_2^{-\bullet}$ – in both oxidative degradation and signal transduction in photosynthetic organisms. We further describe microscopic techniques for ROS detection and controlled generation. Reaction schemes elucidating formation, decay, monitoring and signaling of ROS in cyanobacteria and eukaryotes are discussed. The discussion further targets the rapidly growing field of regulatory effects of ROS on nuclear gene expression.

The excess of ROS under unfavorable stress conditions causes a shift in the balance of oxidants/antioxidants towards oxidants, which leads to the intracellular oxidative stress. Formation of ROS (the production rate) as well as decay of ROS (the decay rate) with the latter one determining the lifetime of ROS finally determines the actual concentration distribution of the ROS pool. The activity of antioxidant enzymes, superoxide dismutase (SOD), catalase, peroxidases, and several others, as well as the content of low molecular weight antioxidants, such as ascorbic acid, glutathione, tocopherols, carotenoids, anthocyanins, play a key role in regulation of the level of ROS and products of lipid peroxidation (LP) in cells (Apel and Hirt, 2004; Biel et al., 2009; Pradedova et al., 2011; Kreslavski et al., 2012b).

The exact mechanisms of neutralization and the distribution of ROS have not all been clarified so far. Especially the involvement of organelles, cells and up to the whole organism, summarizing the complicated network of ROS signalling are still far from being completely understood (Swanson and Gilroy, 2010; Kreslavski et al., 2013a; Schmitt et al., 2014a).

It is obvious that ROS exert deleterious effects. Oxidative destruction by ROS is known and has been studied for decades. However, ROS also act as important signaling molecules with regulatory functions. ROS were found to play a key role in the transduction of intracellular signals and in control of gene expression and activity of antioxidant systems (Apel and Hirt, 2004; Desikan et al., 2001, 2003; Mori and Schroeder, 2004; Galvez-Valdivieso and Mullineaux, 2010; Foyer and Shigeoka, 2011; Schmitt et al., 2014a). Being implicated in reactions against pathogens, (e.g. by respiratory bursts) and by the active participation in signaling, ROS have a protective role in plants (Bolwell et al., 2002; Dmitriev, 2003).

ROS contribute to acclimation and protection of plants, regulate processes of polar growth, stomatal activity, light-induced chloroplast movements, and plant responses to biotic and abiotic environmental factors (Mullineaux at al., 2006; Pitzschke and Hirt, 2006; Miller et al., 2007; Swanson and Gilroy, 2010; Vellosillo et al., 2010). In animals, recent studies have established that physiological H_2O_2 signaling is essential for stem cell proliferation, as illustrated in neural stem cell models, where it can also influence subsequent neurogenesis (Dickinson and Chang, 2011). The following book chapters will describe generation and decay of ROS and their monitoring in cells (chapter 3). Additionally the rapidly growing field of regulatory effects and pathways of ROS will be described (chapter 4) although a complete description of the multitude of roles of ROS from nonphotochemical quenching (NPQ) to genetic signaling is impossible. However, these two chapters provide an overview about the existing knowledge aiming to include the most important original literature and reviews. The following two book chapters are based on the review of (Schmitt et al., 2014a), however, they are significantly broadened to connect the research to overall top down and bottom up signaling in hierarchic networks as it is the aim of this book.

3.1 Generation, Decay and Deleterious Action of ROS

Photosynthetic organisms growing under variable environmental conditions are often exposed to different types of stress like harmful irradiation (UV-B or high-intensity visible light), heat, cold, high salt concentration and also infection of the organisms with pathogens (viruses, bacteria) (Gruissem et al., 2012). Under these circumstances, the balance between oxidants and antioxidants within the cells is disturbed. This imbalance leads to enhanced population of ROS including singlet oxygen ($^1\Delta_gO_2$), superoxide radicals ($O_2^{-\bullet}$ or $HO_2^{\bullet}$), hydrogen peroxide (H_2O_2) and hydroxyl radicals ($HO^{\bullet}$). Other highly reactive oxygen species like atomic oxygen or ozone are either not formed or play a role only under very special physiological conditions and will not be considered here. In this sense, the term ROS is used in a restricted manner. In addition to ROS, also reactive nitrogen- and sulfur-based species play an essential role in oxidative stress (OS) developed within the cells (Fryer et al., 2002; Benson, 2002; Blokhina and Fagerstedt, 2010). However, this interesting subject is beyond the scope of this book.

Under optimal conditions, only small amounts of ROS are generated in different cell compartments. However, exposure to stress can lead to a drastic increase of ROS production and sometimes to inhibition of cell defense systems (Desikan et al., 2001; Nishimura and Dangl, 2010). Plants developed a network of reaction cascades interacting with ROS controlled by ROS and controlling ROS. The elucidation of this interaction network is aimed to be conducted in the following two chapters (chap. 3 and chap. 4) (Biel et al., 2009; Schmitt et al., 2014a).

Rapid transient ROS generation can be observed and is called "oxidative burst" (Bolwell et al., 2002). In this case, a high ROS content is attained within time periods from several minutes up to hours. Oxidative bursts occur during many plant cell processes, especially photosynthesis, dark respiration and photorespiration. Studies using advanced imaging techniques, e.g. a luciferase reporter gene expressed under the control of a rapid ROS response promoter in plants (Miller et al., 2009), or a new H_2O_2/redox state-GFP sensor in zebrafish (Niethammer et al., 2009; see chapter 3.2, *"Monitoring of ROS"*), revealed that the initial ROS burst triggers a cascade of cell-to-cell communication events that result in formation of a ROS wave. This wave is able to propagate throughout different tissues, thereby carrying the signal over long distances (Mittler et al., 2011). Recently, the auto-propagating nature of the ROS wave was experimentally demonstrated. Miller et al. (2009) used local application of catalase or an NADPH oxidase inhibitor to show that a ROS wave triggered by different stimuli can be blocked at distances of up to 5–8 cm from the site of signal origin. The signal requires the presence of the NADPH oxidase (the product of the *RbohD* gene) and spreads throughout the plant in both the upper and lower directions.

3.1.1 Direct $^1\Delta_gO_2$ Generation by Triplet-triplet Interaction

The ground state of the most molecules including biological materials (proteins, lipids, carbohydrates) has a closed electron shell with singlet spin configuration. These spin state properties are of paramount importance, because the transition state of the two electron oxidation of a molecule in the singlet state by $^3\Sigma_g^- O_2$ is "spin-forbidden" and, therefore, the reaction is very slow. This also accounts for the back reaction from the singlet to the triplet state. This situation drastically changes through two types of reactions which transform $^3\Sigma_g^- O_2$ into highly reactive oxygen species (ROS): i) Electronic excitation leads to population of two forms of singlet O_2 characterized by the term symbols $^1\Delta_g$ and $^1\Sigma_g^+$. The $^1\Sigma_g^+$ state with slightly higher energy rapidly relaxes into $^1\Delta_gO_2$ so that only the latter species is of physiological relevance (type I). ii) Chemical reduction of $^3\Sigma_g^- O_2$ (or $^1\Delta_gO_2$) by radicals with non-integer spin state (often doublet state) leads to formation of $O_2^{-\bullet}$, which quickly reacts to $HO_2^\bullet$ and is subsequently transferred to H_2O_2 and $HO^\bullet$ (*vide infra*) (type II). In plants, the electronic excitation of $^3\Sigma_g^- O_2$ occurs due to close contact to chlorophyll triplets that are produced during the photoexcitation cycle (Schmitt et al., 2014a) (see eq. 63, Figure 52, Figure 53). Singlet oxygen is predominantly formed via the reaction sensitized by interaction between a chlorophyll triplet (^{3}Chl) and ground state triplet $^3\Sigma_g^- O_2$:

$$^3\mathrm{Chl} + {}^3\Sigma_g^- O_2 \rightarrow {}^1\mathrm{Chl} + {}^1\Delta_gO_2 \qquad (63)$$

^{3}Chl can be populated either via intersystem crossing (ISC) of antenna Chls or via radical pair recombination in the reaction centers (RCs) of photosystem II (PS II) (for reviews, see Vass and Aro, 2008; Renger, 2008; Rutherford et al., 2012; Schmitt et al., 2014a). Alternatively, $^1\Delta_gO_2$ can also be formed in a controlled fashion by chemical reactions, which play an essential role in programmed cell death upon pathogenic infections (e.g. by viruses).

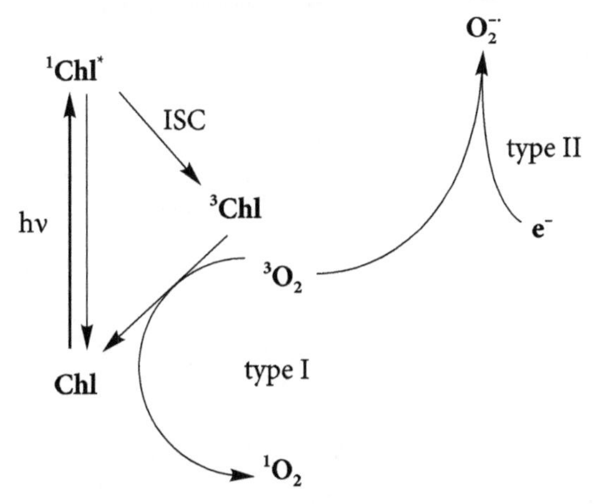

Figure 52. Production of ROS by interaction of oxygen with Chlorophyll triplet states (type I) to $^1O_2$ or chemical reduction of oxygen to $O_2^{-\bullet}$ (type II).

Figure 53 schematically illustrates the pattern of one-electron redox steps of oxygen forming the ROS species $HO^\bullet$, H_2O_2 and $HO_2^\bullet/O_2^{-\bullet}$ in a four-step reaction sequence with water as the final product. The sequence comprises the water splitting, leading from water to $O_2 + 4H^+$ and the corresponding mechanism *vice versa* of the ROS reaction sequence. The production of $^1\Delta_g O_2$ is a mechanism next to that.

The singlet state $^1\Delta_gO_2$ has a long intrinsic lifetime decaying during phosphorescence emission with maximum at 1268 nm. However, it can decay via both radiative and especially non-radiative routes. The lifetime of $^1\Delta_gO_2$ is especially speeded up from its natural half time of 72 min. by collisions with other molecules and its high reactivity (Mattila et al., 2015).

Rapid non-radiative decay of both singlet forms of oxygen $^1\Sigma_g^+ O_2$ and $^1\Delta_gO_2$ occurs via (i) conversion of excitation energy of $^1\Sigma_g^+ O_2$ and $^1\Delta_gO_2$ to vibrational and rotational energy and (ii) a charge-transfer mechanism and (iii) electronic energy transfer (Mattila et al., 2015).

Electronic energy transfer of $^1\Sigma_g^+ O_2$ and $^1\Delta_gO_2$ and a singlet ground state of a quenching molecule produces a triplet state of both the quencher and oxygen.

The final lifetime of $^1\Delta_gO_2$ in aqueous solution is about 3.5 μs (Egorov et al., 1989). But in cells the broad variety of reactive protein residua, lipids and other ROS together with the high reactivity of $^1\Delta_gO_2$ finally reduces the lifetime to a regime in the order of 200 ns reported for $^1\Delta_gO_2$

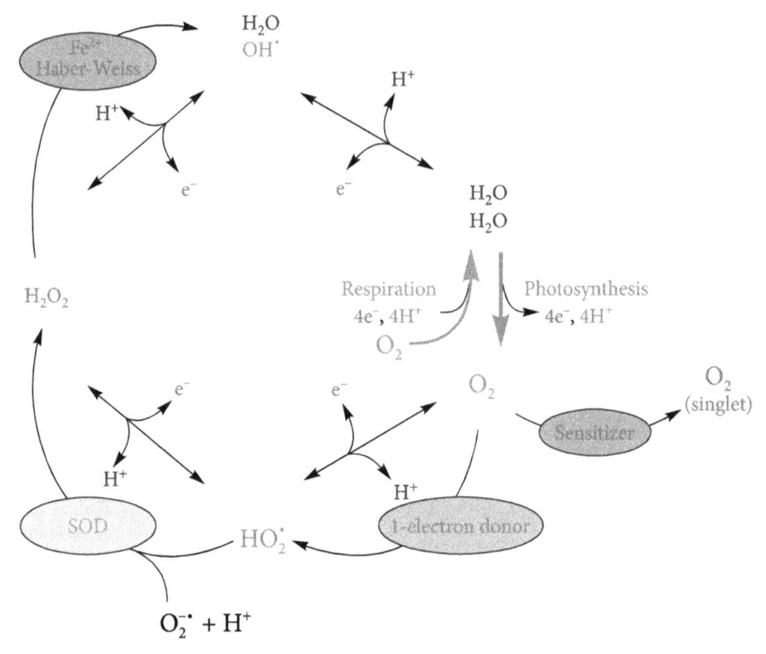

Figure 53. Scheme of ROS formation and water redox chemistry (water-water cycle) according to (Schmitt et al., 2014a). Image reproduced with permission.

in cells (Gorman and Rogers, 1992) so that the species can diffuse not much more than 10 nm under physiological conditions (Sies and Menck, 1992), thus permitting penetration through membranes (Schmitt et al., 2014a). But also distances up to 25 nm have been reported (Moan, 1990) suggesting that $^1\Delta_gO_2$ can permeate through the cell wall of *E. coli*. The singlet oxygen chemistry significantly depends on the environment, solvent conditions and the temperature (Ogilby and Foote, 1983). Higher values of up to 14 μs lifetime and 400 nm diffusion distance in lipid membranes suggest that $^1\Delta_gO_2$ can indeed diffuse across membranes of cell organelles and cell walls (Baier et al., 2005). But as most proteins are prominent targets (Davies, 2003) with reaction rate constants in the range of 10^8–10^9 $M^{-1}s^{-1}$ the potential of $^1\Delta_gO_2$ to work directly as a messanger is rather limited (Wilkinson et al., 1995). Among the canonical amino acids, only five (Tyr, His, Trp, Met and Cys) are primarily attacked by a chemical reaction with $^1\Delta_gO_2$, from which Trp is unique by additionally exhibiting a significant physical deactivation channel that leads to the ground state $^3\Sigma_g^-O_2$ in a similar way as by quenching with carotenoids. The reaction of $^1\Delta_gO_2$ with Trp primarily leads to the formation of peroxides, which are subsequently degraded into different stable products. One of these species is N-formylkynurenine (Gracanin et al., 2009). This compound exhibits optical and Raman spectroscopic characteristics that might be useful for the identification of ROS generation sites (Kasson and Barry, 2012). The reactivity of Trp in proteins was shown to markedly depend on the local environment of the target (Jensen et al., 2012). Detailed mass spectrometric studies revealed that a large number of oxidative modifications of amino acids are caused by ROS and reactive nitrogen species (Galetskiy et al., 2011).

The wealth of studies on damage of the photosynthetic apparatus (PA) by $^1\Delta_gO_2$ under light stress and repair mechanisms is described in several reviews and book chapters on photoinhibition (Li et al., 2012; Vass and Aro, 2008; Adir et al., 2003, Allakhverdiev and Murata, 2004; Nishiyama et al., 2006; Murata et al., 2007; Li et al., 2009; Goh et al., 2012, Allahverdiyeva and Aro, 2012). Such high reactivity leads to an extensive oxidation of fundamental structures of PS II where oxygen is formed in the water-oxidizing complex. $^1\Delta_gO_2$ is directly involved in the direct damage of PS II (Mishra et al., 1994, Hideg et al., 2007; Triantaphylidès et al., 2008, Triantaphylidès and Havaux, 2009; Vass and Cser, 2009), destroying predominantly the D1 protein, which plays a central role in the primary processes of charge separation and stabilization in PS II. The resulting photoinhibition of PS II (Nixon et al., 2010) leads to dysfunction of D1 and high turnover rates during the so called D1-repair cycle. D1 by far exhibits the highest turnover rate of all thylakoid proteins and underlies complex regulatory mechanisms (Loll et al., 2008).

Carotenoids play a pivotal role in ^{3}Chl suppression and quenching (Frank et al., 1993; Pogson et al., 2005) but also direct depletion of $^1\Delta_gO_2$. In addition, NPQ developed under light stress also reduces the population of ^{3}Chl in antenna systems as well as PSII of plants (Härtel et al., 1996; Carbonera et al., 2012, Ruban et al., 2012). The interaction between $^1\Delta_gO_2$ and singlet ground state carotenoids does not only lead to photophysical quenching, but also to oxidation of carotenoids by formation of species that can act as signal molecules for stress response (Ramel et al., 2012). Likewise, lipid (hydro)peroxides generated upon oxidation of polyunsaturated fatty acids by $^1\Delta_gO_2$ can act as triggers to initiate signal pathways, and propagation of cellular damage (Galvez-Valdivieso and Mullineaux, 2010; Triantaphylidès and Havaux, 2009). Further mechanisms of $^1\Delta_gO_2$ generation and decay together with detailed reaction scemes of $^1\Delta_gO_2$ are shown in (Mattila et al., 2015). Detailed studies of the damage of the PA by $^1\Delta_gO_2$ are additionaly found in (Li et al., 2012; Allakhverdiev and Murata, 2004; Nishiyama at al., 2006; Allakhverdiyeva and Aro, 2012; Goh et al., 2012; Nishiyama at al., 2006).

3.1.2 The $O_2^{-\bullet}/H_2O_2$ System

In biological organisms, the four-step reaction sequence of ROS as shown in Figure 53 is tamed and energetically tuned at transition metal centers, which are encapsulated in specifically functionalized protein matrices. This mode of catalysis of the "hot water redox chemistry" avoids the uncontrolled formation of ROS. In photosynthesis, the highly endergonic oxidative water splitting ($\Delta G° = +237.13$ kJ/mol, see Atkins, 2014) is catalyzed by a unique Mn_4O_5Ca cluster of the water-oxidizing complex (WOC) of photosystem II and energetically driven by the strongly oxidizing cation radical $P680^{+\bullet}$ (Klimov et al., 1978; Rappaport et al., 2002) formed via light-induced charge separation (for review, see Renger, 2012).

Correspondingly, the highly exergonic process in the reverse direction is catalyzed by a binuclear heme iron-copper center of the cytochrome oxidase (COX), and the free energy is transformed into a transmembrane electrochemical potential difference for protons (for a review, see Renger and Ludwig, 2011), which provides the driving force for ATP synthesis (for a review, see Junge, 2008). In spite of the highly controlled reaction sequences in photosynthetic WOC and respiratory COX, the formation of ROS in living cells cannot be completely avoided.

The general so called "water-water cycle" as shown in Figure 53 is at most responsible for the subsequent formation of $O_2^{-\bullet}$ or $HO_2^{\bullet}$, H_2O_2 and $HO^{\bullet}$. Among all ROS, the $O_2^{-\bullet}/H_2O_2$ system is one of the key elements in cell signaling and other plant functions. $O_2^{-\bullet}$ and H_2O_2 are assumed to initiate reaction cascades for the generation of "secondary" ROS as necessary for long-distance signaling from the chloroplasts to or between other cell organelles (Baier and Dietz, 2005; Sharma et al., 2012; Bhattacharjee, 2012).

The initial step in formation of redox intermediates of the H_2O_2/O_2 system in all cells is the one-electron reduction of O_2 to $O_2^{-\bullet}$ (see Figure 53). $O_2^{-\bullet}$ and H_2O_2 are mainly formed in chloroplasts, peroxisomes, mitochondria and cell walls (Corpas et al., 2012; Bhattacharjee, 2012). Enzymatic sources of $O_2^{-\bullet}$ / H_2O_2 generation have been identified such as cell wall-bound peroxidases, aminooxidases, flavin-containing oxidases, oxalate and plasma membrane NADPH oxidases (Bolwell et al., 2002; Mori and Schroeder, 2004; Svedruzic at al., 2005). In particular, sources of ROS in the apoplast are oxidases bound to the cell wall, peroxidases, and polyamino oxidases (Minibayeva et al., 1998, 2009). Recent studies have brought strong evidence that O_2 by direct reduction by plastohydroquinone is possible (Khorobrykh et al., 2015). As the direct two electron reduction by $^1\Delta_gO_2$ delivers a stoichiometric amount of H_2O_2 it is assumed without great doubt in the literature that $O_2^{-\bullet}$ is the source for formation of H_2O_2 (see Figure 53) (Khorobrykh et al., 2015).

It was proposed for a long time that the major source of $O_2^{-\bullet}/H_2O_2$ production in chloroplasts is the acceptor side of photosystem I (PS I) (Asada, 1999, 2006). So the exact mechanism of O_2 reduction might still remain a matter of discussion. It was assumed that O_2 mainly is reduced by transfer of electrons from reduced ferredoxin (Fd) to O_2 via ferredoxin-thioredoxin reductase (Gechev et al., 2006) although this assumption was challenged since a long time (Asada et al., 1974; Golbeck and Radmer, 1984). New findings showed that reduced Fd was only capable of low rates of O_2 reduction in the presence of $NADP^+$ with contribution to the total O_2 reduction not exceeding 10% (Kozuleva and Ivanov, 2010; Kozuleva et al., 2014). NADPH oxidase (NOX) is considered to be involved into ROS production both in animal and plant cells (Sagi and Fluhr, 2006) according to the reaction $NADPH + 2O_2 \rightarrow NADPH^+ + 2O_2^{-\bullet} + H^+$.

Under conditions of limited NADPH consumption due to impaired CO_2 fixation rates via the Benson-Bassham-Calvin cycle in photosynthetic organisms, some components of the electron transport chain (ETC) will stay reduced and can perform $^3\Sigma_g^- O_2$ reduction to $O_2^{-\bullet}$. It is suggested that H_2O_2 formation takes place in the plastoquinone (PQ) pool, but with a low rate (Ivanov et al., 2007), studies on mutants of *Synechocystis* sp. PCC 6803 lacking phylloquinone (*menB* mutant) show the involvement of phylloquinone in O_2 reduction (Kozuleva et al., 2014).

Very recent studies revise the assumption that PS I is a predominant source of $O_2^{-\bullet}$ and at all challenge the assumption of a one electron transfer from the PS I acceptor side (Prasad et al., 2015). Prasad et al. (2015) showed with a novel catalytic amperometric biosensor that formation of superoxide can occur via reduction of molecular oxygen by semiplastoquinone. In fact (Prasad et al., 2015) determined the reduction by Q_B^- as the major source of $O_2^{-\bullet}$ challenging the opinion that also PS I is involved. EPR spin-trapping data obtained using the urea-type herbicide DCMU showed the involvement of plastosemiquinone in $O_2^{-\bullet}$ production. From these findings and concomitant measurements of H_2O_2 in accordance with the reaction scheme shown in Figure 53 the authors (Prasad et al., 2015) also conclude that H_2O_2 in PSII membrane is exclusively formed by the dismutation of $O_2^{-\bullet}$.

Recent literature suggests very short lifetimes for $O_2^{-\bullet}$ radicals in the μs regime (1 μs half-life is published in (Sharma et al., 2012), while 2–4 μs are found in (Gechev et al., 2006) – which is about one order of magnitude longer than that of $^1\Delta_gO_2$ (*vide supra*). $O_2^{-\bullet}$ radicals are rapidly transformed into H_2O_2 via the one-electron steps of the dismutation reaction catalyzed by the membrane-bound Cu/Zn-superoxide dismutase (SOD) (see Figure 53) (Asada, 1999, 2006). Three forms of SODs exist in plants containing different metal centers, such as manganese (Mn-SOD), iron (Fe-SOD), and copper-zinc (Cu/Zn-SOD) (Bowler et al., 1992; Alscher et al., 2002), from which Cu/Zn-SOD is the dominant form. The non-enzymatic $O_2^{-\bullet}$ dismutation reaction is very slow (Foyer and Noctor, 2009; Foyer and Shigeoka, 2011). Earlier literature suggested generally a low reactivity of $O_2^{-\bullet}$ radicals indicating that the exact mechanisms of the $O_2^{-\bullet}$ reaction pathways in living cells might need further elucidation (see Halliwell and Gutteridge, 1985 and references therein). In earlier studies, Halliwell (1977) pointed out that $O_2^{-\bullet}$ is a moderately reactive nucleophilic reactant with both oxidizing and reducing properties. The negative charge of the $O_2^{-\bullet}$ radical leads to an inhibition of its electrophilic properties in presence of molecules with many electrons, while molecules with a low electron number might be oxidized. $O_2^{-\bullet}$ oxidizes enzymes containing [4Fe–4S] clusters (Imlay, 2003), while cytochrome c is reduced (McCord et al., 1977). Among the amino acids, mainly histidine, methionine, and tryptophan can be oxidized by $O_2^{-\bullet}$ (Dat et al., 2000). These radicals interact quickly with other radicals due to the spin selection rules. For example, superoxide interacts with radicals like nitric oxide and with transition metals or with other superoxide radicals (dismutation). As an example, Fe(III) is reduced by $O_2^{-\bullet}$, then H_2O_2 interacts with Fe^{2+} (Fenton reaction), in effect forming $HO^\bullet$, which is the most reactive species among all ROS (see also Figure 53). This reaction is particularly mentioned due to its importance for the generation of highly reactive $HO^\bullet$ from long-lived H_2O_2 which might act as long distance messenger. Further information about various reaction rate constants of $O_2^{-\bullet}$ at different conditions, concentrations and pH are found in (Rigo et al., 1977; Fridovich, 1983; Löffler et al., 2007)

H_2O_2 might be the most prominent cell toxic scavenged by a broad series of enzymes including catalases, peroxidases and peroxiredoxins. Within the chloroplasts, H_2O_2 is reduced to H_2O by ascorbate (Asc) via a reaction catalyzed by soluble stromal ascorbate peroxidase (APX)

(Asada, 2006; Noctor et al., 1998) or APX bound to the thylakoid membrane (t-APX). As shown in Figure 54, the Asc oxidized to the monodehydroascorbate radical (MDHA) is regenerated by reduction of MDHA either directly by Fd or by NAD(P)H catalyzed by MDHA reductase (MDHAR). The MDHA radical always decays partially into dehydroascorbate (DHA), which is reduced by DHA reductase (DHAR). In that step, reduced glutathione (GSH) is oxidized to glutathione disulfide (GSSG). The reduction of GSSG to GSH occurs from NAD(P)H by glutathione reductase (GR) (Noctor and Foyer, 1998; Asada, 2006).

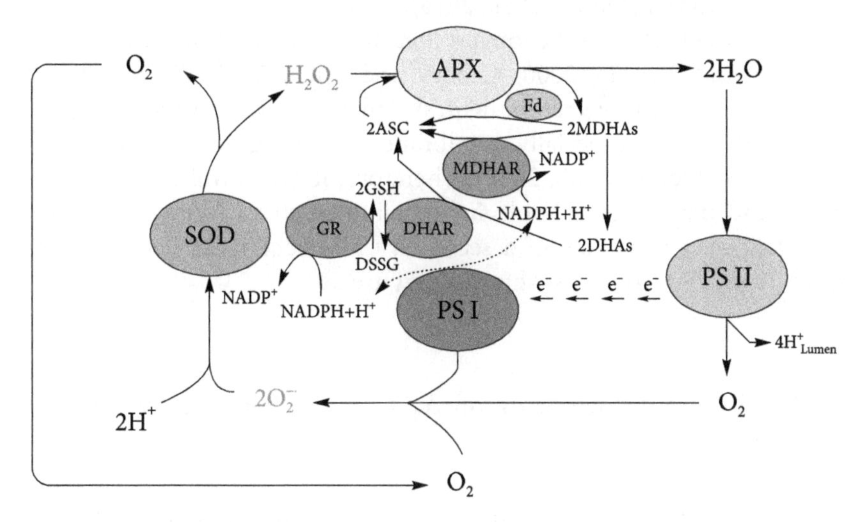

Figure 54. Scheme of pseudocyclic "H_2O–H_2O" electron transport according to (Schmitt et al., 2014a). Image reproduced with permission.

Assuming O_2 reduction to $O_2^{-\bullet}$ at the acceptor side of PS I, followed by dismutation of $O_2^{-\bullet}$ by SOD, and the reduction of H_2O_2 by t-APX one O_2 molecule is finally reduced to two H_2O molecules. This four-electron reduction process counterbalances the oxidation of two H_2O molecules to one O_2 molecule at the donor side of PS II so that no net change in the overall turnover of O_2 is obtained, as is schematically illustrated in Figure 54. The exact role of PS I and PS II might be a matter of discussion. However, this "water-water cycle" is referred to as pseudocyclic electron transport (for details, see Foyer and Shigeoka, 2011; Asada, 1999, 2006) driven by H_2O_2. It has to be kept in mind that this pseudocyclic electron transport can be coupled to the formation of a transmembrane pH difference, ΔpH.

Figure 53 and Figure 54 indicate that H_2O_2 could also be generated by oxidation of two H_2O molecules. Formation of H_2O_2 has long been reported to take place at a disturbed water-oxidizing complex (WOC) under special circumstances (Ananyev et al., 1992; Klimov et al., 1993; Pospisil, 2009). However, under physiological conditions, this process is negligible if taking place at all. Accordingly, H_2O_2 production at PS II should occur via the reductive pathway at the acceptor side under conditions where the PQ pool is over-reduced as proposed by (Ivanov et al., 2007; Prasad et al., 2015). Detailed reaction schemes for the reactions of H_2O_2 are found in (Mattila et al., 2015).

H_2O_2 is the ROS with the longest lifetime, which is in the order of 1 ms (Henzler and Steudle, 2000; Gechev and Hille, 2005). This is mostly supported as the molecule is neutral and therefore can pass lipophilic regions of the cell especially membranes including water channels like aquaporins (Bienert et al., 2007). Therefore, it can travel over large distances and play a central role in signaling of stress (see chapter 4) or undertakes even the role of a secondary photochemical electron transporter from PS II to PS I as shown in Figure 54.

3.1.3 H_2O_2 and Formation of $^1\Delta_gO_2$ and Other Reactive Species like $HO^\bullet$

Even at high light, no substantial amounts of $^1\Delta_gO_2$ and H_2O_2 are accumulated, when the electron flow through the pseudocyclic ETC (*vide supra*) increases and sufficient amounts of $NADP^+$ are present in the cell. At high light intensity and conditions of saturating CO_2 assimilation, the rate of electron flow increases. This leads to its redistribution, i.e. the rate of electron flow to $NADP^+$ decreases and, concomitantly, the rate of electron transfer through the pseudocyclic electron transport increases (Asada, 1999).

H_2O_2 can also participate in the control of $^1\Delta_gO_2$ formation, when an excess of H_2O_2 induces oxidation of the primary electron acceptor of PS II, thus leading to activation of the electron transport. As a result, production of $^1\Delta_gO_2$ is diminished due to reduced probability of ^3Chl population. Accordingly, pseudocyclic electron transport can function as a relaxation system to permit a decline of $^1\Delta_gO_2$ generation (Galvez-Valdivieso and Mullineaux, 2010). Such effects can result in autoinhibited reaction patterns and lead to spatiotemporal oscillations of the ROS distribution e.g. ROS waves.

The steady-state level of cellular H_2O_2 depends on the redox status of the cell (Karpinski et al., 2003; Mateo et al., 2006). Light-induced ROS generation in plants is mainly determined by the physiological state of the PA (Foyer and Shigeoka, 2011; Asada, 1999). Under physiological conditions, the H_2O_2 content in the cell is usually less than 1 μM. At elevated concentration, H_2O_2 inhibits several enzymes by oxidative cross-linking of pairs of cysteine residues. At about 10 μM, H_2O_2 inhibits CO_2 fixation by 50%, which is mainly due to the oxidation of SH groups of Benson–Bassham–Calvin cycle enzymes (Foyer and Shigeoka, 2011). H_2O_2 can block the protein synthesis in the process of PS II repair (Nishiyama et al., 2001, 2004, 2011; Murata et al., 2012). This effect of H_2O_2 has been analyzed in the cyanobacterium *Synechocystis* sp. PCC 6803. It was shown that the translation machinery is inactivated with the elongation factor G (EF-G) being the primary target (see chap. 4.1). Due to that oxidation the protein de novo synthesis is completely blocked via the stop of protein translation. This process has been studied in deep detail and it is understood today mainly as a protective mechanism that avoids an expensive de novo synthesis of proteins in a highly oxidizing environment. Further details on this general type of H_2O_2 signalling are found in chapter 4.

Elimination of H_2O_2 is tightly associated with scavenging of other ROS in plant cells. Both, H_2O_2 production and removal are precisely regulated and coordinated in the same or in different cellular compartments (Karpinski et al., 2003; Foyer and Noctor, 2005; Mateo et al., 2006; Ślesak et al., 2007; Pfannschmidt et al., 2009). The mechanisms of H_2O_2 scavenging are regulated by both, non-enzymatic and enzymatic antioxidants.

The biological toxicity of H_2O_2 appears through oxidation of SH groups and can be enhanced, if metal catalysts like Fe^{2+} and Cu^{2+} take part in this process (Fenton reaction) (see above and Figure 2). The enzyme myeloperoxidase (MPO) can transform H_2O_2 to hypochloric acid (HOCl), which has high reactivity and can oxidize cysteine residues by forming sulfenic acids (Dickinson and Chang, 2011) (see Figure 55).

Thus, H_2O_2 takes part in formation of reactive species like $HO^{\bullet}$ via several pathways.

$$\text{HOCl (enzyme MPO)} \leftarrow H_2O_2 \rightarrow H_2O \text{ (catalase, peroxidase)}$$

$$\downarrow Fe^{2+}, Cu^{2+}$$

$$[HO]^{\bullet}$$

Figure 55. Formation of hypochloric acid and $HO^{\bullet}$ from H_2O_2 and it´s detoxification by enzymes.

Both $O_2^{-\bullet}$ and H_2O_2 are capable to initiate the peroxidation of lipids, but since $HO^\bullet$ is more reactive than H_2O_2, the initiation of lipid peroxidation is mainly mediated by $HO^\bullet$ (Bhattacharjee, 2012; Miller et al., 2009).

Different defense systems have been developed to protect cells from deleterious effects of ROS. The underlying response mechanisms are either leading to diminished generation or enhanced scavenging of ROS. *De novo* synthesis of antioxidant enzymes (SOD, catalase, ascorbate peroxidase, glutathione reductase) and/or activation of their precursor forms take place and low-molecular antioxidants (ascorbate, glutathione, tocopherols, flavonoids) are also accumulated (Foyer and Noctor, 2005; Hung et al., 2005).

The antioxidant defense system contains many components (Pradedova et al., 2011). Essentially, three different types are involved: i) systems/compounds preventing ROS generation, primarily by chelating transition metals which catalyze $HO^\bullet$ radical formation, ii) radical scavenging by antioxidant enzymes and metabolites, and iii) components involved in repair mechanisms. Treatment of mature leaves of wheat plants with H_2O_2 was shown to activate leaf catalase (Sairam and Srivastava, 2000).

3.1.4 The $HO^\bullet$ Radical

The $HO^\bullet$ radical is the most reactive species known in biology. $HO^\bullet$ is isoelectronic with the fluorine atom and characterized by a midpoint potential of +2.33 V at pH 7 (for comparison, the normal reduction potential of fluorine is +2.85 V). In cells, the extremely dangerous $HO^\bullet$ radical can be formed by reduction of H_2O_2 via the Haber–Weiss reaction (Haber and Weiss, 1934) catalyzed by Fe^{2+} (Kehrer, 2000) (see Figure 55). $HO^\bullet$ radicals immediately attack proteins and lipids in the immediate environment of the site of production, thus giving rise to oxidative degradation (Halliwell, 2006). Cells cannot detoxify $HO^\bullet$ radicals and, therefore, a protection can only be achieved by suppression of H_2O_2 formation in the presence of Fe^{2+} using metal binding proteins like ferritins or metallothioneins (Hintze and Theil, 2006). On the other hand, $HO^\bullet$ radicals can be produced in programmed cell death as part of defense mechanisms to pathogenic infections (Gechev et al., 2006).

It has to be mentioned that the $HO^\bullet$ radical is not the only possible product of the reaction between H_2O_2 and Fe^{2+}. New calculations on the electronic structure and *ab initio* molecular dynamics simulations have shown that the formation of the ferri-oxo species $[Fe^{IV}(O^{2-})(H_2O)_5]^{2+}$ is energetically favored by about 100 kJ/mol compared to the generation of the $HO^{\bullet}$ radical (Yamamoto et al., 2012). Therefore, in future mechanistic studies, the species $[Fe^{IV}(O^{2-})(H_20)_5]^{2+}$ should be taken into account for mechanistic considerations on the oxidative reactions of H_2O_2 in the presence of Fe^{2+}.

3.2 Monitoring of ROS

The simultaneous formation of several ROS complicates the monitoring of formation, decay and degradation action of individual species. Especially the discrimination between $^1\Delta_gO_2$ and $H_2O_2/O_2^{-\bullet}$ is of high interest. Due to the fact there is a rising need for highly specific ROS probes able for discrimination between different ROS molecules. In the following, methodologies will be briefly summarized, with especial emphasis on application to cellular systems according to (Schmitt et al., 2014a).

$^1\Delta_gO_2$ can be directly monitored via its characteristic phosphorescence with a maximum around 1268 nm (Wessels and Rodgers, 1995; Mattila et al., 2015) and even singlet oxygen microscopy within the visible spectrum has been reported (Snyder et al., 2004). However, the detection of $^1\Delta_gO_2$ concentration and their time-dependent profiles in biological systems is difficult (for a discussion, see Li et al., 2012), because of the very low quantum yield of the emission, which ranges from 10^{-7} to 10^{-4} depending on the solvent (Schweitzer and Schmidt, 2003). In plant tissue the direct monitoring of $^1\Delta_gO_2$ luminescence is highly distorted by the concomitant Chl phosphorescence and both, Chl and $^1\Delta_gO_2$ cannot easily be discriminated due to the low intensity of both signals. For the direct activation of $^1\Delta_gO_2$ the decay time of $^1\Delta_gO_2$ can be used as a monitor however it strongly depends on the environmental conditions with typical values of 70 ns in living cells (Mattila et al., 2015) while other sources report 200 ns for the lifetime of $^1\Delta_gO_2$ in cells (Gorman and Rogers, 1992). In water typical values of 3,5 μs are found (Egorov et al., 1989) while the natural decay time in vacuum measures 72 min (Mattila et al., 2015).

Generally, detailed analyses require the use of suitable probe molecules as described in 3.2.1 for exogenic fluorescent probes and in 3.2.2 for spin traps generally suitable for ROS carrying own spin like $^1\Delta_gO_2$. In chapter 3.2.3 novel trends of genetically encoded ROS sensors are summarized. Additionally chapter 3.2.4 mentions electrochemical redox electrodes that have been shown to deliver reliable results especially for the detection of $H_2O_2/O_2^{-\bullet}$ using electrodes covered with horseradish peroxidase (Prasad et al., 2015).

3.2.1 Exogenic ROS Sensors

Two types of exogenous probe molecules are typically employed for monitoring of ROS: spin traps, which interact with ROS giving rise to EPR-detectable species (Hideg et al., 1994, 2011; Zulfugarov et al., 2011), and fluorophores, which change their emission properties due to interaction with ROS (*vide infra*).

The use of fluorophores offers a most promising tool because it permits the application of recently developed advanced techniques of time- and space-resolved fluorescence microscopy for *in vivo* studies (see Shim et al., 2012; Schmitt et al., 2013, 2014a, and references therein).

Two different approaches can be used: a) addition of exogenous fluorescence probes, which penetrate into the cell and change their fluorescence properties due to reaction with ROS, and b) expression of ROS-sensitive fluorescent proteins, mostly variants of the green fluorescent protein (GFP), which act as real-time redox reporters for the use in intact cyanobacteria, algae and higher plants. The latter ones are separately described in chapter 3.2.3 (Schmitt et al., 2014a).

Table 3 gives an overview on exogenous fluorophores that typically change their optical properties due to interaction with ROS and additionally mentions the most important properties as ROS specificity, localizability and typical application schemes as well as corresponding references for successful applications.

Most of the mentioned ROS fluorophores are not truly specific to certain ROS. However, as exogenous dyes typically respond in a certain oxidative potential range, appropriate mixtures can permit assays that are selective in a certain range of oxidative potentials (e.g. when both dyes show fluorescence or only one dye shows fluorescence, but the other not). These assays are more selective than those utilizing just a single dye.

Permeability across membranes is necessary at most for the applicability of exogenous ROS-sensing fluorophores (Table 3). Generally, the water/octanol partition coefficient could be utilized to quantify membrane permeability of the probes. For a quantitative analysis, it is necessary to know the reaction mechanism in detail, as well as possible interfering side effects and the cellular localization of these dyes. A good overview on detailed chemical reaction schemes is given by (Mattila et al., 2015).

Table 3. Compilation of ROS-sensitive exogenous fluorescence probes.

Compound/reference	Specificity Further information/localizability
CM-H2DCFDA (Dixit and Cyr, 2003)	Unspecific Permeates into animal cells, requires the presence of cellular esterases. Not easily applicable in plants
Singlet oxygen sensor green (SOSG) (Flors et al., 2006)	Highly specific to singlet oxygen Successfully used for detection of $^1\Delta_gO_2$ in *A. thaliana* leaves
3,3'-diaminobenzidine (DAB) (Thordal-Christensen et al., 1997; Fryer et al., 2002)	Specific to H_2O_2 in presence of peroxidase (and other heme-containing proteins) Generates a dark brown precipitate which reports the presence and distribution of hydrogen peroxide in plant cells. Permeates into plant cells.
Aminophenyl fluorescein (APF)	APF is a cell permeable indicator that can be used to detect hydroxyl radicals ($HO^•$), peroxynitrite ($ONOO^-$) and hypochlorite (OCl^-) production in cells. Shows limited photooxidation.
hydroxyphenyl fluorescein (HPF)	Specific to hydroxyl radical and peroxynitrite. Minor sensitivity to other ROS. HPF is cell permeable.
nitroblue tetrazolium (NBT) (Maly et al., 1989; Thordal-Christensen et al., 1997)	Specific to superoxide and with slightly reduced reactivity to hydrogen peroxide
Proxyl fluorescamine (Cohn et al., 2008)	Specific to hydroxyl radicals and superoxide Complementary use as spin trap
Hydroethidine (dihydroethidium) (Gomes et al., 2005)	Unspecific Binds specifically to DNA, marking the nucleus
DPPP (diphenyl-1-pyrenylphosphine) (Gomes et al., 2005)	Unspecific Lipophilic, detects ROS in lipids, blood plasma, tissues and food
MCLA (2-methyl-6-(4-methoxyphenyl)-3,7-dihydroimidazo[1, 2-a] pyrazin-3-one, hydrochloride) (Godrant et al., 2009)	Specific to superoxide or singlet oxygen
Trans-1-(2'-Methoxyvinyl)Pyrene	Highly specific to singlet oxygen

Some of these ROS probes can be tuned regarding their properties inside the cell by enzymatic reactions. For instance, the comercially available 2',7'-dichlorodihydrofluorescein diacetate, acetyl ester (H_2DCF-DA), a fluorescein-based dye, which is virtually non-fluorescent in the reduced state, becomes fluorescent after oxidation and concomitant splitting of the acetate groups by cellular esterases as 2',7'-dichlorofluorescein (DCF). H_2DCF-DA is widely used in (nonphotosynthetic) animal cells.

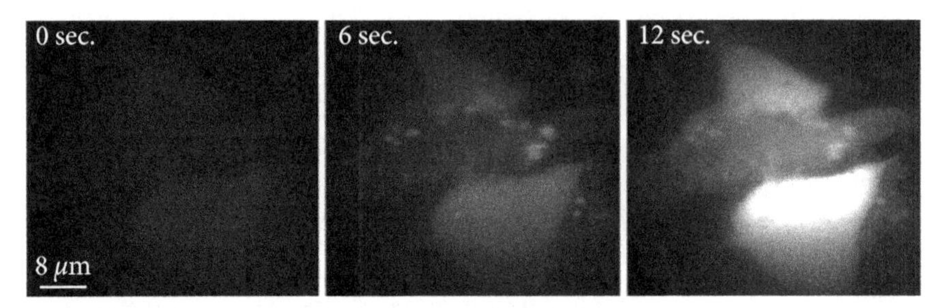

Figure 56. Increase of DCF fluorescence due to ROS production upon exposure of Chinese hamster ovary (CHO) cells to 440–480 nm light. The image shows the ROS content by intensity of the emission of DCF in three different cells.

Figure 56 illustrates the application of CM-H_2DCFDA in monitoring the development of ROS production upon exposure of CHO cells to 440–480 nm light in phosphate buffered saline (PBS). After staining the cells with CM-H_2DCFDA, the fluorescence of the indicator strongly rises upon illumination of the cells with light of 440 nm - 480 nm wavelengths due to the light-induced production of ROS and subsequent photooxidation in presence of oxidative compounds. It can be seen that after 6 sec, the lower cell has a higher cytosolic redox potential (higher fluorescence yield) than the upper two cells that show a less intense luminescence. On the other hand, these cells exhibit white "dots" indicating "hot spots" of accumulated DCF and/or higher local ROS activity (Schmitt et al., 2014a).

DCF was used to monitor the time course of the ROS production (determined from DCF fluorescence) between WT and the *hy2* and *hy3* mutants of *A.thaliana* (Kreslavski et al., 2013b) as it was shown that preillumination of the leaves *of A.thaliana* with red light reduced the inhibitory effect of UV radiation on the PSII activity in WT but not in the *hy3* and *hy2* mutants, which are lacking the phytochrome apoprotein and chromophore, respectively. These findings suggested that the active form of a phytochrome is involved in the protection of the PA against UV-A.

However, as Figure 57 indicates no significant kinetic differences of ROS production between WT and both mutants were observed.

CM-H_2DCFDA is sensitive to ROS only in the living cell environment (*in vivo*) which enables the generation of dyes not only sensitive to ROS but also indicating that the ROS are produced inside the cell. Such studies are necessary especially to avoid side effects due to generation of ROS by the applied dyes, monitoring of ROS outside the cells in solution due to unspecific localisation and/or photooxidation of the dye by illumination (Vitali, 2011; for a detailed description see also Dixit and Cyr, 2003).

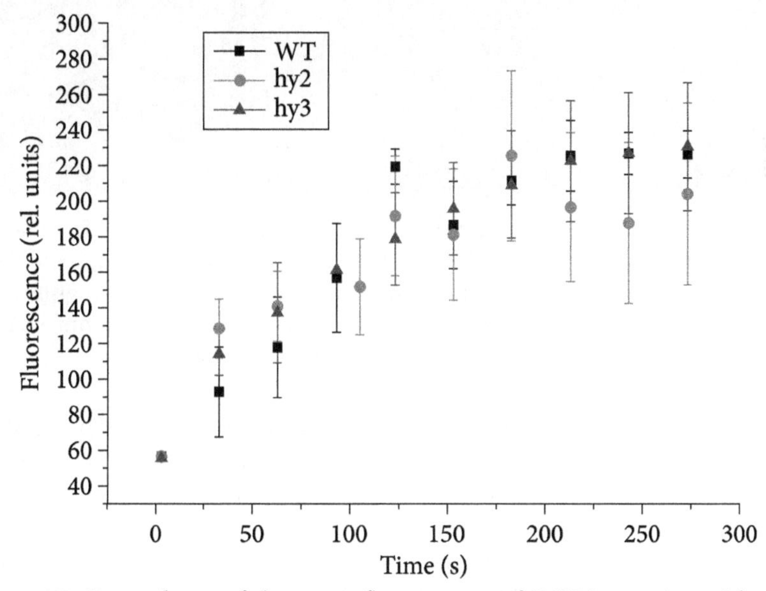

Figure 57. Dependence of the green fluorescence of DCF increasing with time of continuous irradiation with UV-A (360/40 nm). The DCF signal indicates the production of ROS in leaves of WT (black squares) and two phytochrome deficient mutants *hy2* (red circles) and *hy3* mutant (blue triangles) during illumination.

DCF can be used in plant cells of *A.thaliana* leaves for measuring the ROS production (mainly H_2O_2) upon illumination with UV-A. Figure 58 shows the highly fluorescent DCF after incubation of leaves of *A. thaliana* to PBS containing 500 μM H_2DCFDA. The leaves were exposed to H_2DCFDA solution for 1–2 h before starting the UV-A irradiation experiments. It can be seen that after irradiation with 360 nm UV-A light at an intensity of 250 W/m^2 for 10 minutes areas which contained microscopic damages exhibit strong DCF emission (Figure 58, left panel) while the Chl *a* emission at 680 nm appears reduced in the same areas (middle panel) due to photobleaching. The simultaneous reduction of the Chl *a* fluorescence together with enhanced DCF fluorescence becomes evident in the overlay image (Figure 58, right panel) where Chl *a* emission is recolored in red and DCF in green. DCF studies together with observation of Chl *a* bleaching show the interaction of ROS and Chl *a* (Kreslavski et al., 2013b)

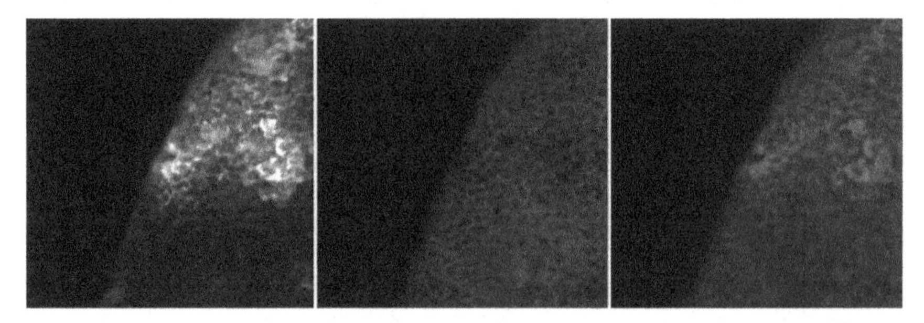

Figure 58. Fluorescence of a section of a 26-d-old *A.thaliana* leaf. The fluorescence was emitted from 2',7'-dichlorofluorescein (DCF) (excited at λ_m = 470 nm) after irradiation of the leaf with UV-A (360 nm; I – 250 Wm^{-2}) registered at 530 nm (left panel) in comparison to the Chl *a* fluorescence at 680 nm (middle panel). The overlay shows both (right panel) after recoloration.

Incubation of *A.thaliana* leaves in a PBS buffer with a final apparent concentration of the polyaromatic hydrocarbon (PAH) naphthalene (Naph) of 100 mg l^{-1} showed severe damage of the cell membrane upon illumination with UV-A as shown in Figure 59. Similar results were obtained for pea leaves (Kreslavski et al., 2014b). UV-radiation in combination with toxic compounds like PAHs lead to generation and accumulation of ROS as visible after staining with DCF.

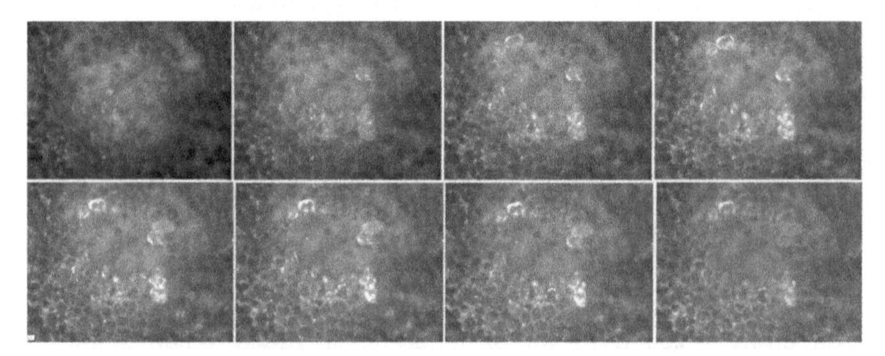

Figure 59. DCF fluorescence in leaves of *A. thaliana* incubated with Naph during illumination with UV-A.

Figure 60 clearly shows the accumulation of ROS in the cell membrane while the red Chl fluorescence is emitted from inside the chloroplasts (see Figure 61).

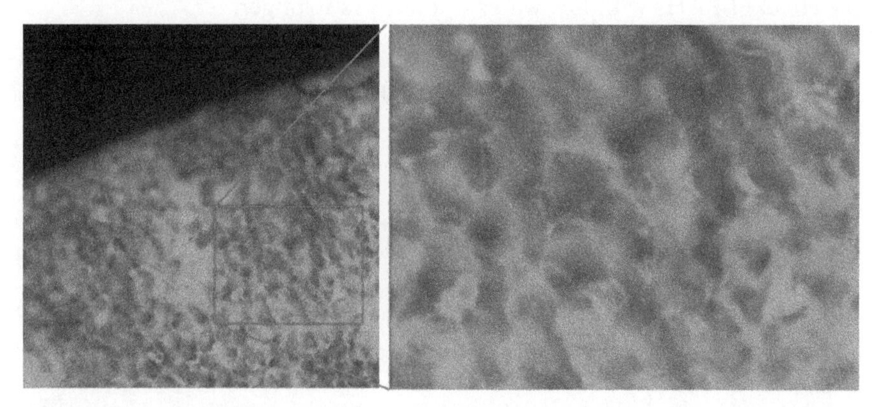

Figure 60. Localization of the green fluorescence of DCF in the cell membrane after illumination with UV light.

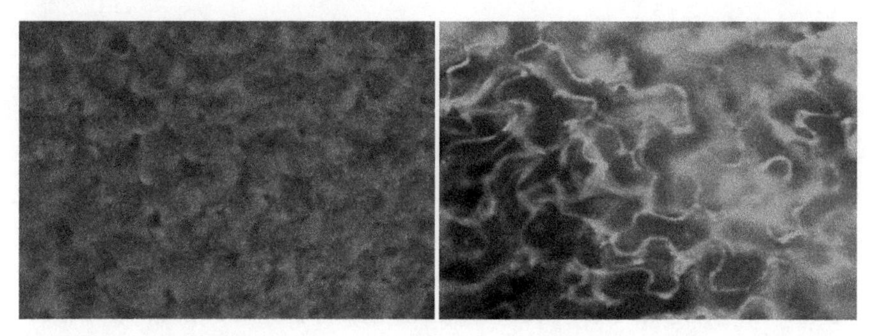

Figure 61. Localization of the green fluorescence of DCF in the cell membrane after illumination with UV light (right side) in contrast to fluorescence emitted from Chl at 680 nm (left side).

In leaves of *A. thaliana* treated with Naph, ROS waves with a temporal frequency of 20 minutes and a "wavelength" of several hundreds of micrometers were observed (Figure 62). Such a behavior is in line with wave-like closure and opening of stomata as observed in green plants under stress conditions.

In pea leaves the reduction of PSII activity at the presence of Naph is accompanied by transient generation of H_2O_2 as well as swelling of thylakoids and distortion of cell plasma membranes (Kreslavski et al., 2014b). It could be shown that Naph-treated leaves of *Arabidopsis thaliana* show enhanced DCF fluorescence in the cell membrane. The comparison of short term and long term exposure to different PAHs revealed that at short term exposure, the PAHs with high water solubility lead to the strongest reduction of PS II activity while after long term exposure the effect of PAHs with low water solubility is stronger.

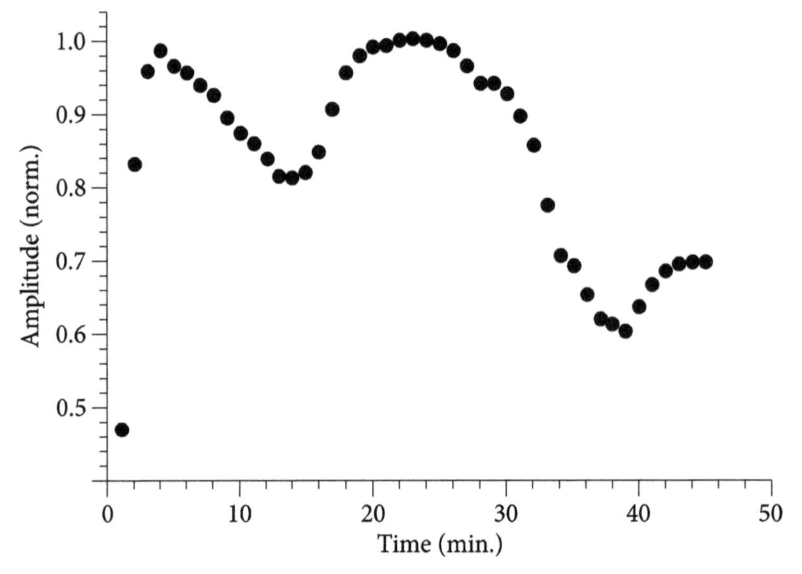

Figure 62. Temporal intensity variation of the DCF fluorescence emitted from a single cell of A.thaliana after incubating the leaves with Naph and illuminating with UV-A over 45 minutes.

Figure 63 illustrates the use of DanePy (Hideg et al., 1998, 2000), see Table 4 in imaging ROS production in *Arabidopsis thaliana* leaves. Figure 63 upper panel illustrates an area of the leaf tip exposed to a photosynthetic photon flux density (PPFD) of 600 μmol m^{-2} s^{-1} for 60 min. The white oval in frame A shows the illuminated area, frames B and C show the images of fluorescence emission from the DanePy before and after high light treatment, respectively. Fluorescence quenching occurs within the area of the leaf tip exposed to the high light. This fluorescence quenching in frame C reflects the formation of non-fluorescent DanePyO due to reaction of $^1\Delta_gO_2$ with DanePy.

Figure 63, lower panel shows ROS sensing in *A.thaliana* with nitroblue tetrazolium (NBT) which is sensitive to H_2O_2 (Maly et al., 1989; Thordal-Christensen et al., 1997; see Table 3). In that case the top half of the leaf was exposed to 600 μmol m^{-2} s^{-1} for 60 min. The purple coloration indicates the formation of insoluble formazan deposits due to reaction of NBT with superoxide.

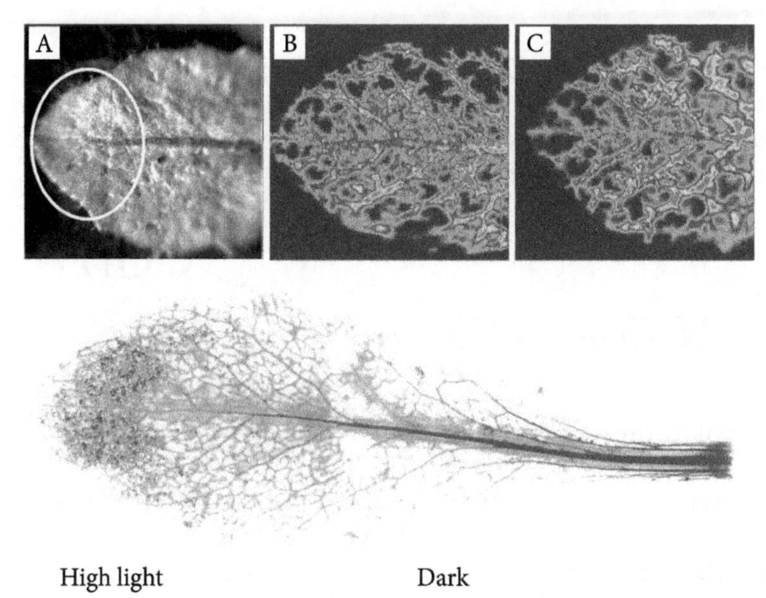

Figure 63. Top panel: Fluorescence emission from an *Arabidopsis thaliana* leaf which was infiltrated with 40 mM dansyl-2,2,5,5,-tetramethyl-2,5-dihydro-1H-pyrrole (DanePy). Bottom panel: Image of an *Arabidopsis thaliana* leaf infiltrated with 6 mM nitroblue tetrazolium (NBT) (adapted from Fryer et al., 2002). Image reproduced with permission.

As DanePy is detectable by fluorescence quenching due to ROS formation it is more often used as a typical spin trap which is highly selective for singlet oxygen.

3.2.2 Spin Traps

While fluorescent ROS-sensing dyes respond to their target molecules without further spectroscopic signal structure, which impedes the selectivity of the otherwise highly sensitive fluorescence technique, the detection of the electron paramagnetic resonance (EPR) with spin traps enables a more selective technique for ROS monitoring. Since some ROS species are radicals, the application of spin traps appears sound to focus on spin-carrying ROS. Therefore, spin traps are widely applied to EPR-detectable ROS species like superoxide and hydroxyl radicals (Hideg et al., 1994, 1998, 2000; Zulfugarov et al., 2011). Fluorescent spin traps for ROS detection like DanePy which is quenched in presence of $^1\Delta_g$ O_2 are suitable for an optical measurement of the interaction between ROS and spin trap molecules (Hideg et al. 1998, 2000, 2001).

The decisive advantage of spin traps is their specific response to the external magnetic field. Even one spin trap that can bind different ROS typically exhibits specific EPR resonance for different bound ROS which makes the applications of spin traps to the most specific approach for selective ROS sensing (see Figure 64).

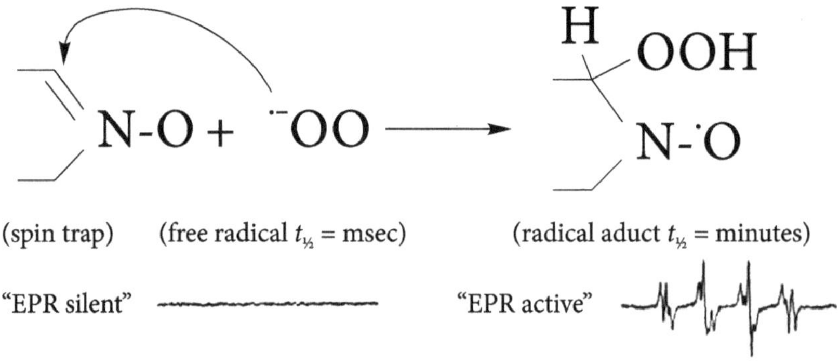

(spin trap) (free radical $t_{½}$ = msec) (radical aduct $t_{½}$ = minutes)

"EPR silent" "EPR active"

Figure 64. Principal scheme for the function of a spin trap and generation of the corresponding specific EPR signal.

Table 4 gives an overview on spin traps used for ROS imaging in plants. Disadvantages of spin traps are the rather large spectroscopic effort for EPR measurements and the impossibility of microscopic applications. Therefore spin traps are highly specific in their target molecules but do not easily carry information about the localization of ROS.

The detection of ROS in cyanobacteria faces additional difficulties because their accessibility to EPR and fluorescent spin traps is limited. An alternative technique is chemical trapping by ROS scavengers like histidine. Recently, it was shown that chemical trapping by histidine is suitable to monitor singlet oxygen generation in *Synechocystis* sp. PCC 6803 (Rehman et al., 2013).

3.2.3 Genetically Encoded ROS Sensors

Fluorescence proteins, in particular the green fluorescent protein (GFP) and its variants are widely used tools to study a large variety of cellular processes (Tsien, 2008). They are used as novel biosensors for the local chemical environment in cells and cell organelles. Highly resolved fluorescence nanoscopy (Klar et al., 2000; Westphal et al., 2005) was boosted by the development of photo-switchable derivatives of GFP (Andresen et al., 2005, 2007; Hofmann et al., 2005; Dedecker et al., 2007, Eggeling et al., 2007; Brakemann et al., 2011).

Table 4. Spin traps suitable for imaging ROS.

Compound	Specificity Further information/localizability
DMPO (Davies, 2002)	Spin trapping of $^1O_2$, superoxide and hydroxyl radicals, Transient EPR spectra specific for trapped radicals but spontaneous decay of DMPO-superoxide adduct with 45 sec. half lifetime
alpha-phenyl N-tertiary-butyl nitrone (PBN) (Davies, 2002)	EPR spectra rather unspecific for trapped radicals.
3,5-Dibromo-4-nitrosobenzenesulfonic acid (DBNBS) (Davies, 2002)	Used for H_2O_2 sensing, specific EPR spectra
5-Diisopropoxyphosphoryl-5-methyl-1-pyrroline-N-oxide (DIPPMPO) (Zoia and Argyropoulosm, 2010)	Used in mitochondria, strongly applied for detecting superoxide
TEMPO-9-AC (Cohn et al., 2008)	Fluorogenic spin trap specific for hydroxyl radicals and superoxide
BODIPY® 665/676 (Pap et al., 1999)	Sensitive fluorescent reporter for lipid peroxidation
DanePy (Hideg et al., 1998, 2000)	Specific to $^1O_2$ Fluorescent spin trap – Fluorescence is quenched in presence of $^1O_2$

GFP can directly be targeted or fused to specific target proteins for precise sub-cellular localization and analysis *in vivo*.

The application of genetically encoded fluorescent proteins (see Figure 65) in fluorescence microscopy provided new insights into the complicated and fascinating world of life on the microscopic scale. Superresolution microscopy was developed by Stephan Hell who was awarded the Nobel Prize in chemistry in 2014. As fluorescent proteins are specific markers that can be fused as tags to selected proteins, it became possible to follow the dynamics of a certain protein or enzyme in the living cell with minimal disturbance of the cell and its metabolism. Novel microscopic techniques like laser switching contrast microscopy (LSCM) allows for the imaging of the dynamics of optically switchable proteins in single cell compartments. We recently present an application for the monitoring of diffusive properties of single molecules of the photoswitchable fluorescent protein Dreiklang (Schmitt et al., 2016; Junghans et al., 2016), see Figure 66.

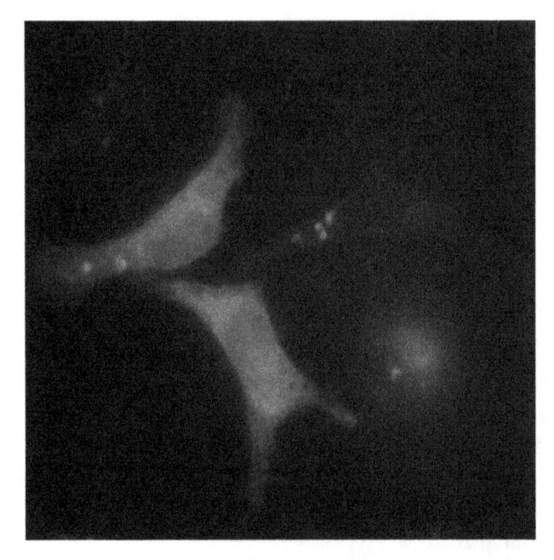

Figure 65. CHO cells expressing red fluorescent protein (RFP) in the cytosol and green fluorescent protein (GFP) localized in cell membranes.

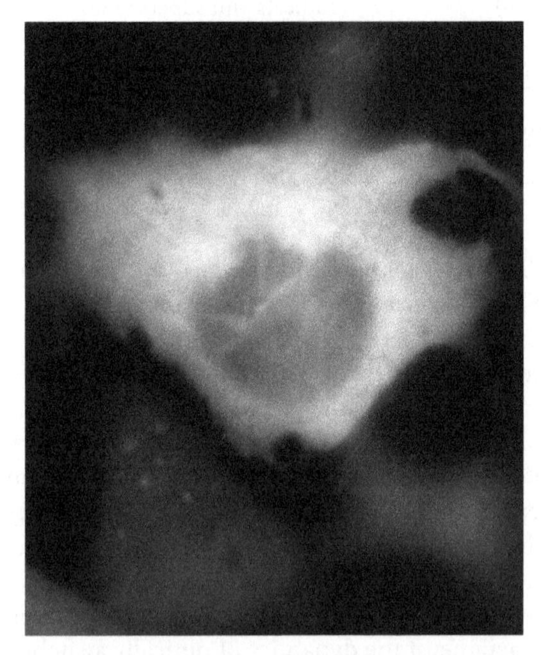

Figure 66. High resolution image of a CHO cell expressing Dreiklang (Brakemann et al., 2011) after OFF photo-switching in the cell nucleus. The image shows a sum of 20 images captured after off-switching of Dreiklang by a 405 nm laser due to diffusion of the single molecules into the laser focus (Schmitt et al., 2016). Image reproduced with permission.

Genetically encoded ROS sensors are therefore one approach to overcome problems regarding the specificity of localization when monitoring ROS. To exploit the potential of GFP for sensing the local chemical matrix, extensive studies have been undertaken to develop GFP-based *in vivo* sensors by targeted mutations and generalized approaches like directed evolution. The optical properties of these biosensors depend on selective binding of protons, oxygen atoms, water molecules and/or cofactors or are induced by electron transfer (Heim et al., 1995; Yang et al., 1996; Brakemann et al., 2011; Kremers et al., 2011).

Fluorescent proteins which are sensitive to the microenvironment like pH (Miesenböck et al., 1998; Campbell and Choy, 2001; Hanson et al., 2002; Bizzarri et al., 2009; Schmitt et al., 2014b), ROS (Ostergaard et al., 2001; Schwarzländer et al., 2009; Belousov et al., 2006) or NADH (Hung et al., 2011; Tejwani et al., 2017; Wilkening et al., 2017) are used as standard tools for the selective imaging of physiological parameters and their dynamics. Often the GFP-based ROS sensor variants contain pairs of redox-active cysteines forming a disulfide bridge as redox switch. These proteins can be selectively expressed as fluorescence markers, fused to specific target proteins or to organelle-specific targeting sequences, thus enabling a specific and localized monitoring (and manipulation) of ROS at a molecular level (for a review, see Swanson et al., 2011).

Progress in engineering of ROS-sensitive fluorescence proteins led to the development of several derivatives of GFP containing the mentioned redox-active cysteines forming a disulfide bridge as redox switch (Jimenez-Banzo et al., 2008). One example is roGFP (Hanson et al., 2004; Schwarzländer et al., 2008). Derivatives of the yellow fluorescent protein (YFP) have also been described, which are modified by introduction of redox-active cysteines in constructs termed $rxYFP^{149}{}_{202}$ (Ostergaard et al., 2001) or HyPer (Belousov et al., 2006). Chromophore transformations in red-fluorescent proteins offer tools for designing suitable red-shifted probes, which are advantageous for imaging studies due to the strong absorption in the green spectral range, in which chlorophylls exhibit only very low absorption. Excitation with longer wavelengths also leads to reduced autofluorescence (for a review, see Subach and Verkhusha, 2012).

The disulfide bridge in the oxidized rxYFP leads to a distortion of the typical beta-barrel structure of GFP derivatives, thus changing the fluorescence properties of rxYFP (Ostergaard, 2001). The mitochondrially-targeted redox sensitive GFP termed roGFP-mito does not specifically react in response to a certain species of ROS, but it is used to selectively label mitochondria in plants (Schwarzländer et al., 2009).

In an alternative approach, the H_2O_2-sensitive probe HyPer was constructed by fusing the regulatory domain of the H_2O_2-sensitive transcription factor OxyR from *E.coli* to a cyclically permuted YFP (Belousov et al., 2006). For applications of the genetically encoded ROS sensors in studies on ROS effects, see (Maulucci et al., 2008; Meyer and Dick, 2010; Mullineaux and Lawson, 2009).

Table 5 gives an overview on genetically encoded fluorescence proteins and their basic properties of selectivity and applicability in plants.

Table 5. Genetically encoded fluorescence proteins applicable for ROS monitoring.

Compound/reference	Specificity Further information/localizability
rxYFP (Ostergaard et al., 2001)	Unspecific
roGFP (Schwarzländer et al., 2008, 2009)	Unspecific, applied to label plant mitochondria
HyPer (Belousov et al., 2006)	H_2O_2 sensitive by fusing the regulatory domain of the H_2O_2-sensitive transcription factor OxyR to YFP, not yet expressed in plant cells
GFP redox sensor (Niethammer et al., 2009)	Specific to H_2O_2, successfully applied in Zebrafish larvae to detect H_2O_2 patterns after wounding

The application of fluorescence markers for ROS sensing is generally complicated by photobleaching. In addition, fluorophores often act as $^1\Delta_gO_2$ sensitizers themselves. This problem is especially important for GFP derivates as ROS sensors. However, the generation of new GFP mutants that produce reduced amounts of ROS is a promising approach to overcome this problem, which again votes for the importance of developing improved genetically encoded fluorescence proteins for ROS sensing for future studies.

3.2.4. Electrochemical Biosensors

Spectroscopic techniques are always a challenge in photosynthetic research. Therefore more and more approaches aim in generally overcoming the need of probes and spectroscopy. One direction uses electrochemical biosensors to monitor the redoxchemistry of ROS as electric signals. Novel developments present electrochemical biosensors for real time monitoring of H_2O_2 generation at the level of sub-cellular organelles (Prasad et al., 2015). Electrochemical biosensors typically utilize enzymes to achieve ROS turnover, producing a current that can then be measured at the surface of an electrode (Pohanka and Skladai, 2008). For H_2O_2 sensing combined enzymes such as horseradish peroxidase (HRP) are used and novel materials like carbon nanotubes and polymeric matrices are used as electrodes to improve sensor sensitivity and specificity (Enomoto et al., 2013). A detailed review on typical electrochemical biosensor architectures is found e.g. in (Pohanka and Skladai, 2008). More recently novel types of electrochemical sensors have been developed and used for highly specific monitoring of ROS. For a review on electrochemical ROS sensors see also (Calas-Blanchard et al., 2014).

Electrochemical biosensors are produced with highly elaborated techniques like photolithography which is now used in the fabrication of lab-on-a-chip electrochemical biosensors (Enomoto et al., 2013).

3.3 Signaling Role of ROS

During evolution ROS were steadily forming a constraint for the growth of both, autotrophic and heterotrophic life forms as living plants that always needed to adapt to the ROS containing environment. Therefore in billions of years of evolutionary acclimation not only the damaging effects of ROS were defeated but ROS developed a strong signaling role as the trigger mechanism for the processes that help to acclimatize to high ROS levels or even need ROS as cofactors. ROS have not only damaging but also a signaling role (Hung et al., 2005; Mubarakshina et al., 2010; Zorina et al., 2011; Kreslavski et al., 2012b, 2013a; Schmitt et al., 2014a). Figure 67 gives a general overview on the molecular generation and signaling of ROS and the acclimation of the photosynthetic apparatus (PA) (Zorina et al., 2011). The response of cells starts with the perception of stress by sensors or the response of sensors to ROS that are formed under stress conditions (Kanesaki et al., 2010).

Typically, cascades of mitogen-activated protein kinases (MAPK), other transcription factors (TFs), Ca^{2+}, phytohormones and other compounds function as sensors and/or transducers (Kaur and Gupta, 2005; Jung et al., 2009). In the following, characteristic examples will be described for the signaling action of $^1\Delta_gO_2$, H_2O_2 and $O_2^{-\bullet}$ in cells, separately discussed for cyanobacteria (chap. 4.1) and plants (chap. 4.2).

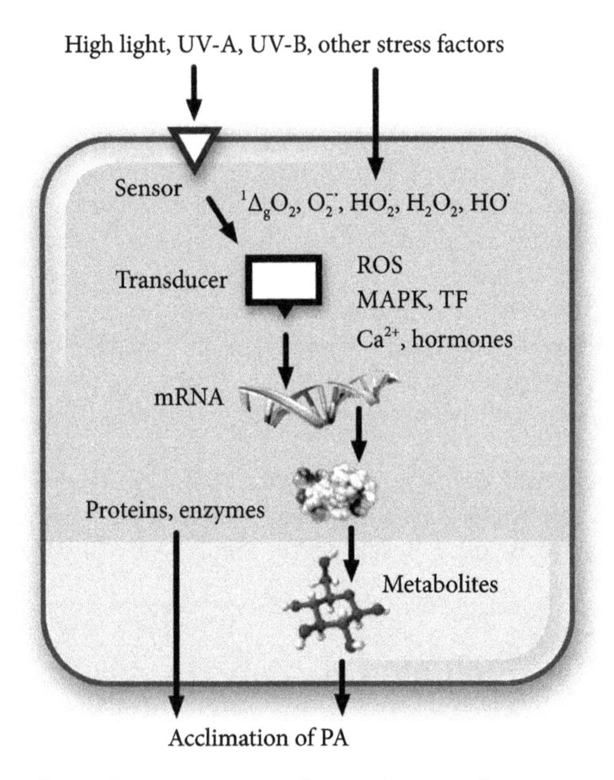

Figure 67. Scheme for perception and transduction of stress signals and formation of ROS as signal molecules for genetic signaling supporting the acclimation of cells to stress conditions (adapted from Zorina et al., 2011). Image reproduced with permission.

In plant cells, ROS are produced in different organelles, predominantly in chloroplasts and peroxisomes, while the contribution from mitochondria is smaller (Foyer and Noctor, 2005). Imaging of oxidative stress in leaves of *Arabidopsis thaliana* revealed that $^1\Delta_gO_2$ and $O_2^{-\bullet}$ are primarily located in mesophyll tissues, while H_2O_2 was predominantly detected in vascular tissues (Fryer et al., 2002), see Figure 63 in chap. 3.2.1.

Another important approach to study effects of different ROS next to the selection of special assay conditions is given by the use of mutant strains that differ in the generation of individual ROS and/or the content of protection systems/enzymes. This point is of high relevance for studies on specific signaling pathways (see chap. 4 "ROS signaling in coupled nonlinear systems").

Detailed studies of such systems have been conducted in cyanobacteria. The important aspects of superoxide and hydrogen peroxide signaling in cyanobacteria have been treated separately (chap. 4.1). As mentioned above, cyanobacteria serve as efficient models for studying the molecular mechanisms of stress responses. The genes of these cells can be easily knocked out or overexpressed which permit straightforward approaches to investigate the genetic aspects of signaling. This enabled the intensive studies of the potential stress sensors and signal transducers in cyanobacteria (Los et al. 2010; Kanesaki et al., 2010; Zorina et al., 2011, Kreslavski et al., 2013a).

Resulting from a long evolutionary adaption process, systems of perception and transduction of stress signals, as well as the hormonal regulation system (see Figure 67) work in close coordination. Their interaction was fine-tuned during billions of years of evolution.

In the cytoplasm of plant cells, low temperature, drought and salinity cause an increased concentration of Ca^{2+}. In this case, calcium channels may serve as multifunctional sensors that perceive stress-induced changes in the physical properties of cell membranes (see Figure 67). The discovery of such multifunctional sensory systems is important to understand perception and transmission of stress signals. Apparently, changes in membrane fluidity, regardless of the nature of the stress effect are a signal that is perceived by sensory histidine kinases or ion channels localized in the membranes (Kanesaki et al., 2007; Zorina et al., 2011; Los et al., 2013; Kreslavski et al., 2013a).

It is known that ROS are produced in all cell compartments and their formation is necessary for the functioning of photosynthetic organisms (Suzuki and Mittler, 2006). Certain ROS are considered as signaling molecules and regulators of expression of some chloroplast and nuclear genes (Schmitt et al., 2014a; Kreslavski et al., 2013a; Minibayeva et al., 1998; Minibayeva and Gordon, 2003; Desikan et al., 2001, 2003; Hung et al., 2005; Galvez-Valdivieso and Mullineaux, 2010; Mubarakshina et al., 2010; Dickinson and Chang, 2011). A new view on the effects of ROS as signaling molecules first appeared in the study of hormone signaling and the regulation of expression of genes involved in plant protection from pathogen infections (Chen et al., 1993), conditions under which interactions of ROS with salicylic acid and nitric oxide play a crucial role in regulation of the response to infection (Wilson I.D. et al., 2008; Kreslavski et al., 2013a; Vallad and Goodman, 2004).

One of the key points in understanding of the effect of ROS on photosynthesis was the discovery of the formation of the superoxide anion and hydrogen peroxide in the pseudo-cyclic electron transport (See Figure 53 and Figure 54), which does not lead to the reduction of $NADP^+$, but to the absorption of O_2 (Asada, 1999). In addition, it was shown that the activation of plasma membrane redox-systems and the increased formation of ROS in the apoplast is one of the universal reactions of plant cells to stress (Kreslavski et al., 2013a; Minibayeva et al., 1998, 2009; Minibayeva and Gordon, 2003; Dickinson and Chang, 2011).

It was found that the main generators of ROS in the apoplast of root cells are the cell wall peroxidases (Minibayeva et al., 2009; Minibayeva and Gordon, 2003). Apparently, the release of ROS from cells followed by a switch of peroxidase/oxidase modes of extracellular peroxidases form the basis for the fast response of plant cells to stress.

In addition to ROS, the stress signaling functions may be attributed to some metabolites, whose formation is initiated by ROS, for example, the products of lipid peroxidation (LP). The primary subjects for peroxidation in living cells are unsaturated fatty acids that constitute major components of phospho- and glycolipids of biological membranes.

ROS regulate the processes of polar growth, the activity of stomata, light-induced movement of chloroplasts and plant responses to the action of biotic and abiotic environmental factors (Pitzschke and Hirt, 2006; Miller et al., 2007; Swanson and Gilroy, 2010).

Signaling by ROS may be realized through changes in potential of the redox-sensitive cell systems and through phosphorylation/dephosphorylation cycles of signaling proteins (transcription factors, etc). The accumulation of redox-active compounds such as ROS within the chloroplast is associated with the rate of photosynthetic electron transport. Redox-sensitive thioredoxin or PQ may act as sensors of changes in redox properties under stress conditions (Figure 54). Signals generated from modulation in the activity of ETC may also lead to changes in gene expression (Vallad and Goodman, 2004).

Although many things have been ruled out from the mechanisms of action of ROS as signal molecules, there are still many gaps in understanding the complete network of these regulatory events. The sensor(s) of H_2O_2 in higher plants remain largely unknown (Galvez-Valdivieso and Mullineaux, 2010; Mubarakshina et al., 2010; Kreslavski et al., 2012b). There is no information about specific proteins that convert a signal about an increase in the intracellular ROS levels to a biochemical response in the cells. It is not known exactly which particular ROS play a signaling role in the chloroplast and other cellular compartments and how different signaling pathways respond to an increase in the level of different types of ROS. Knowledge of the mechanisms of regulation of these signaling pathways may help to construct biochemical pathways and to produce genetically engineered plants with enhanced stress resistance.

High light, especially high doses of UV-A or UV-B lead to the damage of the PA. Plastoquinones (the primary and secondary plastoquinone, Q_A and Q_B, respectively) as well as the D1 and D2 proteins are amongst the primary targets of UV radiation (Strid et al., 1994; Babu et al., 1999; Asada, 2006; Carvalho et al., 2011). The Mn_4CaO_5 cluster of PSII is also vulnerable to damage by UV irradiation (Ohnishi et al., 2005, Najafpour et al., 2013, 2015, 2016). However, UV light-induced damage can also depend on the additional interaction with light in the visible region. Or – more generally spoken – the light induced signaling after interaction of visible light with certain sensors leads to the activation of protection mechanisms against UV-A and UV-B.

Red light (RL) of low intensity can alleviate the negative effect of UV radiation on plants and their PA (Lingakumar and Kulandaivelu, 1993; Qi et al., 2000, 2002; Biswal et al., 2003; Sicora et al., 2003; Kreslavski et al., 2012a,b, 2013a,b,c, 2014a,b). Recent studies have shown that low intensity RL pulses activate the phytochrome system, which triggers protective mechanisms against UV-radiation (Kreslavski et al., 2013a, b). However, many details of this protective action of RL acting via the phytochrome system on PA have not been clarified so far.

The phytochrome system plays an important role in plant growth and PA development. This concept is in agreement with recent studies on mutant *Arabidopsis* strains with deficiencies in different types of phytochromes, which demonstrated that deletion of phytochromes is critical for plant development (Strasser et al., 2010; Zhao et al., 2013). Even if light capable of driving photosynthesis is available, normal seedling greening and plant development is impossible if phytochromes are absent (Strasser et al., 2010; Zhao et al., 2013). The effects of phytochrome deficiency on photosynthetic parameters have been investigated in previous studies, including the impact on PSII activity (Kreslavski et al., 2013b) and Chl (a+b) content (Strasser et al., 2010; Zhao et al., 2013).

Protective effects against UV are caused by the RL-induced formation of the far-red-absorbing active form of phytochrome and/or enhancement of phytochrome biosynthesis as a result of RL illumination (Kreslavski et al., 2012a, 2013b,c). It was suggested that this protective effect is due to decreased Chl degradation and higher stability of the PSII, as well as higher photochemical activity and a reduced damage of thylakoid membranes (Lingakumar and Kulandaivelu, 1993; Biswal et al., 2003; Kreslavski et al., 2004). On the other hand, a decreased phytochrome level can reduce the resistance of the PA. For example, *hy2* mutants of *Arabidopsis* show a decreased level of PhyB and other pytochromes due to reduced biosynthesis of the phytochrome chromophore, phytochromobilin (Parks and Quail, 1991). This *hy2* mutant also showed decreased UV-A resistance of PSII, as determined from delayed luminescence emission (Kreslavski et al., 2013b). It was also shown that the resistance of PA in *Arabidopsis* WT increased after preillumination with RL, whereas in the *hy2* mutant the PSII resistance to UV-A did not change upon the same treatment. It was suggested that the PA resistance to UV radiation depends on the ratio of pro- and antioxidant compounds, which can be affected by PhyB and other phytochromes (Kreslavski et al., 2013b). The role of different phytochromes for the UV resistance of PSII has not been studied so far.

PhyB, one of the key phytochromes in green plants, is involved in the synthesis of photosynthetic pigments, chloroplast development (Zhao et al., 2013), as well as in the synthesis of some photosynthetic proteins and stomatal activity (Boccalandro et al., 2009). It is also known that an increased PhyB content can enhance the resistance of the photosynthetic machinery to environmental stress (Thiele et al., 1999; Boccalandro et al., 2009; Carvalho et al., 2011; Kreslavski et al., 2004, 2012a,c, 2013b,c). In particular, transgenic cotton plants, in which the phytochrome B (*PhyB*) gene of *Arabidopsis thaliana* was introduced, showed more than a two-fold increase in the photosynthetic rate and more than a four-fold increase in transpiration rate and stomatal conductance (Rao et al. 2011). In addition, the increase of PhyB content in transgenic potato plants (Dara-5 and Dara-12), which are superproducers of PhyB, led to enhanced resistance of the PA to high irradiance (Thiele et al., 1999). It can be suggested that the increased resistance results from higher Chl content or enhanced stomatal conductance.

4
ROS Signaling in Coupled Nonlinear Systems

Today it is well established that ROS exert important functions in signaling pathways within the cells of both plants and animals. The mode of signaling under the participation of ROS depends on the nature of stress. In response to different types of stress, ROS can act in a dual manner: i) by functioning as signal molecules which induce molecular, biochemical, and physiological responses leading to development of adaptive mechanisms and improving the tolerance of the organisms to stress (acclimation) or ii) by inducing reaction sequences that eventually cause programmed cell death (Galvez-Valdivieso and Mullineaux, 2010; Mittler et al., 2011; Vranova et al., 2002; Kreslavski et al., 2007, 2011; Jaspers and Kangasjärvi, 2010; Los et al., 2010; Schmitt et al., 2014a).

In general, significant differences exist in the response to abiotic (light, draught, cold, heat etc.) and biotic (infection by viruses and bacteria) stress. This includes also the type of ROS molecules involved. In case of abiotic stress, ${}^1\Delta_gO_2$ is often formed in addition to $O_2^{-\bullet}$ and H_2O_2, while biotic stress mainly leads to enzymatic generation of $O_2^{-\bullet}$ and H_2O_2 which are used as a defense mechanism against biotic stress (Laloi et al., 2004). Different types of ROS give rise to specific signaling, as shown in animal cells (Klotz et al., 2003).

ROS have several advantages in acting as signal molecules (Miller et al., 2009; Mittler et al., 2011): i) Cells are able to rapidly generate and scavenge different forms of ROS in a simultaneous manner, thereby permitting rapid response to stress. ii) The subcellular localization of ROS signals can be strongly controlled within cells, i.e. a spatial control of ROS accumulation exists in a highly specific manner. iii) ROS can be used as rapid long-distance auto-propagating signals to be transferred throughout the plant, as recently reported for *Arabidopsis thaliana*, in which ROS signals propagate at rates of up to 8.4 cm/min (Miller et al., 2009, Schmitt at al., 2014a). iv) ROS are tightly linked to cellular homeostasis and metabolism.

Most probably, the mechanism of stomatal closure and it spatiotemporal patterns result from an underlying ROS signaling mechanism. Therefore, it is proposed that ROS are implemented in very general signaling schemes that influence the expression of genes and consequently the molecular biology of green plants. Additionally, ROS are responsible for macroscopic long range effects that are directly observable on the cellular level like stomatal closure. It is a trigger for adaption of the whole cell metabolism and, in case of biotic stress, actively produced with respect to long range interaction as found for $O_2^{-\bullet}$ and H_2O_2 (in contrast to singlet oxygen which has a much shorter lifetime) to be used as oxidezing defense molecules against the biotic stressors.

A comparison with ROS signaling in animal cells revealed that the communication of mitochondria in heart cells occurs via ROS-induced waves. An abrupt collapse or oscillation of the mitochondrial energy state is assumed to be synchronized across the mitochondrial network by local ROS-mediated interactions (Zhou et al., 2010; Zhou and O'Rourke, 2012). This model is based on the idea that a depolarization of the electrical potential difference across the coupling membrane is specifically mediated by $O_2^{-\bullet}$ via its diffusion and the $O_2^{-\bullet}$-dependent activation of an inner membrane anion channel. This is in agreement with experimental data. This mode of ROS-induced ROS release mechanisms in animal cells can also be used in plants for propagation of cell-to-cell ROS signaling over long distances (Miller et al., 2009, Mittler et al., 2011). The concept of a transient ROS burst occurring in selected cells can be further extended to the more general concept of a ROS wave propagating in time and space as response to different types of stress (Schmitt et al., 2014a).

4.1 Signaling by Superoxide and Hydrogen Peroxide in Cyanobacteria

Various mechanisms are involved in the signal function of ROS. At first, ROS-induced modifications of proteins can lead to changes of either structure or activity or both, in particular via oxidation of thiol groups.

Illustrative examples are the suppression of CO_2 fixation and the blockage of the elongation factor EF-G in cyanobacteria and iron-containing clusters in enzymes (Spadaro et al., 2010). The oxidation of EF-G represents a rather general signaling scheme. Such a reaction chain as represented by the EF-G oxidation can be understood as a chemical inactivation process that is switched on and off by the oxidative potential. High oxidative potential in cyanobacteria leads to the oxidation of the two residues 105Cys and 242Cys in EF-G, and subsequent formation of a disulfide bridge between the two cysteine residues blocks the elongation of translation (Kojima et al., 2007). Replacement of these conserved cysteine residues by serine makes EF-G insensitive to ROS (Kojima et al., 2009). The mechanism of translation blockage under the influence of oxidative stress via post-translational redox regulation of the elongation factor state is a universal way of cell protection against ROS. Thus, EF-G is a primary target for ROS action and a key regulator of the translation efficiency (Nishiyama et al., 2011; Murata et al., 2012). This H_2O_2-induced blockage of the translation machinery interrupts the repair of photodamaged PS II, thus eventually leading to the disappearance of PS II and, consequently, the interruption of the linear electron transport chain. Studies on the effect of other stress factors (heat, drought, salinity) on photoinhibition have shown that the suppression of PS II repair determines the PS II sensitivity of cyanobacteria to environmental conditions (Allakhverdiev and Murata, 2004; Murata et al., 2007, 2012; Nishiyama et al., 2011).

The two Cys residues oxidatively linked to an S-S bridge by H_2O_2 are highly conserved in EF-G of cyanobacteria and of chloroplasts in algae and higher plants. Therefore, it seems very likely that ROS induces similar effects in the chloroplasts of plant cells. The translation of the D1 protein in chloroplasts is also regulated by redox components at both initiation and elongation steps (Zhang et al., 2000). A marked difference to cyanobacterial D1 is the possibility to phosphorylate D1 in plants. This property permits a regulation process of the circadian rhythm of degradation and metabolism. In this context, it is important to know the extent of how ROS affect the phosphorylation pattern of D1 in plants. This question remains to be answered in future studies.

Depending on the lifetime, different types of ROS molecules can either directly act as signal molecules or generate signal chains by formation of oxidation products.

It must be emphasized that fundamental differences exist between prokaryotic cyanobacteria and eukaryotic plants. In cyanobacteria, the photosynthetic and respiratory electron transport reactions take place in the intracytoplasmic (thylakoids) and cytoplasmic membrane, respectively, and close interactions exist between both photosynthesis and respiration (Peschek, 2008). On the other hand, eukaryotic plant cells contain semi-autonomous organelles (chloroplasts, mitochondria, peroxisomes, nucleus) with specific functional activities. This differentiation requires a more complex signaling system for "cross-talk" between these organelles. As a consequence, the mechanisms of "handling" stress-induced ROS and the modes of protection are markedly different between cyanobacteria and plants, and even within the plant kingdom. Therefore, a generalized and unified scheme cannot be presented at the moment and only selected characteristic examples of signaling are presented.

In cyanobacteria and plants, the $O_2^{-\bullet}$ radical is assumed to be predominantly produced at the acceptor side of PS I (Asada, 1999). The lifetime of $O_2^{-\bullet}$ is mainly determined by the presence of SOD and does not exceed a few microseconds in cells (Gechev et al., 2006). The signaling function of $O_2^{-\bullet}$ has been investigated by analyses of gene expression using DNA microarrays (Scarpeci et al., 2008) and studies on $O_2^{-\bullet}$ accumulation in plants deficient in Cu/Zn-SOD (Rizhsky et al., 2003). The results are in favor of a signaling role of this radical but details of the pathway(s) are not well known today.

$O_2^{-\bullet}$ can react with NO under formation of peroxinitrite. This species is likely to be synthesized in chloroplasts, where it can fulfill signaling functions (Foyer and Shigeoka, 2011). Under normal pH conditions, $O_2^{-\bullet}$ is deprotonated in animal cells (pH = 7.4 in blood cells) due to its pK_a value of 4.8. However, at sufficiently low pH values (e.g. sometimes existing in the thylakoid lumen, see Joliot and Joliot, 2005), $O_2^{-\bullet}$ anion radicals become protonated and the neutral hydroperoxyl radical ($HO_2^{\bullet}$) can cross membranes (Sagi and Fluhr, 2006) (see chapter 3.1). The formation of H_2O_2 occurs mainly via the formation of $O_2^{-\bullet}$ followed by SOD-catalyzed dismutation (Asada, 1999, 2006) and in the process of photorespiration (Foyer and Noctor, 2009). H_2O_2 is markedly less reactive than $^1\Delta_gO_2$ (Halliwell and Gutteridge, 1985) (see chap. 3.1) and characterized by a much longer lifetime in the order of 1 ms (Henzler and Steudle, 2000; Gechev and Hille, 2005). Therefore, H_2O_2 is a most promising candidate to function as an intra- and intercellular messenger (Vranova et al., 2002; Hung et al., 2005; Bienert et al., 2007; Foyer and Shigeoka, 2011; Mittler et al., 2011). Numerous results on H_2O_2

signaling were reported for both, prokaryotic cyanobacteria and eukaryotic plants.

Eubacteria, including cyanobacteria, actively use characteristic two-component systems of signal perception and transduction (Kreslavski et al., 2013a).

Such two-component regulatory systems are typically composed of a sensory histidine kinase (Hik) and a response regulator form the central core of the phosphate signaling system in cyanobacteria (Los et al., 2010; Kreslavski et al., 2013a). The sensory histidine kinase perceives changes in the environment with its sensory domain. A subsequent change of its conformation often leads to autophosphorylation of the conservative histidine residue in a Hik from a donor ATP molecule from which a phosphate group is then transferred to the conserved aspartate in a receiver domain of the response regulator protein (RRP). After phosphorylation, the RRP also changes its conformation and gains (positive regulation) or loses (negative regulation) the ability to bind to DNA. The RRP usually binds the promoter region(s) of genes for proteins that are involved in the stress signal network or are linked to stress protection pathways (Kreslavski et al., 2013 a).

Hik33 of *Synechocystis* is the multisensory protein, which perceives cold, salt, and oxidative stresses. The mechanisms by which Hik33 recognizes the stresses are still not fully clear. It is assumed that changes in the physical mobility of membrane lipids and changes in the surface charge on the membrane, associated with changing mobility, are activators for Hik33. Activation may be also caused by depolarization of the cytoplasmic membrane upon cold stress or due to changes in charge density of the membrane surface under stress (Nazarenko et al., 2003; Kreslavski et al., 2013a).

Sensory histidine kinases are also important for the functioning of genes involved in photosynthesis and/or regulated by high light intensity. Experiments with the *Synechocystis* mutant deficient in Hik33 (this mutant is also named DspA, see Hsiao et al., 2004) revealed that low or moderate light intensity causes retardation in growth and decrease in photosynthetic oxygen evolution in mutant cells, compared to wild-type cells, under photoautotrophic conditions. The addition of glucose neutralized these differences. However, mutant cells were more sensitive to light intensity and quickly died under strong light (Hsiao et al., 2004; Kreslavski et al., 2013a).

The defense of bacteria against oxidative stress and adaptive regulation mechanisms have been thoroughly analyzed in *Escherichia* (*E.*) *coli* and *Bacillus* (*B.*) *subtilis.* In eubacteria (heterotrophic, autotrophic, and chemotrophic), two global regulators, OxyR and PerR, are involved in the control of gene transcription induced by H_2O_2 addition (Zheng et al., 1998) (see Figure 69). Both these regulators have active thiol groups and can directly recognize changes in the redox state of the cytoplasm. The ferric uptake repressor (Fur) type protein PerR was found to be the central regulator of inducible stress response (Herbig and Helmann, 2001; Mongkolsuk and Helmann, 2002). In the cyanobacterium *Synechocystis* sp. PCC 6803, a gene (*srl*1738) encoding a protein similar to PerR was identified as being induced by H_2O_2 (Li et al., 2004) in methylviologen treated cells upon illumination (Kobayashi et al., 2004). It was concluded that the Fur-type protein Slr 1738 functions as a regulator in inducing the potent antioxidant gene *sll*1621, which encodes for a peroxiredoxin.

The regulator OxyR is absent in *Synechocystis* sp. PCC 6803. Studies on *Synechocystis* sp. PCC 6803 incubated for 20 min with 0.25 mM H_2O_2 proved how several histidine kinases can serve as H_2O_2 sensors (Kreslavski et al., 2013a). Mutations of genes Hik34, Hik16, Hik41, and Hik33 encoding histidine kinases led to blockage of the H_2O_2-induced gene expression (Zheng et al., 1998; Kanesaki 2007). Peroxidases were found to control 26 of 77 genes induced by H_2O_2. The histidine kinase Hik34 was shown to regulate the expression of the gene *htpG* under oxidative stress. This kinase was characterized as regulator of gene expression due to heat (Suzuki et al., 2005), salt, and hyperosmotic stress (Shoumskaya et al., 2005). In addition, Hik34 is subjected to autoregulation in the presence of H_2O_2. The pair of histidine kinases Hik16-Hik41 regulates the genes *sll0967* and *sll0939* with unknown functions not only in response to H_2O_2 but also under salinity and hyperosmotic stress. Hik33 controls 22 genes; among them are *ndhD2* encoding NADH dehydrogenase, three *hli* (high-light-inducible) genes, *pgr5* encoding ferredoxin-plastoquinone reductase, the genes *nblA1* and *nblA2* involved in phycobilisome degradation, and others. It should be noted that ROS induce also the expression of the genes *hspA, dnaJ, dnaK2, clpB1, ctpA* and *sigB*. These genes were activated by mechanisms without the involvement of histidine kinases, although Hik34 is acting as repressor of genes encoding heat shock proteins (Zheng et al., 1998; Kanesaki et al., 2002; Suzuki et al., 2005; Los et al., 2010).

In addition to histidine kinases, the transcription factor PerR participates in the response of *Synechocystis* sp. PCC 6803 to ROS. PerR is involved in the regulation of only six genes, of which four encode proteins with unknown function. Based on evidence for both PerR and Hik33 being components in the control of the induction of *nbl* gene expression due to oxidative stress, it seems possible that PerR interacts with a two-component regulatory system (Zheng et al., 1998; Kanesaki et al., 2002; Suzuki et al., 2005; Los et al., 2010).

4.2 Signaling by Singlet Oxygen and Hydrogen Peroxide in Eukaryotic Cells and Plants

In plants cells, $^1\Delta_gO_2$ is known to function predominantly as a plastid ROS signal which activates nuclear gene expression (Li et al., 2012). Because of its high reactivity, $^1\Delta_gO_2$ has a very short lifetime in cells (about 200 ns, see Gorman and Rodgers, 1992), other sources report times down to 70 ns (Prasad et al., 2015). For a report on markedly longer lifetime in cells, see (Skovsen et al., 2005). The short lifetime leads to a rather limited diffusion radius. However it seems as if the diffusion radius is about 10 nm, especially as longer lifetime is reported in Lipid membranes and therefore singlet oxygen is able to cross cell membranes (Schmitt et al., 2014a). However, for long range signaling, the involvement of additional components is necessary as the diffusion radius of $^1\Delta_gO_2$ is surely limiting. An important pathway is signal transfer from the site of formation within the chloroplast through the cytosol to the nucleus, which is termed chloroplast-to-nucleus signaling (Kreslavski et al., 2012b, Schmitt et al., 2014a).

Generally like in cyanobacteria, so also in plants, two types of fundamentally different responses to $^1\Delta_gO_2$ stress are known: i) development of increased tolerance and ii) induction of programmed cell death.

A $^1\Delta_gO_2$ signaling pathway in *C.reinhardtii* was shown to give rise to gene expression that leads to increased tolerance to ROS (acclimation). This phenomenon comprises enhanced expression of genes for ROS protection and detoxification, e.g. of a glutathione peroxidase-homologous gene (*gpxh/gpsx*), and also the expression of a Δ-class glutathione-S-transferase gene (*gsts*) greatly increases (Fischer et al., 2012). The effect on components participating in signaling was analyzed in a $^1\Delta_gO_2$-resistant mutant (SOR 1). The results obtained revealed the involvement of reactive electrophilic species that are formed by $^1\Delta_gO_2$-induced lipid peroxidation (Fischer et al., 2012). It was found that the *SOR1* gene encodes a leucine zipper transcription factor, which controls the expression of numerous genes of stress response and detoxification. It was inferred from these results that reactive electrophilic species play a key signaling role in acclimation of *C.reinhardtii* cells to $^1\Delta_gO_2$ stress (Kreslavski et al., 2012b).

In many cases, $^1\Delta_gO_2$ signaling induces programmed cell death, in particular under biotic stress. Much information on the genetic control of this phenomenon has been gathered from investigations on the *A. thaliana* mutant *flu1* which is defective in the feedback control of the Chl biosynthesis pathway. This mutant, which accumulates the photosensitizer protochlorophyllide in the dark, generates $^1\Delta_gO_2$ within the first minute of illumination after a dark-to-light shift (op den Camp et al., 2003). The $^1\Delta_gO_2$ formation taking place in the vicinity of the thylakoid membrane (Przybyla et al., 2008) can be manipulated by altering the degree of light exposure and the preceding dark period. In contrast to wild-type plants, the $^1\Delta_gO_2$ production in *flu1* is not associated with excess excitation of PSII (Mullineaux and Baker, 2010). The studies on the *flu1* mutant revealed that $^1\Delta_gO_2$ can trigger the activation of programmed cell death and that two chloroplast-located proteins, EXECUTER1 and 2 (EXE1 and EXE2), control this process (Przybyla et al., 2008; Wagner et al., 2004; Lee et al., 2007) (see Figure 71). EXE1 and EXE2 act as suppressors (Wagner et al., 2004; Lee et al., 2007), but their mode of function in signaling of the $^1\Delta_gO_2$-induced programmed cell death is not yet resolved.

As a consequence of the special mode of $^1\Delta_gO_2$ formation in the *flu1* mutant, a cell death in their leaves can be induced either due to direct oxidative destruction (necrosis) under a large excess of ROS or at a slower rate of $^1\Delta_gO_2$ formation via signaling the activation of a programmed cell death pathway.

On the basis of data obtained on *C.reinhardtii* cells, $^1\Delta_gO_2$ was inferred to be able to leave chloroplasts directly into the cytosol and even to reach the nucleus, thereby inducing the expression of nuclear gene *gpxh*, which encodes glutathione peroxidase (Fischer et al., 2007). Since the fraction of "mobile" $^1\Delta_gO_2$ is extremely small, a direct effect of $^1\Delta_gO_2$ was manifested only under high light and so far observed only in cells of this microalga (Fischer et al., 2007). It appears much more likely that oxidation products of special molecules are formed, which act as second signal messengers and are transferred via the cytosol and to the nucleus. This idea is confirmed by experimental data obtained on both, cells of the unicellular green alga *C.reinhardtii* and multicellular leaves of the higher plant *A.thaliana*. In the latter, plant ß-cyclocitral was shown to be formed by the oxidation of ß-carotene under ROS stress and identified as a stress signal that acts as a second messenger in $^1\Delta_gO_2$ signaling (Ramel et al., 2012). Likewise, oxidation of polyunsaturated fatty acids due to interaction with $^1\Delta_gO_2$ in the lipid fraction of thylakoid membranes leads to formation of reactive electrophilic species, which are able to exit into the cytosol (Galvez-Valdivieso and Mullineaux, 2010). Via autocatalytic cascades, lipoperoxide radicals can result in generation of $^1\Delta_gO_2$ in the cytosol (Flors et al., 2006) and trigger the EXE1/EXE2-mediated pathway of programmed cell death (Wagner et al., 2004).

The enzymatic peroxidation of lipids is catalyzed by lipoxygenases. These enzymes play an essential role in response to pathogen infection and wounding (Feussner and Wasternack, 2002; Overmyer and Brosché, 2003; Hoeberichts and Woltering, 2003). Specific lipoxygenase pathways lead to formation of lipoxide species which are likely to be different when induced by chemically different ROS like $^1\Delta_gO_2$ versus $O_2^{-\bullet}/H_2O_2$. Studies on the *flu1* mutant of *A. thaliana* revealed that 70 genes are up-regulated by $^1\Delta_gO_2$ but not by $O_2^{-\bullet}/H_2O_2$, the latter being formed at PS I via the methylviologen mediation reaction (op den Camp et al., 2003).

The signaling pathway(s) of $^1\Delta_gO_2$ leading to cell apoptosis tightly interact(s) with other signaling pathways involving hormones and other ROS. The $^1\Delta_gO_2$ species activates the signaling pathways controlled by salicylic and jasmonic acid resulting in changes of the expression of numerous genes which are related to anti-stress defense systems. An example of regulatory interaction is the decrease in cell injury and death induced by $^1\Delta_gO_2$ due to its conversion into H_2O_2 (Laloi et al., 2007).

This effect is possible because the cell is able to scavenge $^1\Delta_gO_2$ via the increase of the amount of lipid-soluble antioxidants and also the acceleration of reduction of photodamaged D1 protein in the PS II reaction center. This pathway counteracts cell apoptosis along the EXE1 and EXE2 pathways (Mullineaux and Baker 2010).

H_2O_2 stress in plants induces the expression of many chloroplast and nuclear genes (Figure 68) (Foyer and Noctor, 2009; Bechtold et al., 2008). It was found that several genes are down-regulated, while others are up-regulated (Vandenabeele et al., 2004). In particular, H_2O_2 was shown to activate several genes encoding antioxidant and signaling proteins: ascorbate peroxidase (APX), glutathione reductase, catalase, mitogen-activated protein kinase (MAPK), and phosphatases (see Figure 54, Figure 68, Figure 71) (Mullineaux et al., 2000; Vranova et al., 2002). H_2O_2 of chloroplast origin can serve as a redox signal which triggers the expression of the gene encoding cytoplasmic APX2 (Davletova et al., 2005). Likewise H_2O_2 of extracellular/plasmamembrane origin has been shown to be important for APX2 expression (Bechtold et al., 2008; Galvez-Valdivieso et al., 2009). Furthermore, H_2O_2 is also involved in inducing the expression of some light-responsive genes.

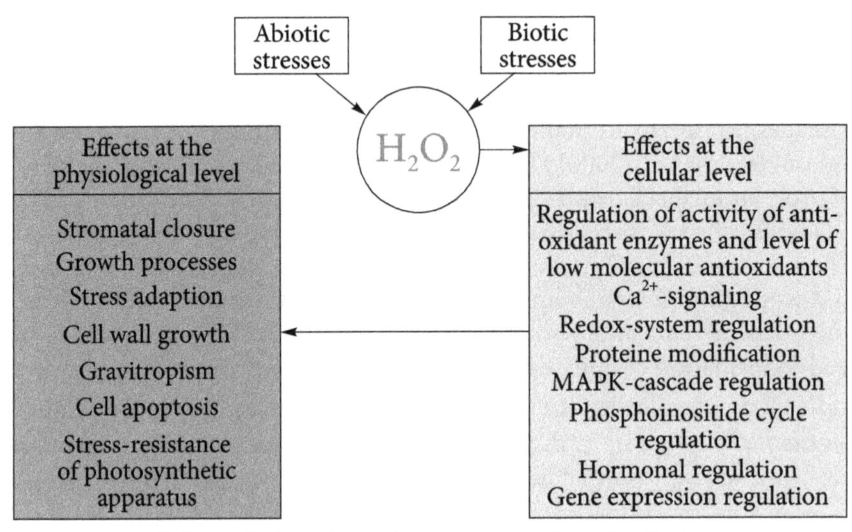

Figure 68. Cellular and physiological processes regulated by H_2O_2.

Plant treatment with H_2O_2 stimulates the expression of the gene *APX2* and of genes *ZAT10* and *ZAT12*, which encode transcription factors (Davletova et al., 2005; Rossel et al., 2007). Both factors *ZAT10* and *ZAT12* mediate different subsets of the high light-inducible or -repressible gene set, including genes coding for *APX2* and *APX1*, respectively.

It has been suggested that H_2O_2 molecules produced in the chloroplasts can exit the organelles by diffusion, likely via water channels (aquaporins), and induce signaling processes in the cytoplasm (Mubarakshina et al., 2010), i.e. triggering the MAPK cascade (Pfannschmidt et al., 2009), by which nuclear genes are activated in the cell, in particular the gene encoding cytoplasmic APX (Apel and Hirt, 2004; Yabuta et al., 2004; Vranova et al., 2002). H_2O_2, produced on the cytoplasmic membrane or in the apoplast, can also play a signaling role, possibly by functioning together with abscisic acid (Bechtold et al., 2008; Yabuta et al., 2004).

A new genetic approach for analyzing the signaling effect of H_2O_2 in plants has been reported (Maruta et al., 2012). This method is based on chemically inducible RNAi. It has been shown that silencing the expression of ascorbate peroxidase bound to the thylakoid membrane (t-APX) in *A. thaliana* leaves leads to both, an increase of the fraction of oxidized proteins in chloroplasts and to effects on the expression of a large set of genes. Among these, the transcription levels of a central regulator of cold acclimation are suppressed and the levels of salicylic acid (SA) increase together with the response to SA. The results reveal synergistic and antagonistic effects of H_2O_2, when chloroplasts are exposed to high light.

Another striking feature is the finding that growth of *A.thaliana* plants under short-day illumination gives rise to a diminished expression of several genes which are involved in sensing and hormone synthesis (Thimm et al., 2004). It was found that the level of ROS production is higher by a factor of about two in leaves from short-day (8 h light) tobacco than in leaves from long-day (16 h light) plants. Based on these results, an unknown regulatory protein was proposed to exist which changes the relative extent of cyclic and pseudo-cyclic photosynthetic electron transport, thereby affecting the ROS content in chloroplasts (Michelet and Krieger-Liszkay, 2012). These findings suggest that light sensor(s) participate in this phenomenon.

The expression of ROS sensitive genes was shown to depend on diurnal and circadian conditions, thus illustrating a role of the biological clock in transcriptional regulation of these genes. Likewise H_2O_2, generation and scavenging exhibit a diurnal rhythm. These findings indicate that an important functional relation exists between ROS signaling and circadian output, which provides a mechanistic link for plant response to oxidative stress (Lai et al., 2012). The components involved and the underlying mechanism(s) of these mutual interactions of signal networks are not yet resolved and represent challenging topics for future research.

4.3 ROS and Cell Redox Control and Interaction with the Nuclear Gene Expression

The discovery of multisensory systems is important for understanding the basic question how perception and transmission of stress signals operate in large cell cultures and tissue. As is shown in the example of the multisensory protein Hik33 in *Synechocystis* one protein is able to react to a variety of different stress signals (chap. 4.2). Hik33 can sense Redox stress but additionally salt and cold stress and all lead to the induction of similar responsive patterns. Multisensory systems become more complicated in plants than in cyanobacteria. Apparently, changes in membrane fluidity, regardless of the nature of the stress effects, are a signal that is perceived by sensory histidine kinases or ion channels localized in the membranes (Los et al., 2013). The ROS signals are transduced by rather macroscopic structures like membranes that regulate single molecules like Hik signaling directly to the molecular fundament of each cell, the genome. In such way, involving histidine kinases as one example ROS signaling covers a large hierarchy from top down and bottom up mechanisms involving the genome, the proteome, cell tissues and the whole organism. Bottom up ROS can oxidize EF-G (see chapter 4.1) and translation of genes is stopped. Top down the ROS level determines membrane rigidity and therefore the activity of histidine kinases.

Structural modifications give rise to detachment of weakly bound enzymes from the cell wall, as shown for the isoforms of peroxidases (Minibayeva and Gordon, 2003). Concomitantly, ROS modulate the activity of antioxidant enzymes, e.g. in case of catalase and ascorbate peroxidases (Shao et al., 2008). Likewise, also the redox state of cell components acting as antioxidants or being involved in signaling (glutathione system, ascorbate system, plastoquinone pool, thioredoxin, etc.) is prone to changes by ROS. Furthermore, the activity of ion channels can be affected, going along with variations in the concentration of relevant ions like Ca^{2+} in the cytosol.

The by far most important regulatory control in acclimation of organisms to different stress factors is the modulation of gene expression. ROS in general and H_2O_2 in particular play an important role in cell signaling pathways and are involved in the regulation of gene expression (Apel and Hirt, 2004; Laloi et al., 2007). Studies using DNA microarrays (Gadjev et al., 2006; Scarpeci et al., 2008) revealed that an increase of the ROS concentration affects the expression of a rather large number of genes. This response can sometimes comprise up to one third of the entire genome. Experiments performed with the unicellular green alga *Chlamydomonas* (*C.*) *reinhardtii* showed that H_2O_2 and $^1\Delta O_2$ interact with different targets leading to activation of specific promoters (Shao et al., 2007).

With respect to regulatory and signaling effects of ROS including the function of second messengers, the general control of cellular processes by the ambient redox conditions should be pointed out clearly as a generalized working framework of the cell chemistry rather than ROS taking the role of "isolated" signal molecules (Foyer and Noctor, 2005; Pfannschmidt et al., 2009; Shao et al., 2008; Buchanan and Luan, 2005). Several models have been proposed for this redox control which includes oxidation/reduction of thiol groups, iron-sulfur centers, hemes, and flavin (Foyer and Noctor, 2005; Vranova et al., 2002). The redox homeostasis in cells is mainly controlled by the presence of large pools of the thiol buffer glutathione and of $NADPH/NADP^+$ and also by high concentrations of ascorbic acid (Foyer and Noctor, 2005). The fraction of reduced glutathione is normally higher than 90% (Noctor et al., 2002). Concomitant with these hydrophilic compounds, tocopherols can function as a lipophilic redox buffer system (for the antioxidant efficiency of different tocopherol species, see (Krumova et al., 2012). These systems protect lipids and other membrane components of chloroplasts by physical scavenging and chemical interaction with ROS (Krumova et al., 2012).

The main sources of ROS in plant cells are still the chloroplasts, where ROS are produced as the so called "waste product" during photosynthesis. Within the chloroplasts primarily the components of the photosynthetic ETC produce ROS (Shao et al., 2007). Several possible sources are known for chloroplast signals, which can affect gene expression in the nucleus of the plant cells (Galvez-Valdivieso and Mullineaux, 2010; Buchanan and Luan, 2005). These sources include the biosynthetic pathway of tetrapyrrole compounds, changes in the redox state of photosynthetic ETC components (e.g. the PQ pool) and ROS generation. All these pathways induced by different signals are interconnected and therefore often considered as tightly coupled. At present, only rather limited information is available on the exact mechanisms of the transduction of signals from the chloroplast to the nucleus due to the accumulation of redox-sensitive compounds and ROS in the chloroplasts (Fey et al., 2005a,b; Pogson et al., 2008; Shao et al., 2008, Pfannschmidt et al., 2009; Kreslavski et al., 2011; Schmitt et al., 2014a).

One of the key sensors in adaptation of the ETC to light is the redox state of the PQ pool which regulates the phosphorylation of light-harvesting complexes II (LHC II) (Vener et al., 1998) and also acts as a signal for regulation of the expression of a set of plastid and nuclear genes (Pfannschmidt et al., 2003), such as *Lhcb*, *petE*, *APX2*, and *ELIP2* encoding light-harvesting complex proteins, plastocyanin, ascorbate peroxidase 2 and early light inducible protein, respectively. Likewise, the expression of SOD is affected (Shaikhali et al., 2008). It should be noted that the expression of *Lhcb* genes is only partially controlled by the redox state of PQ, because additional factors are involved like ATP synthesis and the electric potential difference across the thylakoid membrane (Yang et al., 2001).

The redox state of PQ is proposed to induce two signaling pathways, which are initiated under the influence of high and low light and subsequently activate the expression of plastid and nuclear genes (Fey et al., 2005a,b; Pfannschmidt et al., 2009). ROS arising from reactions at the acceptor side of PSII could be one type of signals which trigger the regulation of these pathways (Ivanov et al., 2007). Under high light stress of *A. thaliana*, the PQ pool was shown to be oxidized by both ${}^{1}\Delta_{g}O_{2}$ and less electron input from PS II due to the effect of NPQ with implications on redox signaling (Kruk and Szymańska, 2012). Furthermore, a plastid terminal oxidase (PTOX) leading to slow PQH_2 oxidation is probably involved in a ROS-triggered signal transduction cascade (Trouillard et al., 2012). The underlying mechanism of the specific role of PTOX in acclimation of plants to high light remains to be clarified.

The expression of genes encoding PSI proteins (psaD and psaF) is also affected (Pogson et al., 2008). Changes in the redox state of components on the PSI acceptor side contribute to the regulation of nuclear and chloroplast genes (Shaikhali et al., 2008). This effect is primarily related to the redox state of thioredoxin, which depends on the rate of electron transport from ferredoxin (Scheibe et al., 2005). The redox states of thioredoxin, glutathione and glutaredoxin act as signals for regulation of stress-responsive genes (Mullineaux and Rausch, 2005; Schürmann and Buchanan, 2008).

Several examples of regulation by thiols, in particular chloroplast gene translation and transcription have been described in recent reviews (Oelze et al., 2008). In addition to signaling modes where ROS generated in response to different types of stress act either directly as signal molecules, or, via generation of second messengers (e.g. oxidation products of Cars and lipids, *vide supra*), also pathways operating in the opposite direction are established in plants. This does not only account for producing ROS as defensive molecules but it also accounts for the active production of ROS as messenger molecules. In both these cases, ROS are produced as second messengers or as reactive species in response to stress, e.g. in the defense to biotic infection.

Special proteins are involved in the development of cell response to changes of the redox state. These proteins are encoded by so-called reporter genes. Investigations on redox signaling between chloroplasts and nucleus have been focused on the induction of genes of cytosolic ascorbate peroxidases APX1 and APX2, genes *ZAT10* and *ZAT12* encoding zinc-finger transcription factors, and also gene *ELIP2* encoding the early light-induced chlorophyll-binding protein ELIP2. Both transcription factors, ZAT10 and ZAT12, favor the induction of gene clusters related to activation of the photosynthetic ETC under high light by switching on the expression of genes *APX1* and *APX2* (Davletova et al., 2005; Pogson et al., 2008). The expression of genes *APX2*, *ZAT10*, and *ZAT12* is stimulated by treatment with H_2O_2 (Karpinski et al., 1999; Davletova et al., 2005; Pogson et al., 2008) It seems reasonable to assume that H_2O_2 regulates the expression of these genes, thus acting via direct or indirect effects of the redox state signaling from chloroplast through cytosol to the nucleus. Kinetic experiments revealed that the redox signal arising from the response to a change in light quality (spectral composition) is transmitted within about 30 min from the chloroplast to the nucleus (Zhang et al., 2000).

Studies on high-light effects in *A. thaliana* plants showed that the nuclear genes encoding cytosolic peroxidases APX1 and APX2 were activeted in 15–20 min. It was also found that the activation of these genes is part of the systemic response to superfluous light exposure (Karpinski et al., 1999). The induction of chloroplast gene expression occurred also in the range of 15–20 min in response to changes of the redox state in the organelles induced by changes in light quality (Pfannschmidt et al., 2009). These kinetic results on signal transduction suggest that some components of the signal cascade triggered by stress-induced changes of ROS concentration are present in the cells already under optimum conditions and do not need to be synthesized in response to stress. This idea explains the correspondence of the kinetics of signal transduction in response to ROS appearance and of the transduction of other intracellular stimuli (Pfannschmidt et al., 2009).

A genetic screen aimed at identifying $^1\Delta_gO_2$-responsive genes (Baruah et al., 2009) led to the proposal of the gene named "pleiotropic response locus 1" (*PRL1*) acting as a point of convergence of several different signaling pathways, thus integrating various intra- and extracellular signals. Under pathogen-induced stress, the gene "enhanced disease susceptibility 1" (*EDS1*) plays a role in development of the hypersensitive reaction and in mediating EXE1/EXE2-regulated cell death induced by $^1\Delta_gO_2$ (Ochsenbein et al., 2006). The EDS1 protein has been shown to be required for the resistance to biotrophic pathogens and the accumulation of SA. SA likely enhances the plant defenses by inducing the synthesis of pathogen-related proteins (Mullineaux and Baker, 2010). EDS1 seems to play a pivotal role in a mutually antagonistic system, integrating ROS signals from chloroplasts in cells suffering from photooxidative stress (Straus et al., 2010).

A general problem in identifying different $^1\Delta_gO_2$-induced signal pathways and their (synergistic) interplay has to be mentioned. The effect on the gene expression pattern is expected to depend on the nature of the nearest neighborhood of $^1\Delta_gO_2$ formation, if one accepts that signaling directly by $^1\Delta_gO_2$ can take place only at a site very close to its generation. This would also imply that the signal pathway comprises the participation of oxidation products of Cars, lipids and other molecules acting as second messengers which can induce different genetic responses. The $^1\Delta_gO_2$ site differs in WT plants and in mutants like *flu1* and, also, if $^1\Delta_gO_2$ is generated by using exogenous sensitizers (Hideg et al., 1994; Krieger-Liszkay, 2005). Therefore, different types of second messenger species are likely to be formed in mutant studies in contrast to wild type studies. Thus, it is difficult to gather straightforward conclusions on the mechanism of $^1\Delta_gO_2$ signaling from studies performed under different assay conditions and using different sample material including single gene mutants. This important problem needs to be further addressed in forthcoming studies.

ROS can be involved in several signaling pathways by modulating the activity of different components like MAPKs and phosphatases, transcription factors, and calcium channels (Pei et al., 2000; Mori and Schroeder, 2004; Pfannschmidt et al., 2009; Kreslavski et al., 2011; Pogson et al., 2008; Kovtun et al., 2000; Gupta and Luan, 2003). Heterotrimeric G-proteins may also participate in the signaling pathways initiated by ROS (Joo et al., 2005). An effect of ROS on the activity of Ca^{2+} channels was shown to arise for both abiotic stress and plant-pathogene interaction (Demidchik et al., 2003).

Very few details have been resolved so far on the nature of steps which link various pathways in coordinating ROS signaling. One piece of the puzzle is the finding that MAPKs are involved in transducing signals derived from ROS generated by sources in chloroplasts (Liu et al., 2007).

One mode of ROS-induced signaling is given by the activation of transcription factors containing SH groups like OxyR in eubacteria and/or iron-sulfur clusters (Zheng et al., 1998). Formation of S-S bridge(s) by H_2O_2 is expected to change the structure of OxyR, thereby inducing the transition from the inactive into the active form, as is schematically illustrated in Figure 69.

But another possibility of ROS signaling changes of the subcellular distribution of these factors is seen in the case of yeast. Yeast cells express the protein Yap1, which is functionally homologous to the transcription factor OxyR in eubacteria. Yap1 can regulate the transcription of specific genes in response to changes of the redox state of the cell (Liu et al., 2005). The inactive form of Yap1 is localized in the cytoplasm. H_2O_2 oxidizes Yap1 via the peroxidase Gpx3 (Delaunay et al., 2000) under formation of disulfide bonds between neighboring cysteines, thus leading to conformational changes, which enables the transport of this Yap1 form to the nucleus, where it induces the expression of genes encoding for components of antioxidant defense system(s). Mutants lacking Yap1 were shown not to be able to induce antioxidant defense upon treatment with H_2O_2 (Liu et al., 2005).

In analogy to the thiol-based sensor OxyR of bacteria, Yap1 is part of a relatively simple regulatory loop, where ROS induce the expression of certain antioxidative enzymes. Although a gene homologous to *OxyR* is absent in higher plants, attempts to complement mutations of the gene *OxyR* in *E. coli* by using the expression gene library of *A.thaliana* identified the *AnnAt1* gene encoding annexin as to be capable of restoring a functional defect in the *OxyR* bacterial mutant (Gidrol et al., 1996). Recently identified signaling proteins undergoing thiol modulation (modification) in plants include a protein tyrosine phosphatase (Dixon et al., 2005) and a histidine kinase ETR1, which is involved in ethylene signaling (Desikan et al., 2005).

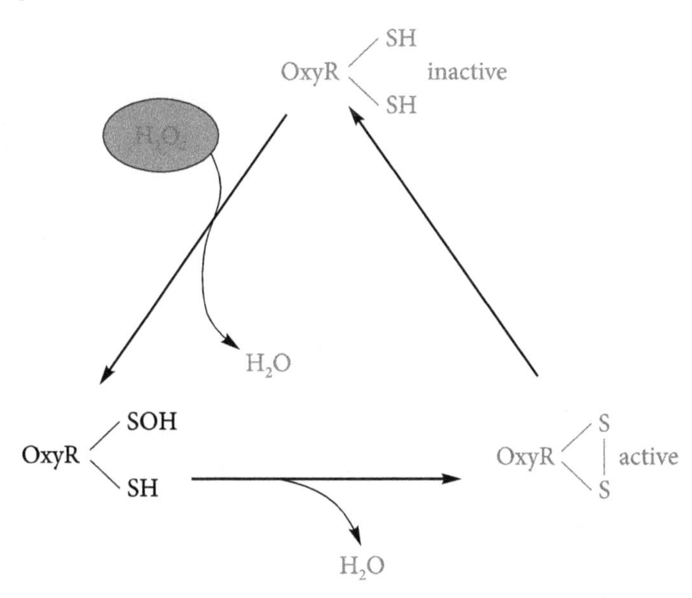

Figure 69. Hypothetical scheme of regulation of bacterial transcription factor OxyR activity. The inactive form contains thiol groups (SH). Under the influence of H_2O_2, the thiol group is oxidized with the formation of an SOH group and then rapid formation of a disulfide bond occurs and OxyR transits into its active form.

Based on these findings, it seems reasonable to assume that analogous mechanisms also exist in plant cells, but they are likely to be more complex. Figure 70 presents a proposed simplified scheme for redox-sensitive sensors (RsS) acting as primary sensors of H_2O_2 signal transduction.

The signal can be transmitted directly from H_2O_2 or via RsSs to the MAPK cascade and/or to transcription factors (Neill et al., 2002; Pogson et al., 2008; Pfannschmidt et al., 2009). The conformation and activity caused by reversible oxidation of cysteine residues of regulatory proteins, which are involved in gene expression at different developmental stages, offer a simple and elegant mechanism for regulation of transcription and translation systems under oxidative stress. As a result, transcription of nuclear genes required for ROS scavenging is activated. By oxidation of the translation elongation factor G (EF-G) in chloroplasts and by blocking translation of new proteins, H_2O_2 can also regulate gene expression on the level of translation, in particular (see chap. 4.1 for more details) (Nishiyama et al., 2011; Murata et al., 2012).

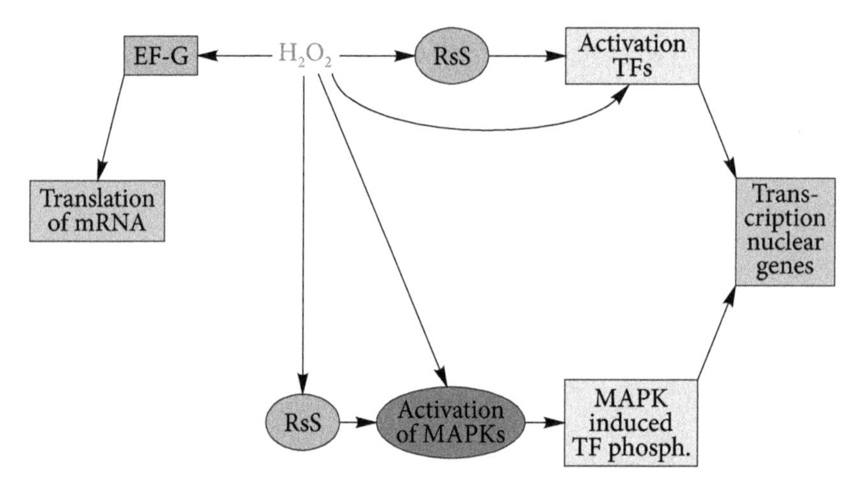

Figure 70. Effects of H_2O_2 on processes of transcription and translation. RsS, MAPK and TF are redox-sensitive sensor(s), MAP-kinase and transcription factor(s), respectively.

Figure 71 presents a hypothetical scheme of pathways of photosynthetic redox signal transduction in plants. It summarizes selected mechanisms described in the chap. 4.1, 4.2 and 4.3 in a general picture that aims to denote the complex networking of different species to establish ROS-iniated and ROS-mediated signaling pathways between cell organelles.

4.4 ROS as Top down and Bottom up Messengers

In chapters 4.1–4.3 we have found a broad series of ROS signaling that might comprise both bottom up or top down signaling (see Figure 71). Indeed some of the effects described were formerly believed as unknown top down signaling procedures, however, as we will point out in the following, bottom up and top down might occur concomitantly and it is not easy to comprise only one direction in a complicated communicating network.

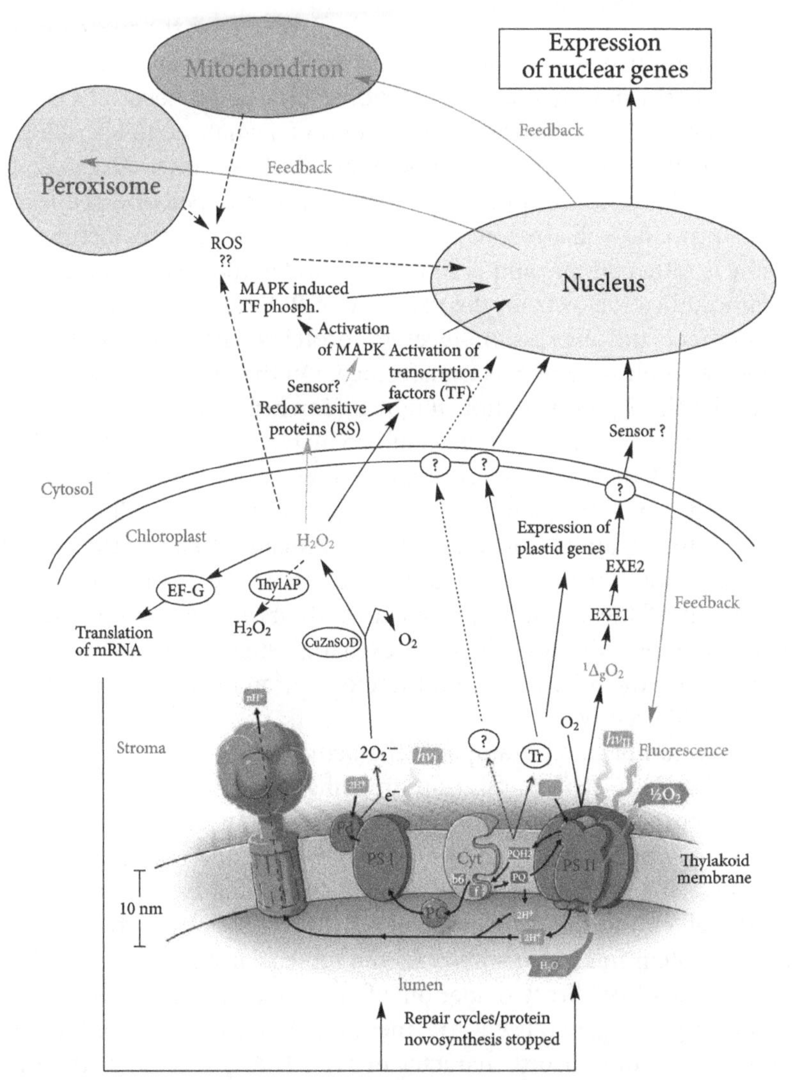

Figure 71. Hypothetical scheme of pathways of selected photosynthetic redox signal transduction in plants initiated at the thylakoid membrane. For the sake of simplicity, other cell organells (nucleus, mitochondrion, peroxisome) are symbolized by colored ovals. Interrupted arrows designate hypothetic pathways of signal transduction. The question marks designate unknown components of signal transduction pathways. Solid lines designate signal transduction pathways with some experimental confirmation. Dotted lines designate experimentally established signal transduction pathways in chloroplast ETC and in the stroma. The abbreviations RS, MAPK and TF denote redox-sensitive protein(s), MAP-kinase and transcription factor(s), respectively according to (Schmitt et al., 2014a). Image reproduced with permission.

To reconcile some examples we might first look at the mechanism of EF-G oxidation which stops the translation of RNA and therefore the overall protein novosynthesis in cyanobacteria under severe ROS stress (see chapter 4.1). As the primary oxidation of a single complex, the EF-G compartment of ribosomes, the activity of ribosomes is stopped. The activity of the ribosomes in gene translation is stopped and the level of certain proteins will decrease according to their average lifetime. The scheme is rather simple and is sometimes understood as a typical bottom up communication as from the presence of ROS molecules in the microenvironment the expression level of proteins changes and therefore finally the whole organism is transformed. On the other hand the original trigger for the EF-G oxidation might be intense sunlight and concomitant ROS that disactivate the protein production machinery. In such way the change of the microenvironment follows a macroscopic command from top down. If the plant could make the voluntary decision to expose itself to intense sunlight or to bring its organism into a situation of ROS overproduction most people would agree that we have a top down communication of the whole organism towards its molecular composition. It is comparable with a decision to exchange single elements to enforce a variation on the microscale and therefore enforcing the whole structure to behave differently.

A more complicated example is the activation of histidine kinases, as Hik33; this is a typical reaction to enhanced membrane rigidity as a consequence of a higher ROS levels. Here the macroscopic change of a structure induces the change of a single molecule configuration (see chap. 4.3). We acknowledge that transmission of stress signals operates in large cell cultures and tissues, and Hik33 is understood to be a multisensory protein in *Synechocystis.* As the molecular answer to membrane rigidity instead of direct oxidation of molecular cysteines the reaction scheme appears generalized to "membrane tension" instead of ROS defining the multisensory character of Hik33. Any change of the membrane fluidity, which might be ROS, salt, cold stress or age activates Hik33 and the typical property of a larger structure to react to stress in a common way (changing fluidity of the membrane). The ROS signals are transduced by rather macroscopic structures like membranes that regulate single molecules like Hik signaling directly to the molecular fundament of each cell, the genome. It seems as we have a generalized top down communication where the membrane reacts to an external trigger and induces a change on the moleculare level. However finally the activation of genes is bottom up again. In fact a top down strategy seems more likely be able to induce a bottom up change while a bottom up change will change if induced by a macroscopic signal. So top down and bottom up strategies somewhat go hand in hand and ROS signaling covers a large hierarchy from top down and bottom up mechanisms involving the genome, the proteome, cell tissues and the whole organism.

Other examples arise as structural modifications give rise to detachment of weakly bound enzymes from the cell wall, as shown for the isoforms of peroxidases (Minibayeva and Gordon, 2003). On the other side again, ROS modulate the activity of antioxidant enzymes, e.g. in case of catalase and ascorbate peroxidases (Shao et al., 2008). Likewise, also the redox state of cell components acting as antioxidants or being involved in signaling (glutathione system, ascorbate system, plastoquinone pool, thioredoxin, etc.).

Bottom up ROS can oxidize EF-G (see chapter 4.1) and translation of genes is stopped. Top down the ROS level determines the change of macroscopic structures and therefore the activity of histidine kinases. However that is only a single aspect as the existence of ROS signalling itself might be understood as a top down phenomenon.

In higher animals ROS level change due to voluntary activation of ROS. Animals that carry a free will might just make the decision to raise their ROS level. The body can be activated and activity, active movement, i.e. "sport" conducted as a consequence of the free will, can raise the ROS level severely and therefore activate changes on the molecular level down to gene expression in both ways, by direct oxidation and by transmitted activation of structural changes that initiate the activity changes of multi-sensory proteins.

But should we therefore always understand top down as a consequence of free will? It seems more likely that a pattern interaction in hierarchical systems contributes to phenomena of both qualities: top down and bottom up signaling. Indeed all these interaction schemes follow a long evolutionary adaption process that adds up to a higher reproduction probability under given constraints. So in both directions, bottom up as well as top down, we find the most important regulatory control in acclimation of organisms to different stress factors – the modulation of gene expression (Apel and Hirt, 2004; Laloi et al., 2007). As mentioned above, this response can sometimes comprise up to one third of the entire genome.

Redox-sensitive enzymes serve as a molecular "switch" by undergoing reversible oxidation and reduction reactions in response to redox changes within the cells. ROS can oxidize the redox-sensitive enzymes directly or indirectly under the participation of low-molecular redox-sensitive molecules like glutathione or thioredoxin, from which the latter interacts with ferredoxin (Foyer and Noctor, 2005; Shao et al., 2008). In this way, the whole cell metabolism can be tuned. On the other hand, redox-sensitive signaling proteins function in combination with other components of signaling pathways, including MAPKs, phosphatases, transcription factors, etc. (Foyer and Noctor, 2005; Shao et al., 2008, Pfannschmidt et al., 2009). Redox regulators available in the apoplast have been suggested to be among the key ROS sources during stress (Minibayeva et al., 1998, 2009; Minibayeva and Gordon, 2003). It has to be mentioned that the molecular mechanisms that transfer ROS by dedicated oxidation of covalent bonds producing chemical oxidation products are maybe just a minor example of the general regulation that occurs from the overall redox state of the cell. Fundamental concepts may also arise from physical principles like the membrane potential or the decoupling or coupling of photosynthetic subunits by electrostatic interaction, which has been suggested to be the driving mechanism for the coupling state of cyanobacterial light-harvesting complexes and the cell membrane in *A. marina* (see Schmitt et al., 2006; 2007) or in artificial systems consisting of cyanobacterial antenna complexes and semiconductor quantum dots (Schmitt et al., 2010; 2011; Schmitt, 2011).

If we look to the activation of ROS as messenger or defense molecules we find another situation where two directions of a reaction pattern are conducted depending on the general circumstances in which the system is found. As mentioned above ROS is generated in response to different types of stress acting as signal molecules, but via generation of second messengers (e.g. oxidation products of Cars and lipids, *vide supra*), also pathways operating in the opposite direction are established in plants. This does not only account for producing ROS as defensive molecules but it also accounts for the active production of ROS as messenger molecules.

Therefore we have again something that might be understood as being top down and bottom up regarding the point of view from where we look at it. Active production of ROS as single molecules initiated by external stimuli or by free will is top down as it deactivates single cells (for example biotic pathogens) as a reaction to an externally introduced cell damage or even as a consequence of free will and a decision of the whole organism (if it carries consciousness, like animals and humans and is able to conduct "decisions"). Bottom up the signal molecules oxidize MAPKs and therefore directly influence gene expression.

In the following we will try to establish this kind of understanding of top down and bottom up networking in another, more generalized point of view.

Light is the most important signal in regulating a vast majority of processes in living organisms, as reflected by numerous light sensors and biological clocks. But light has to be understood as both, a top down messenger (as it is conducted from the solar radiation field onto the earth surface) and additionally as a bottom up messenger (as it induces conformational changes on the molecular level that conduct molecular reaction schemes changing finally the basic chemical composition in cells).

Light is at first the unique Gibbs free energy source for the existence of living matter though the process of photosynthesis. Light should generally be understood as the most basic prerequisite of life as it carries a reduced entropy as compared to the thermal equilibrium of the planet. When light is transferred to a planetary surface and reemitted as heat radiation and in such way carrying away the most entropic radiation, the black body radiation, from the planetary surface, it can take up entropy and therefore allow for the growth of complex structures and locally less entropic systems.

On the other hand, light at high intensities also leads to stress giving rise to the deleterious process of photoinhibition in photosynthetic organisms (Adir et al., 2003; Allakhverdiev and Murata, 2004; Nishiyama et al., 2006; Murata et al., 2007; Vass and Aro, 2008; Li et al., 2009; Goh et al., 2012; Allahverdiyeva and Aro, 2012). Imbalances in the redox state of components of the electron transfer chain (ETC) lead to dangerous ROS production. Therefore, suitable sensors are required to permit efficient adaptation to illumination conditions which vary in time (diurnal, seasonal rhythm) and space (e.g. plants in different altitudes of a tropical rain forest or bacteria in different water depth and living environment).

4.4.1 Stoichiometric and Energetic Considerations and the Role of Entropy

The statements of Jennings as published in (Jennings et al., 2005; 2006; 2007) suggest the possibility that the second law of thermodynamics might stand in contradiction to the primary photochemistry of plants. Such an opinion seems to be rather challenging, if not to say wrong as we will seek to show in the following.

We will try to briefly discuss the basic underlying thermodynamic concepts and their relation to photosynthesis. In fact, it is hard to prove that photosynthesis does not proceed with negative entropy formation, but it is equally difficult to prove the opposite. As for all thermodynamic processes that might violate the second law of thermodynamics, no plants are observable that grow with a universe overall negative entropy formation. It turns out that plants grow in local nonequilibrium situations in the stationary overall equilibrium of the energy and entropy flux from the sun that interacts with the earth and leads to a high entropy production on the earth surface. We strongly believe and make suggestions for a proof that no process violates the second law of thermodynamics. The second law of thermodynamics mainly reveals that unlikely conditions relax to the most probable distribution. The appearance of plants is just a kind of nonequilibrium dynamics in such process that leads from a local strong nonequilibrium to the equilibrated situation. Plants slow down the pathway of a system into its most probable state. This aspect of physics and living matter is shortly treated in the following subsection.

The photosynthetic formation of sugar from carbon dioxide and water as denoted by chemical equation 1 is highly endergonic and therefore needs an input as driving force, the Gibbs free energy of the absorbed photons. The endothermic character is shown by the difference of the enthalpy (H) which is $\Delta H = +2808$ kJ/mol. The reduction of entropy (S) can be calculated to $\Delta S = -259.1$ J/(mol*K) (see Müller, 2005). Therefore at room temperature (T) (295 K) one gets $T\Delta S = -76.4$ kJ/mol.

Generally a process occurs spontaneously if the change of Gibbs free energy (G) is negative (i.e. if the Gibbs free energy of the product compounds is smaller than the Gibbs free energy of the reaction products, $\Delta G < 0$):

$$\Delta G = \Delta H - T\Delta S. \tag{64}$$

Without the photon contribution the resulting Gibbs free energy of the photosynthetic reaction described in chemical equation 1 calculates to $\Delta G = +2884.4$ kJ/mol according to equation 64 indicating the endergonic character of the reaction. The equilibrium of the chemical equation 1 is far at the left side of the reaction, i.e. at the side of the chemical educts, carbon dioxide and water.

The analysis of the photosynthetic processes leading to the production of one mol glucose shows that at least 60 photons are absorbed per single molecule glucose that is generated (Häder, 1999; Campbell and Reece, 2009).

The absolute minimal value for the energy uptake should at least correspond to the absorption of 4 photons for the oxidation of one water molecule (two turnover cycles of PS I and PS II, each) and therefore to 48 photons per mol glucose. The splitting of $2H_2O$ to $4H^+$ and O_2

is releasing 4 electrons in the photosystem II (PS II) only. These electrons have to be pumped from the PS II via PS I to the place where NAD+ is reduced (see Figure 51 and Renger and Renger, 2008; Renger 2007, 2008, 2008b; Kern and Renger, 2007). PS I and PS II are working hand in hand which doubles the absorbed number of photon quanta per mol glucose. In fact, 8 photons are absorbed to split 2 water molecules, i.e. 48 photons are absorbed per mol glucose (see chemical equation 2).

This leads to a photonic contribution of at least $\Delta H^{phot} = 48 \cdot N_A h\nu$ per mol that is delivered by the photon energy (Jennings et al., 2005). The longest absorbed wavelength in the PS I is about 700 nm and therefore an additional $\Delta H^{phot} = 48 \cdot N_A h\nu \approx 8200$ kJ/mol is involved into chemical equation 1.

For detailed thermodynamic considerations the eq. 64 has to be evaluated for each single chemical step inside a plant. Such accurate thermodynamic analysis of photosynthesis is a very complicated task. It might be the reason for the recent discussions whether or not photosynthesis might violate the second law of thermodynamics (Jennings et al., 2005; 2006; 2007; Lavergne, 2006).

One mol of absorbed red photons with 700 nm wavelength contains an energy of 171 kJ. It is not fully consistently answered in the literature how much entropy a single photon contains or even if it is possible to define the entropy of a single photon. The entropy of the photon ensemble can be described by the entropy of the Planck spectrum and therefore thermal single photons exhibit a probability distribution that carries the corresponding amount of entropy. The absorbed light energy of 8200 kJ/(mol glucose) seems to carry a huge amount of "excess" energy in comparison to the energy consumption $\Delta G = +2884,4$ kJ/mol in chemical equation 2.

Some of the relevant literature suggests that this "excess energy" shows that photosynthesis could be much more efficient. We believe that this point of view is too simple: how the directed transfer of photon energy to the chemical Gibbs free energy of glucose is possible is an important question. During the reaction sequence highly energetic compounds are formed as ATP that is urgently necessary to drive the functional work of the cell. It is not possible to drive the photosynthetic processes without energy dissipation.

The transformation of solar radiation to free energy is possible due to the low entropy per enthalpy (H) ratio S/H of the sunlight. If the volume and pressure do not change during a thermodynamic process, as we assume for the absorption of a photon, then $dH = dU$ and $1/T = S/H$.

The inverse temperature of the solar radiation and therefore $(S/H)^{solar}$ is much lower than the $(S/H)^{earth}$ ratio of thermal photons that are released after absorption on the earth surface. The wavelength maximum (λ_{max}) of the emitted black body radiation determines the so called colour temperature (T_C) of radiation which is associated with the S/H ratio according to Wien´s law:

$$\frac{S}{H} = \frac{1}{T_C} = \frac{\lambda_{max}}{2897{,}8\,\mu mK}. \tag{65}$$

Equation 65 shows, that the wavelength maximum (λ_{max}) of electromagnetic radiation is proportional to the entropy per enthalpy ratio. Therefore S/H is much lower for solar light than for thermal radiation of a 280 K black body.

In that sense the plant uses a pool of "negative entropy" from sunlight. There exist different estimations of the effective colour temperature T_C determined by the light spectrum that is available for photosynthetic organisms taking into account scattered light or light absorbed by the earth atmosphere (Müller, 2005), but for all estimations the available light spectrum corresponds to $T_C \gg 1000$ K in marked contrast to the typical entropy content of thermal baths in the sea or the earth atmosphere at $T = 280$ K. The formation of highly organized structures is a dissipative process in accordance to the second law of thermodynamics because the process occurs during the equilibration of sun radiation and earth temperature. The plants grow driven by a fractally local nonequilibrium.

As we have seen the photosynthesis is a quantum process which is driven by absorption of single photons. Entropy is a statistical quantity which is not well defined for single particles.

Generally the equation

$$\frac{\Delta S}{\Delta Q} \geq \frac{1}{T} \tag{66}$$

holds where the equality denotes reversible processes when the uptake of heat goes along an isothermal path while the inequality is characteristic for irreversible processes, i.e. when the system is dissipative and the dynamics is not restricted to a thermodynamic path.

At this point it has to be clearly stated that the photosynthetic process does not occur in the equilibrium. The equations of thermodynamics are only valid assuming local equilibrium conditions. This might not be possible for light absorption and the interaction of excited states with the surrounding environment. The reader should therefore keep in mind that our considerations suggest extensions of formulas that are valid in equilibria conditions in spite of strong nonequilibrium conditions.

Evaluating equation 65 we can make a rough suggestion for the average entropy of a single photon by identifying it's energy with enthalpy and heat: $Q = H = h\nu$.

The equality of heat and enthalpy is fulfilled for an isobar thermodynamic process if the pressure remains constant which can be assumed to be valid for reactions of the photon gas.

From that simple consideration we would suggest

$$S_{phot} \approx \frac{hc}{2.8978 \cdot 10^{-3} mK} = 6.9 \cdot 10^{-23} \text{ J/K} \tag{67}$$

as contribution of the probability distribution of the single photon to the entropy. Interestingly S_{phot} as given by eq. 67 is a fixed value independent from the photon wavelength and qualitatively comparable to the value $S_{phot} \approx k_B = 1.38 \cdot 10^{-23}$ J/K as presented in (Kirwan, 2004).

Kirwan Jr. (2004) suggests a comparable value for the photon entropy as shown here, while Gudkov (1998) takes the viewpoint that "light is a form of high grade energy which carries no thermodynamic entropy". Already Planck calculated the black body radiation spectrum considering the thermodynamic entropy of the radiation field. The thermal black body radiation which is in thermal equilibrium with the environment carries no Gibbs free energy: $\Delta G = 0$ which corresponds to $\Delta H = \Delta Q = T\Delta S$ according to eq. 64.

The conversion of the energy of 1 mol photons in the solar radiation field at T_C = 5000 K into thermal radiation generates about 17 mol photons at T_C = 280 K according to Wien's law $\frac{1}{T_C} = \frac{\lambda_{max.}}{2897{,}8 \mu mK}$ and the fact that $E_{photon} = \frac{hc}{\lambda}$. As about half of the radiation is converted to excited states in the plant while the rest of the incoming spectrum dissipates the generation of 1 mol glucose it at least correlated with the emission of $8 \cdot 48 \approx 400$ mol thermal photons. These photons might carry away an entropy of $\Delta S_{thermal} \approx 400 \cdot N_A \cdot k_B \approx 3.3 \text{ kJ/(mol} \cdot \text{K)}$ which is easily compensating $\Delta S_{glucose} \approx -259.1 \text{ J/(mol} \cdot \text{K)}$ necessary to be reduced for the production of one mol glucose as shown above. $\Delta S_{thermal}$ exceeds $\Delta S_{glucose}$ by a factor of ten. The rather rough and incomplete estimation as presented here might finally lead to the conclusion that a plant can not grow without dissipating a large fraction of its absorbed energy as this is necessary to preserve the second law of thermodynamics.

The necessary rise of the overall entropy when a plant grows is correlated with dissipation of energy into the environment. If a system absorbs photons and then relaxes to a final state emitting more photons than previously absorbed (in particular the production of phonons or any bosons fullfills the necessary increase of entropy) then the final state can be of lower entropy than the initial state in full accordance to the second law of thermodynamics because the environment takes up entropy.

At the moment there is no reason to assume that photosynthesis violates the second law of thermodynamics as done by Jennings et al. (Jennings et al., 2005) and answered by detailed description of the processes which are in line with the second law of thermodynamics by Lavergne (Lavergne, 2006) which let to further comments of Jennings (Jennings et al., 2006, 2007).

Generally entropy is a quantity of an ensemble describing the probability of an ensemble´s state. The most general formulation of this property is given by the Shannon entropy

$$S = -k_B \sum_i p_i \ln p_i \tag{68}$$

calculating the entropy from the single probabilities p_i of each state i that occurs in the thermodynamic equilibrium (k_B is the Boltzmann constant). A more general formulation where $p_i(t)$ is time dependent and the system is not necessarily in an equilibrium is suggested by Haken (1990).

$$S = -k_B \sum_i p_i(t) \ln p_i(t). \tag{69}$$

In eq. 69 $p_i(t)$ might be calculated from rate equations. We will use equation 68 and equation 69 to estimate the probability of a forward in comparison to backward steps in the rate equations as given by eq. 8 (see chap. 1.3). Equation 69 is a generalization of the problem of calculating entropy in the nonequilibrium case which helps us to understand the correlation of dynamics of probabilities and entropy when a photosynthetic complex relaxes after light absorption.

In the thermodynamic equilibrium equation 68 and equation 65 connect the macroscopic observables of thermodynamics (here: temperature T as given in eq. 65) with the microscopic probabilities of certain distributions of space and momentum in an ensemble of states (equation 68). Therefore the combination of equation 68 and 64 is the most important step for a statistical motivation of thermodynamics.

Eq. 8 and 9 (see chap. 1.3) can be derived from eq. 68 if we use the second law of thermodynamics which postulates that the Shannon entropy function of eq. 68 is at maximum in the equilibrium case of a closed system.

Then

$$\delta S = 0 = \delta\left(\sum_i p_i \ln p_i\right)$$
$$= \delta\left(\sum_i p_i \ln p_i - \lambda\left(\sum_i p_i - 1\right) - \beta\left(< E > - \sum_i p_i E_i\right)\right)$$

where λ is an arbitrary Lagrange parameter with $\delta\lambda = 0$ that can be added because $\sum_i p_i = 1$ and β is a Lagrange Parameter that can be added because $< E > = \sum_i p_i E_i$.

Accomplishing the variation one gets:

$$e^{(\lambda-1)} = \frac{1}{\sum_i e^{-\beta E_i}} \Rightarrow p_i = \frac{e^{-\beta E_i}}{\sum_i e^{-\beta E_i}}.$$

Taking eq. 66 (the equality for the reversible processes) and eq. 68 one gets:

$$S = -k_B\left(\sum_i p_i \ln p_i\right) = \frac{Q}{T}$$
$$\Rightarrow -k_B\left(\frac{\sum_i e^{-\beta E_i}\left(-\beta E_i\right)}{\sum_j e^{-\beta E_j}}\right) = \frac{Q}{T} \Rightarrow k_B \beta \langle E \rangle = \frac{Q}{T}.$$

For the thermally equilibrated system without work $\langle E \rangle = Q$ one finds: $\beta = \frac{1}{k_B T}$.

And therefore

$$p_i = \frac{e^{-\frac{E_i}{k_B T}}}{\sum_i e^{-\frac{E_i}{k_B T}}} := \frac{e^{-\frac{E_i}{k_B T}}}{Z} \tag{70}$$

where Z denotes the standard canonical partition function.

From eq. 70 the proposed Boltzmann distribution for state populations according to eq. 9 and for rate constants according to eq. 8 follows directly.

4.4.2 The Entropy in the Ensemble of Coupled Pigments

As mentioned shortly in the last chapter the rate equation formalism as e.g. given by eq. 4 and 38 (see chap. 1.3, 1.4 and 2.2) delivers a thermodynamic approach to excited states that migrate in systems of coupled pigments. In the equilibrated case the probabilities of excited state populations follow the Boltzmann distribution:

$$S = -k_B \sum_i p_i \ln p_i = -k_B \sum_i \left(\frac{e^{-\frac{E_i}{k_B T}}}{\sum_j e^{-\frac{E_j}{k_B T}}} \ln \frac{e^{-\frac{E_i}{k_B T}}}{\sum_j e^{-\frac{E_j}{k_B T}}} \right)$$

$$= -k_B \sum_i \left(\frac{e^{-\frac{E_i}{k_B T}}}{\sum_j e^{-\frac{E_i}{k_B T}}} \left(\frac{-E_i}{k_B T} - \ln \sum_j e^{-\frac{E_j}{k_B T}} \right) \right) \tag{71}$$

$$= k_B \ln Z + k_B \sum_i p_i \frac{E_i}{k_B T} = k_B \ln Z + \frac{\langle E \rangle}{T}.$$

In the dissipative (nonequilibrium) situation the solution of eq. 38 for the time dependent excited state population can be used to estimate the entropy dynamics:

$$S(t) = -k_B \sum_i p_i(t) \ln p_i(t) = -k_B \sum_i \left(\frac{N_i(t)}{\sum_j N_j(t)} \ln \frac{N_i(t)}{\sum_j N_j(t)} \right)$$

$$= -k_B \sum_i \left(\frac{\sum_j A_{ij} e^{-k_j t}}{\sum_k \sum_j A_{kj} e^{-k_j t}} \ln \frac{\sum_j A_{ij} e^{-k_j t}}{\sum_k \sum_j A_{kj} e^{-k_j t}} \right). \tag{72}$$

In the following eq. 72 will be shortly evaluated. For that purpose we choose the most simple system in this context, i.e. two coupled states as shown in Figure 72. In the initial moment only the energetically higher state is excited and after excitation the system relaxes. For the sake of simplicity we assume that the excited states cannot decay into the ground state. This is a good approximation for strongly coupled systems where the energy transfer processes and the thermal equilibration occur much faster than the excited state relaxation.

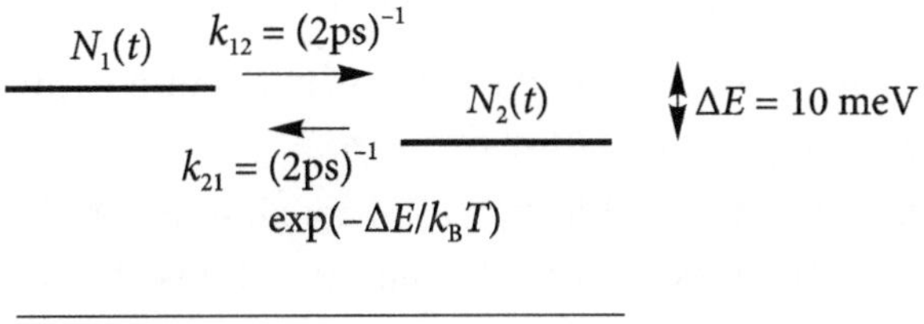

$N_1(t)$ $k_{12} = (2\text{ps})^{-1}$ $N_2(t)$ $\Delta E = 10$ meV

$k_{21} = (2\text{ps})^{-1}$ $\exp(-\Delta E/k_B T)$

Figure 72. Two coupled excited states which are separated by $\Delta E = 10$ meV. The energy transfer from state one to state two has a probability of (2 ps)$^{-1}$. The back transfer probability follows the Boltzmann distribution.

With the formalism described in eq. 39 in chap. 2.2 one can calculate the excited state population of the system shown in Figure 72. Then eq. 72 can be used to calculate the time dependent entropy of the system shown in Figure 72. While a numerical solution of the problem is possibly independent from the complexity of the coupled system the problem given in Figure 72 can be solved analytically and the result denotes to:

$$N_1(t) = \frac{e^{-k_{12}(1+\exp(-\Delta E/k_B T))t}}{1+\exp(-\Delta E/k_B T)} + \frac{\exp(-\Delta E/k_B T)}{1+\exp(-\Delta E/k_B T)},$$

$$N_2(t) = \frac{-e^{-k_{12}(1+\exp(-\Delta E/k_B T))t}}{1+\exp(-\Delta E/k_B T)} + \frac{1}{1+\exp(-\Delta E/k_B T)}.$$

For that system the entropy according to eq. 72 takes the form

$$\begin{aligned} S(t) &= -k_B\left(N_1(t)\ln\left(N_1(t)\right) + N_2(t)\ln\left(N_2(t)\right)\right) \\ &:= -k_B\left(p(t)\ln\left(p(t)\right) + \left(1-p(t)\right)\ln\left(1-p(t)\right)\right). \end{aligned}$$

This entropy function has a maximum of $S^{\max} = k_B \ln 2$ for the equal population probability of $N_1(t) = N_2(t)$:

$$\frac{\partial S(t)}{\partial p(t)} = 1 + \ln(p(t)) - 1 - \ln(1-p(t)) \overset{!}{=} 0 \Rightarrow p(t) = N_1(t) = \frac{1}{2} = N_2(t).$$

The time $t^{\max}$, when this maximum is reached denotes to:

$$t^{\max} = \frac{1}{k_{12}} \frac{-\ln\left(1/2 - 1/2 \cdot \exp\left(-\Delta E / k_B T\right)\right)}{1 + \exp\left(-\Delta E / k_B T\right)}.$$

For $\Delta E \to \infty$ or $T = 0$ we get

$$t^{\max} = \ln 2 / k_{12}.$$

While for $\Delta E = 0$ or $T \to \infty$ the entropy maximum is coincident with the equilibrium $N_1(t) = N_2(t)$ for $t^{\max} \to \infty$.

That means that $S(t)$ exhibits a local maximum for all cases where $\Delta E > 0 \wedge T < \infty$ which is the case for all relaxations that correspond to time directed (dissipative) processes.

The time dependent excited state population for $N_1(t)$ and $N_2(t)$ calculated according to eq. 39 for 300 K and the entropy curves $S(t)$ at different temperatures are shown in Figure 73.

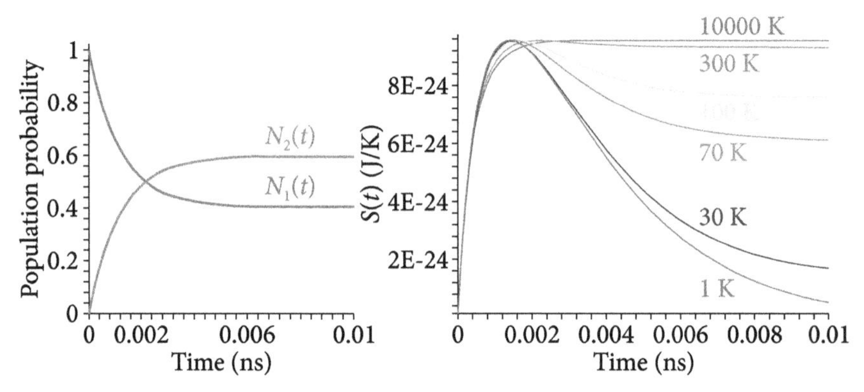

Figure 73. Time dependent population at 300 K of $N_1(t)$ (red curve, left side) and $N_2(t)$ (green curve, left side) and calculation of the entropy of the system shown in Figure 72 according to eq. 72 at 10.000 K (red curve, right side), 300 K (green curve, right side), 100 K (yellow curve, right side), 70 K (light blue curve, right side), 30 K (dark blue curve, right side) and 1 K (magenta curve, right side).

As expected for the high temperature limit the entropy rises monotonously to the maximum that is reached with the equilibrium of the system (Figure 73, right side, red curve for 10.000 K).

The situation is different at lower temperatures. At room temperature (Figure 73, right side, green curve for 300 K) the system entropy reaches the maximum in the time scale near to the inverse transition probability (about 2 ps) fast but decays afterwards to a somewhat lower level.

For very low temperatures the entropy calculates to $\lim_{T \to 0} S(t \to \infty) \to 0$. The difference of the maximal entropy S^{max} and the entropy $S(t \to \infty) := S^{inf}$ in dependency on the temperature is shown in Figure 74, left side. While this difference is rather low at physiological temperatures the relaxation to the thermal equilibrium leads to a strong reduction of the system entropy in comparison to S^{max} at lower temperatures. The system generally goes through an entropy maximum and afterwards the entropy decays significantly to $S^{inf} < S^{max}$ if $k_B T < \Delta E$. Therefore a local entropy maximum is observed if $\Delta E > k_B T$.

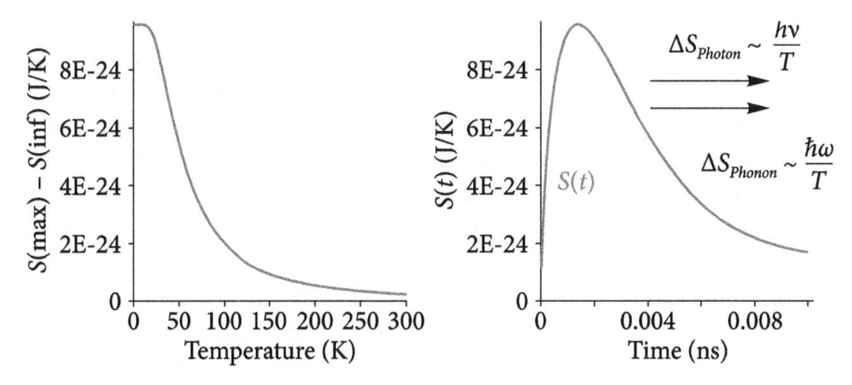

Figure 74. Left side: Temperature dependent difference of the maximal entropy $S^{max} = k_B \ln 2$ and the entropy after full relaxation S^{inf} calculated for the system shown in Figure 73. Right side: schematic cartoon how photons and phonons transfer entropy to the local environment during relaxation of the system shown in Figure 72.

At first glance the local maximum of the entropy function at temperatures $\Delta E > k_B T$ might look as if the system given in Figure 72 could violate the second law of thermodynamics. But such a violation is surely not the case. In fact it should be kept in mind that a relaxation of the pure isolated system as given in Figure 72 could not occur if there would not exist surrounding states that are able to dissipate ΔE.

The environment effectively takes up the value $\frac{1}{1+e^{-\frac{\Delta E}{k_B T}}} \Delta E$ during the dissipation process. This leads to a rise of environmental entropy that is bigger than the reduction of the isolated system's entropy during the transition $S^{max} \rightarrow S^{inf}$. This preserves the second law of thermodynamics and it becomes clear that the system complexity could never arise if this system would not be able to interact with its environment. There is no violation of the laws of thermodynamics. Jennings mentioned that if we analyze a photosynthetic system from the lowest energy limit, we could observe a violation of the second law of thermodynamics (Jennings et al., 2005, 2006, 2007).

This might be true for a single photon of 680 nm wavelength that is absorbed by a plant and drives a single quantum process in the photosynthetic nanomachine. But this is neither the continuous reality of a growing plant nor is it a process that is forbidden due to the thermodynamic laws as long as it counts for a single absorbed photon only which can violate the second law of thermodynamics according to the Jarzynski equality which denotes the probability for a trajectory violating the second law of thermodynamics similar to eq. 8 (see e.g. ref. Crooks, 1998) and references therein for details). The second law of thermodynamics is a pure statistic interpretation of ensembles. It is not a law that can be applied to a single quantum process. As denoted by the Jarzynski equality a single molecular process is allowed to violate the second law of thermodynamics in a transient way. In the time- or the ensemble average the laws of thermodynamics hold.

The calculation of the entropy as given in eq. 72 for the nonequilibrium ensemble enables a suggestion of a nonequilibrium partition function that would enable the calculation of all thermodynamic variables for a full nonequilibrium situation in systems that can be generally described by excited state probabilities. With eq. 73 all the nonequlibrium observables of the ensemble like the time dependent temperature $T(t)$, the time dependent Gibbs energy $G(t)$ etc... can be calculated. However, we used the denotation of the entropy as given in the eq. 71 that is only valid for the equilibrium. To achieve an nonequilibrium situation the expectation value $\langle E \rangle (t)$ of the energy and the "momentary" temperature $T(t)$ would have to be treated time dependently:

$$\begin{aligned} S(t) &= -k_B \sum_i p_i(t) \ln p_i(t) = k_B \ln Z(t) + \frac{\langle E \rangle (t)}{T(t)}, \\ Z(t) &= \exp\left(-\sum_i p_i(t) \ln p_i(t) - \frac{\langle E \rangle (t)}{k_B T(t)} \right) \\ &= \exp\left(-\frac{\langle E \rangle (t)}{k_B T(t)} \right) \prod_i \exp\left(-p_i(t) \ln p_i(t)\right), \end{aligned} \tag{73}$$

$$Z(t) = \exp\left(-\frac{\langle E\rangle(t)}{k_B T(t)}\right)\prod_i \exp\left(\ln\left(p_i(t)^{-p_i(t)}\right)\right),$$

$$Z(t) = \exp\left(-\frac{\langle E\rangle(t)}{k_B T(t)}\right)\prod_i \frac{1}{p_i(t)^{p_i(t)}}.$$

4.5 Second Messengers and Signaling Molecules in H_2O_2 Signaling Chains and (Nonlinear) Networking

Although our current knowledge on the signalling networks of ROS is still rather fragmentary, Figure 71 summarizes some pathways in the signal networks including the interference between various pathways.

It was found that H_2O_2 stimulates a rapid increase of intracellular Ca^{2+} concentration (Kim et al., 2009). The development of oxidative stress controls the activity of several isoforms of calmodulin. In plant cells, on the other hand, the ROS generation in mitochondria is activated by an increase of the Ca^{2+} concentration. Likewise, under certain conditions, the Ca^{2+} concentration depends on the actual ROS level and additionally Ca^{2+} stimulates the formation of ROS in plant cells (Bowler and Fluhr, 2000). These findings indicate that ROS/redox state and calcium-dependent signaling pathways are closely interconnected in a strongly nonlinear way (Yin et al., 2000). Heat hardening, at least a short-term treatment, can also be accompanied by an increase of the ROS content in cells (Dat et al., 1998), i.e. ROS might function in transduction of a temperature signal (Suzuki and Mittler, 2006; Yu et al., 2008). It is also suggested that ROS participate in acclimation of the photosynthetic apparatus to high light under conditions similar to heat hardening (Kuznetsov and Shevyakova, 1999; Allakhverdiev et al., 2007; Kreslavski et al., 2009, 2012a).

Common intermediates were found to participate in mechanism(s) of ROS and phytohormone action (Jung et al., 2009), as is shown by the involvement of the species $O_2^{-\bullet}$ and O_3 in programmed cell death together with ethylene- and jasmonate-dependent metabolic pathways. On the other hand, ROS can function as second messengers in the transduction of hormonal signals, as was shown for the auxin affect on gene expression, where ROS are used as second messengers, which simultaneously regulate activity and expression of glutathione transferase (Tognetti et al., 2012).

H_2O_2 induces the phosphoinositide cycle that switches signaling pathways associated with the secondary messengers, IP3 and diacylglycerol (DAG) (Munnik et al., 1998), whereas phospholipase D was reported to stimulate H_2O_2 production in *A. thaliana* leaves via generation of phosphatidic acid acting as lipid messenger (Sang et al., 2001).

It is well known that exogenous salicylic acid and pathogens induce a burst of ROS generation in plant tissues (Dmitriev, 2003). However, it remains unclear how and to what extent ROS are involved in the improvement of plant stress resistance. Exogenous salicylic acid was found to give rise to enhanced plant cold tolerance (Horváth et al., 2002). This effect is attributed to the inhibition of catalase and related to oxidative stress leading to accumulation of H_2O_2. Studies on two maize genotypes revealed that the cold-resistant line had a molecular form of catalase, which was more severely inhibited by salicylic acid than the catalase of the sensitive line (Horváth et al., 2002). Oxidative stress caused by exogenous salicylic acid depends on the calcium status of the cells and is not manifested in the presence of calcium channel blockers.

When considering the existence of signaling networks, it must be emphasized that different hierarchies of complexity exist in ROS-induced signaling depending primarily on the evolutionary level of the organism. Networking is evidently simpler in prokaryotic than in eukaryotic cells, which contain various cell organelles (chloroplasts, mitochondria, peroxisomes, nucleus, endoplasmatic reticulum) and loci of genomic information (e.g. plastids and nucleus in plants). An even more complex signal network exists in multi-cellular organisms, e.g. in leaves of higher plants with different cell types (mesophyll, bundle sheath, guard cells). The deciphering of the latter type of networks requires detailed analyses, and this topic is just at the beginning to reach a level of deeper understanding.

4.6 ROS-Waves and Prey-Predator Models

As mentioned in chap. 3.2.1 in leaves of *A. thaliana* treated with naphthalene ROS waves were observed that spread over the cell tissue with a temporal frequency of 20 minutes and a wavelength of several hundreds of micrometers (see Figure 62). Such a behavior is in line with wave-like closure and opening of stomata as observed in green plants under stress conditions. A simple explanation for the formation of waves is the relaxation of a nonequlibrium system that contains compounds that inhibit their own production (autoinihibition). The overreduction of the electron transfer chain (ETC) can lead to the production of ROS as electron transfer is blocked and ^{3}Chl accumulates. ROS in turn oxidizes the ETC and an overreduced ETC is known to activate NPQ. So growth of electronic charge in the ETC forms ROS which inhibits the overreduction. Similarily all ROS triggered mechanisms that lead to the depletion of ROS on the molecular level might cause ROS waves with characteristic oscillation frequencies and intensity distributions as well as localizations delivering information on the underlaying mechanisms.

ROS waves can be described by prey-predator models based on rate equations. In a simplified approach, the basic prerequisite for any oscillating reaction demands for processes that are autocatalytic or autoinhibited. If we look at ROS nearly all induced processes are autocatalytic or autoinhibited because they are nonlinear. Higher ROS level activates catalase. For example treatment of mature leaves of wheat plants with H_2O_2 was shown to activate leaf catalase (Sairam and Srivastava, 2000).The higher amount of catalase will reduce the H_2O_2 level and the reduced H_2O_2 level again reduces the catalase activity. Similarly a high concentration of excited singlet states in the light-harvesting complex can form singlet oxygen. The overall sequence of such reactions occurs as an oscillating chemical reaction as observed in simple prey-predator models like the fox and the rabbit as shown in Figure 75.

Figure 75. Fox and rabbit in a simplified prey-predator model.

Both species, fox and rabbit as shown in Figure 75 are autoinhibiting due to the coupling with the other species. The production of rabbits leads to more rabbits, however the reproduction rate is constant. More rabbits lead to the growth of foxes which inhibit the rabbit population. In dependence on the exact distribution of the reproduction and inhibition rates this can lead to oscillations.

To model the system we assume that the rabbit population X can reproduce according to the equation for the feeding of rabbits: $A + X \rightarrow 2X$ by consumption of a constant food supply A. Rabbits can die according to natural death following an equation for the death of rabbits: $X \rightarrow B$. If we assume the rabbits to represent ROS then the first equation describes the production rate and the second the natural ROS scavenging. The fox population Y is fed by rabbits: $nX + Y \rightarrow 2Y$ where a number of n consumed rabbits leads to a reproduction of a fox. This process is the most important one as it describes the autoinhibition of the rabbits by producing foxes. Also the fox might die due to natural death: $Y \rightarrow C$.

The fox might represent the catalase molecules which is activated due to a higher ROS level.

All these equations that are needed to describe the population of foxes and rabbits are characterized by certain rate constants according to the rate formalism presented by eq. 8 in chapter 1.3 and 1.4:

$$
\begin{aligned}
A + X &\xrightarrow{k_1} 2X, \\
X &\xrightarrow{k_2} B, \\
nX + Y &\xrightarrow{k_3} 2Y, \\
Y &\xrightarrow{k_4} C.
\end{aligned}
$$

The exact representation of these equations describes the change of the rabbit X and fox population Y in time according to:

$$
\begin{aligned}
\frac{dX}{dt} &= k_1[A][X] - k_2[X] - nk_3[Y][X], \\
\frac{dY}{dt} &= k_3[X][Y] - k_4[Y].
\end{aligned}
$$

Solving these equations we will find that both, rabbit and fox population can follow an oscillating concentration in time as plot in Figure 76 which is the solution of the rate equation system shown above.

Figure 77 illustrates a time series of propagating ROS on naphthalene treated leaves of *Arabidopsis* that is strongest in the beginning (light areas on the leaf) and decays whereby in some areas the intensity reaches a local minimum and starts to grow again.

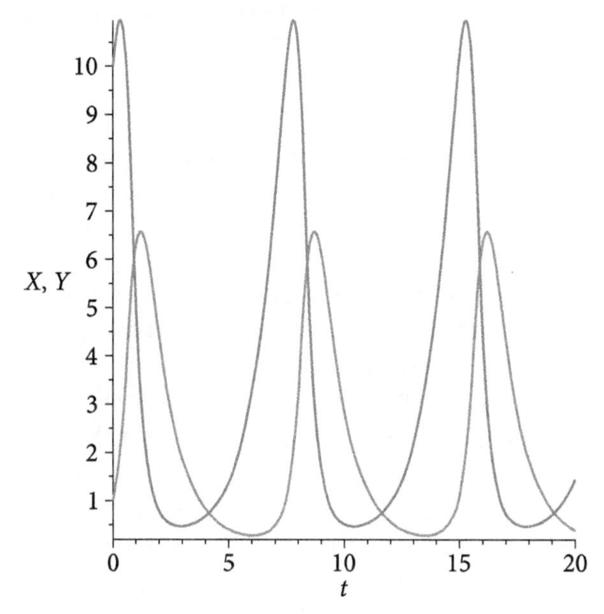

Figure 76. Population of rabbits (red) and foxes (green) as the solution of the equations shown above.

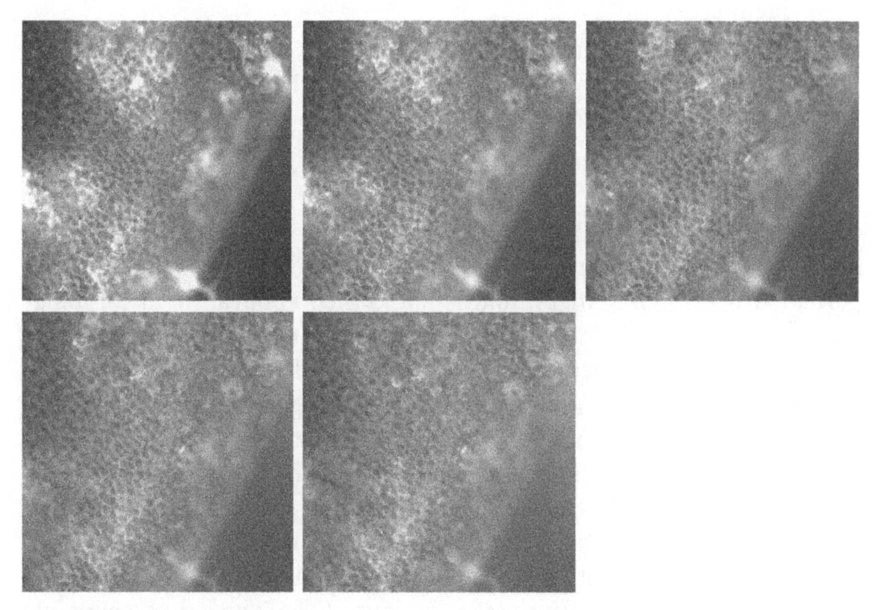

Figure 77. Spread of ROS waves in *A. thaliana* treated with naphthalene under continuous illumination. The temporal development of the sensor intensity (DCF, see Table 3) that is used to monitor ROS is shown in Figure 62.

4.7 Open questions on ROS Coupling in Nonlinear Systems

The enormous work performed during the last decades has clarified the deleterious effects of ROS on photosynthetic organisms. However, this is only one side of ROS functions. The other side is the very important signaling role of ROS in the response of cyanobacteria, alga and higher plants to different forms and conditions of stress. In spite of significant progress achieved during the last decade, our current state of knowledge on this topic is still rather fragmentary.

There are several questions that need to be answered:

1. How do ROS generated in chloroplasts affect the transcription of the chloroplast genome?
2. How can ROS leave the chloroplasts and directly induce a significant expression of genes of the nuclear genome?
3. What is the nature and the mechanistic function of second messengers formed by reaction of ROS with specific molecules like lipids and Cars?
4. What determines the mechanism of ROS wave propagation in plant cells?
5. What is the identity of the primary sensor(s) of ROS (transcription factors and/or protein kinases) and the primary genes responding to oxidative stress?
6. Do ROS induce new signaling pathways by acting as second messengers?

Significant progress in answering these questions is expected from the development of new spectroscopic methods for monitoring ROS, in particular with high spatial resolution, and their application in combination with directed genetic engineering of plants. Among the methods for manipulations at molecular level the targeted ROS production within specific cell compartments and organelles is of high interest.

One important approach towards exploitation of photosynthetic organisms as sustainable sources of biomass is the improvement of the resistance of the cells against environmental stress conditions. This problem targets world food and world energy supply. On the other side plants function as sensitive indicators for the environmental conditions and photosynthetic activity changes in contact to diverse dust pollutants leading to dynamic changes of chlorophyll fluorescence. The full understanding and technical exploitation of these mechanisms has implications on food production. It additionally opens the way to develop rapid alert systems for dust pollutants or, more generally, as reporters for the environmental quality, by monitoring the fluorescence properties of plants like for example lichens (e.g. *Peltigera aphtosa*) which are sensitive to pollutants (Maksimov et al., 2014b).

As the World Health Organization (WHO) just recently pointed out air pollution as the worst environmental threat for human and environmental health, the need for techniques to quantify the air contamination and its impact on plants is pressing. The WHO estimates that seven million people worldwide died due to illnesses linked to air pollution in 2012 alone, according to new data released on March 25th 2014. These shocking developments urgently require new techniques and initiatives that are able to quantify the air pollution and might help to decontaminate air especially in the big cities. Plants with selected and specialised properties might be a solution for these problems.

The genetic transformation of plants according to a deep molecular biological knowledge of all processes that interact with ROS delivers a tool to produce enhanced plants as ROS sensors, ROS scavengers or crop plants with improved resistance to ROS. The analyses of the capability of cyanobacteria and algae for the decontamination of water and air give rise to genetically enhanced ROS scavengers.

5 The Role of ROS in Evolution

Evolution is understood as the process that leads to the variation of phenotypes observed along all species as first described by Charles Darwin in *The Origin of Species* (Darwin, 1859). We define evolution as a development process of lifeforms that is caused by mutation, reproduction and selection as a minimum set. Further important mechanisms as recombination and detailed properties of the gene transfer are neglected as we do not intend to present a reinterpretation of evolution but a small motivation how ROS might be a driving force in the origin of species.

As pointed out ROS plays an important role as an environmental constraint exhibiting selection pressure to organisms which have not developed the necessary systems for ROS scavenging. In the following chapters we will see that possibly the strongest environmental change that ever happened in the ecosphere was the big bang of ROS production after the development of oxygenic photosynthesis (5.1). In chapter 5.2 the role of selection pressure to drive evolution and its impact in conducting leaps in evolution is elucidated. We will shortly illustrate how genetic diversity is of major impact for the adaptivity of organisms to external contraints and changes of the environmental conditions than mutations, which are random products that do not lead to predictable directions of evolution. Selection pressure is a driving force for complexity in systems that follow simple rules. We will now close the circle from the very beginning of this book where the explanation for hierarchical structures was motivated by Steven Wolfram's cellular automata (see chapter 1).

5.1 The Big Bang of the Ecosphere

About 3 billion years ago the atmosphere started to transform from a reducing to an oxidizing environment as evolution developed oxygenic photosynthesis as key mechanism to efficiently generate free energy from solar radiation (Buick, 1992; Des Marais, 2000; Xiong and Bauer, 2002; Renger, 2008; Rutherford et al., 2012; Schmitt et al., 2014a). Entropy generation due to the absorption of solar radiation on the surface of the earth was retarded by the generation of photosynthesis, and eventually a huge amount of photosynthetic and other complex organisms developed at the interface of the transformation of low entropic solar radiation to heat. The subsequent release of oxygen as a "waste" product of photosynthetic water cleavage led to the aerobic atmosphere (Kasting and Siefert, 2002; Lane, 2002; Bekker et al., 2004), thus opening the road for a much more efficient exploitation of the Gibbs free energy through the aerobic respiration of heterotrophic organisms (for thermodynamic considerations, see (Renger, 1983; Nicholls and Ferguson, 2013)).

From the very first moment this interaction with oxygen generated a new condition for the existing organisms starting an evolutionary adaptation process to this new oxidizing environment. ROS became powerful selectors and generated a new hierarchy of life forms from the broad range of genetic diversity in the biosphere. We assume that this process accelerated the development of higher, mainly heterotrophic organisms in the sea and especially on the land mass remarkably.

The efficient generation of biomass and the highly selective impact of ROS lead to a broad range of options for complex organisms to be developed in the oxidizing environment. Important and more complex side effects next to the direct destructive impact of ROS are examples like the fact that the molecular oxygen led to generation of the stratospheric ozone layer, which is the indispensable protective shield against deleterious UV-B radiation (Worrest and Caldwell, 1986). ROS lead to new complex constraints for evolution that drove the biosphere into new directions – by direct oxidative pressure and by long range effects due to environmental changes caused by the atmosphere and the biosphere themselves.

For organisms that had developed before the transformation of the atmosphere the pathway of redox chemistry between water and O_2 by oxygenic photosynthesis was harmful, due to the deleterious effects of ROS. It is assumed that about 90% of all living organisms died due to the impact of ROS in that period of earth history making the big bang of the biosphere, understood as the development of oxygenic photosynthesis, the most deleterious environmental change ever with its subsequent selection pressure. O_2 destroys the sensitive constituents (proteins, lipids) of living matter. As a consequence, the vast majority of these species was driven into extinction, while only a minority could survive by finding anaerobic ecological niches. All organisms developed suitable defence strategies, in particular the cyanobacteria, which were the first photosynthetic cells evolving oxygen (Zamaraev and Parmon, 1980).

5.2 Complicated Patterns Result from Simple Rules but Only the Useful Patterns are Stable

As we had seen from Steven Wolfram´s cellular automata complicated patterns can result from simple rules. Evolution does not need complicated rules to conduct complexity (see Figure 78). The simple fact that for each position only 4 different nucleotides are taken into account when constructing the genetic code leads to an incredible number of possible configurations for a sequence of hundred nucelotides (for 100 nucleotides 4^{100} configurations are thinkable which equals about 10^{25}, i.e. the number of atoms in one pound of carbon). Some people believe that this enormous number is an indication for the fact that the world can not be a product by chance, however these interpretations often neglect a series of further details important to know when the real probability for random developments in evolutionary processes shall be correctly estimated.

The estimation of the possible configurations of a polypeptide chain in proteins and the comparison of this number with the time necessary to conduct the folding of a protein by chance lead to the Levinthal-paradox found by Cyrus Levinthal in 1969. Due to the very large number of degrees of freedom in an unfolded polypeptide chain a number of possible conformations in the order of 10^{100} has to be "tested" until the right configuration is found. Levinthal noted that the time necessary to "find" the right folding would exceed the age of the universe by far. (Levinthal, 1969).

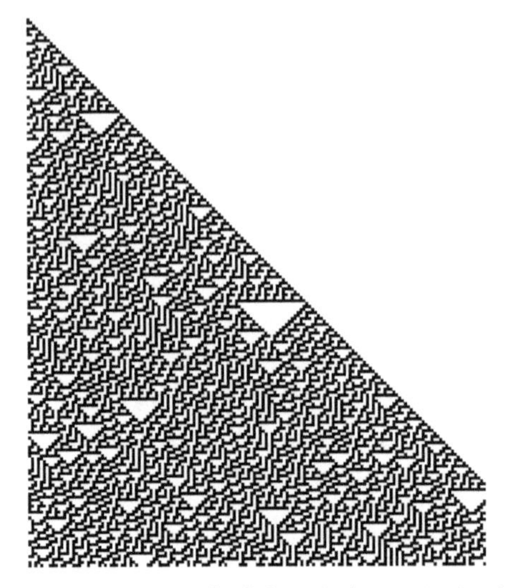

Figure 78. Detail of the Mathematica® calculation of the "rule 30" (see Figure 1 and Figure 2) of Steven Wolfram's cellular automata (left side) in comparison to a pigmented seashell (right side) (Coombes 2009). Image reproduced with permission.

This is true even if conformations are sampled at rapid (nanosecond or picosecond) rates. The "paradox" is that most small proteins fold spontaneously on a millisecond or even microsecond time scale. Of course Levinthal was aware of this fact and he had already suggested favorized pathways for the folding process (Levinthal 1968). He suggested that the paradox can be resolved if "protein folding is sped up and guided by the rapid formation of local interactions which then determine the further folding of the peptide; this suggests local amino acid sequences which form stable interactions and serve as nucleation points in the folding process" (Levinthal, 1969). Today it is known that the protein folding follows an energetic landscape and that in case that a single configuration is stable it is kept in this stable position and a large majority of further configurations is not tested anymore because they become energetically disadvantaged due to the presence of the stable partial configuration that lead the protein into an energetic local minimum. The correct configuration of local interactions determines the further folding of the peptide. The local stable interactions serve as nucleation points. The process is comparable to a skier that has to test a ski slope. The skier could do that following a large number of pathways. However there is a small selection of all possible slopes in the beginning that clearly point to the energetic minimum, i.e. downhill direction. The skier will follow this route especially if he does not use his free will. As soon as he had started and finished already some part of the skiing slope he will not be able to return to the starting point again and test for another route. So the Levinthal paradox is similar to the assumption that a skier would never reach the valley as there are so many possible routes leading from the top to the bottom. But we know that in spite of the fact that such a large number of pathways is possible is possible he will quickly reach the valley. The skier just takes the right path – and we should be aware that the answer to the Levinthal-Paradox is similarly trivial somehow. The image of the "correctly chosen pathway" is representing a selection process: There is a constraint (the energetic landscape, for the skier the gravity) that choses the right path from all possible pathways. The paradox is resolved by a selection process that was unknown or underestimated when the paradox was formulated.

If we look at evolution the Levinthal-paradox is sometimes applied to genes. Some people argue that the age of the universe would not be large enough to develop something as complicated as a human being. However such an assumption is rather baseless and probably does not take into account some constraints evolution is facing. This might be even more true as genetics comprises features which are not mentioned up to now. One very important factor that drives the diversity of genes is reproduction. Each gene tends to reproduce itself due to the double helix character of the DNA. However reproduction can only occur if a minimum stability of the gene is guaranteed as a prerequisite and in consequence each gene which undergoes a mutation that delivers an advantage for the organism hosting that gene is reproduced.

It is a bit misleading to argue that any complex gene will be found at any future time by chance as this is underestimating the power of reproduction and selection that both work on the pool of genetic variation. Assuming an absolute random process that has to occur until the final gene configuration for a complex organism is found relies on the so called infinite monkey theorem. The infinite monkey theorem states that a monkey hitting keys at random on a typewriter keyboard for an infinite amount of time will almost surely type a given text, such as the complete works of William Shakespeare. This argument is surely true. However it can be abused arguing that the time needed to find only one of Shakespeare´s great works might exceed the age of the universe. In an infinite number of time any finite text would be typed – but the time is not infinitely most probable.

The whole concept is misleading and also the actual variants of the theorem including multiple and even infinitely many typists, and the

Figure 79. According to the infinite monkey a monkey hitting keys at random on a typewriter keyboard for an infinite amount of time will almost surely type a given text, such as the complete works of William Shakespeare.

target text varying between an entire library and a single sentence do not respect the important mechanisms of reproduction and selection which are necessary to correctly understand evolution. To correlate the infinite monkey theorem with evolution we would have to introduce two significant variations that respect reproduction and selection in a correct way. Reproduction occurs if any of the mutations happening when the large number of typing monkeys (each single evolving organism and it means each individual of a population should be represented by a single monkey) just finds a short "correct" sequence by chance. This might be a single word or it can be half a page, any change in the text that is not contraproductive meaning that it does not lead to a disadvantage in reproduction will be copied many times. The importance of reproduction is even bigger if a mutation in the text leads to a reproduction advantage for the new mutant. In that case the new, meaningful mutation which, as we should keep in mind, is possibly just one correct word in the large text, will be reproduced in larger number than the other sequences. The new mutant will possibly quickly grow over the whole population. This effect might additionally be stronger if we keep in mind that next to the reproduction capability other selection rules exist like predators or the natural lifespan. If a new predator shows up he might be hungry for the monkey´s texts and just eats them up, except one fragment that contains a specially long text sequence of Shakespeare´s *Romeo and Juliet* and is therefore able to escape. As now there is much space for the diminished population to grow it might be possible that after a short time only this surviving text fragment is copied many times and present everywhere (all others are extinguished) and just in the next round all monkeys start to type on the text that has been largely worked over and contains the correct text sequence in all copies now.

Such a leap of evolution in all monkeys texts does not follow severe mutations but free space for the genetic diversity to grow together with selection pressure. A prominent example for such a leap of evolution is given in the next chapter 5.3.

5.3 Genetic Diversity and Selection Pressure as Driving Forces for Evolution

As indicated in chap. 5.2 the selection is a very strong driving force for evolution. However selection additionally needs genetic diversity. Misleading concepts in history which tried to argue that any evolutionary process might be accelerated by extinguishing actively a part of the genetic diversity are wrong. Hopefully in any case of selection pressure the genetic diversity grows quickly to its old values and above them. Supporting a population´s resistance and evolution always means that the genetic diversity has to increase as we will see in the following.

The concept of the infinite typing monkey showed that next to mutations especially genetic diversity (the number of text variations present when all monkeys are typing) are important if a predator shows up that is extinguishing a part of the text fragments. This picture is a symbol for a new selection pressure that induces directed evolution. Reproduction is strongly correlated with the selection pressure as selection most generally is not represented by an extinction of individuals in the population but by disadvantage in the reproduction speed for a certain genotype. A broad genetic diversity together with new selection mechanisms can easily lead to a leap of evolution as it was for example observed by Stuart et al. from the University of Austin, Texas when they investigated quick evolution of lizard feet (Stuart et al., 2014).

Figure 80. Rapid evolution of a native lizard species caused by pressure from an invading lizard species. The native species had much smaller feet in average (left side) than observed 15 years later after their feet evolved to better grip branches (right side).

The authors observed that Anolis carolinensis, otherwise known as the green anole had made notable physical changes within just 20 generations (15 years). The study discovered that the rapid evolution of a native lizard species was caused by pressure from an invading lizard species, introduced from Cuba. After contact with the invasive species, only the native lizards which were able to perch higher into the trees than others were able to escape. In such way the population began perching higher in trees, and, generation after generation, their feet evolved to become better at gripping the thinner, smoother branches. So only after a few generations the size and phenotype of the lizard´s feet had changed significantly (Stuart et al., 2014). The scientists were astonished by the speed the adaption took place as significant changes occurred within a few months when native lizards shifted to higher perches and showed more sticky scales on their feet. "We did predict that we´d see a change, but the degree and quickness with which they evolved was surprising," said Yoel Stuart, author of the study (Stuart et al., 2014).

This latest study is a well-documented example of what evolutionary biologists call "character displacement", i.e. when similar species in competition with each other evolve differences to take advantage of different ecological niches. Charles Darwin observed a similar example when two species of finch in the Galápagos Islands diverged in beak shape as they adapted to different food sources. Probably this observation was a key element for Darwin to develop his understanding for evolution as the origin of species (Darwin, 1859). The cruel fact that induced the observed character displacement is represented by a new selection pressure due to the invading species.

Similarily the most severe selection pressure ever might have been the "big bang" in the biosphere when the atmosphere changed from reducing to an oxidizing environment as evolution developed oxygenic photosynthesis. The release of oxygen led to the aerobic atmosphere thus opening the road for a much more efficient exploitation of the Gibbs free energy through the aerobic respiration of heterotrophic organisms who had a large advantage and quickly reproduced in large amount while ROS became powerful selectors extinguishing the large majority of the former population.

6 Outlook: Control and Feedback in Hierarchic Systems in Society, Politics and Economics

It is often assumed that if a system is disturbed in the apparently stable but fragile state, it will be prone to fall into chaos. However, regarding our ecosphere, this is obviously not the case on a long time scale. Or, put more precisely, it would be necessary to define chaos quite well to be able to judge opinions like that. However if we assume that chaos is something like the aforementioned big bang in the ecosphere driving a large percentage of all species into extinction in a short time span we might at least realize that in spite of the fact that we disturb our ecosphere quite severely it is not necessarily behaving absolutely chaotic.

Evolution turns out to be a strong mechanism selecting stable structures that survive after a distortion of the environment and start to grow again. This means that we will most probably not eliminate all fish from the planet by overfishing, but we might cause a new form of "fish" to emerge, which is not interesting for our menus and therefore has adapted to the constraint of a mankind that shows the tendency to overfish the oceans. We will not eliminate life in the oceans, but we will shift it into a direction that releases the pressure of the fishermen's nets and fills the ecologic niches that are opened due to overfishing. Actually it is observed that the number of jellyfish is significantly rising, which is most probably caused by overfishing and the extinction of regular fish populations (Gershwin, 2013). When reading Lisa-ann Gershwin's book *Stung! On Jellyfish Blooms and the Future of the Ocean* one may wonder why the jellyfish bloom was not foreseen twenty years earlier. After all, the correlations between overfishing of regular fish, opening of new ecologic niches, growing toxicity of the oceans, global warming and the growth advantages for jellyfish are quite evident.

For most experimental scientists it is of interest to extract the maximum amount of information from a sample. In complex systems, where the dynamics of several coupled states takes place within the same time domain, it is difficult to make clear forecasts. However, often there does not even seem to be any serious attempts to do so. We need a new methodology for simultaneous spectroscopy of multiple parameters over several orders of magnitude in space and time in multiscale hierarchical systems. In biophysics, samples can be imaged with a spatial hierarchy from the (resolution limited) molecular level, involving cell organelles and the cell structure, tissue formation and the whole organism. Fluorescence correlation spectroscopy (FCS) on single molecules up to macroscopic scenes covers a time range from the ps domain up to minutes. Therefore a hierarchy over 14 orders of magnitude in time and 9 orders of magnitude in space are accessible for generalized spatiotemporal correlation spectroscopy. Such deep hierarchical investigations are lacking in many other disciplines.

In economics, for example, one typically finds investigations of processes that happen on the minute time scale up to financial concepts covering five years. However these investigations are not analysed in a concerted scientific approach. More likely the economics study itself is based on a concept of win-loss estimations or equilibria. These concepts might be of extremely high scientific relevance. However nevertheless there is a lack of a hierarchically structured dynamic studies.

Such a temporal hierarchy should cover even more than the mentioned seven orders of magnitude in time. Extending the time scales turns out to be rather successful. This counts for the introduction of high-frequency trading or 60 second pitches for the presentation of start-up ideas on one end of the scale but it should also lead to an extended view on our economic strategy that can cover more than five years and clearly focus on predictable outcomes on long timescales of several generations.

We will refrain at this point from an instruction in how mathematical concepts might be applied to social, economic or political networks: instead we will point to the urgent need to react to predicted outcomes following current strategies.

Inspired by Hermann Haken's *Synergetics* (Haken, 1990) Wei-Bin Zhang wrote *Synergetic Economics*. She focused on problems of time and change in economic systems and was able to deal with the complexity resulting from nonlinearity and leading to instability, bifurcation and chaos in economic evolution. Zhang describes a hierarchy of instabilities in her book and shows how economic systems that develop along such a hierarchy of instabilities evolve structural patterns. Similar to Hermann Haken's synergetics, in Zhang's work external control parameters reduce the degree of freedom in complex systems. However, while synergetics focusses on self-organization, Zhang focuses on the possibilities of chaos that arise from the hierarchy of instabilities. Changes of external parameters can lead to chaos. They may lead to new spatial-temporal patterns of the system, such as oceans filled of jellyfish as a result of overfishing the seas.

Evolution forces the disturbed system to stabilize on the next probable stability path after its distortion. The same thing generally happens in society when the constraints change. This might be an event like a terror attack or the election of a new president who releases a broad series of new laws. In nature the appearance of the biosphere will change if we change a control parameter like the average temperature or the average number of fish in the sea. Between these stabilities there exists a transition time in which a more or less chaotic behavior is observed. In that time the old and the new form coexist with rather quickly changing population density. After some time the stabilization is completed and the new equilibrium holds a quite stable population of both – the old and the new species, but with changed ratio as compared to the former situation.

Politics might be the example in which the transition from one to another structure is at most prone to chaos. Changes in external control parameters which might be the outcome of an election, a revolution or an economic change can lead to chaotic situations that finally reach a new equilibrium. Political systems might also be the most evident structures indicating how top down and bottom up signalling both contribute to the current state of the structure. Elections and peoples' behaviour are surely bottom up. However, the release of laws and the exchange of people in responsible positions are often top down. Both influence the dynamics of the system.

Bibliography

Adir N., Zer H., Shochat S., Ohad I. Photoinhibition – a historical perspective, *Photosynth. Res.*, 2003, Vol. 76, p. 343–370.

Adolphs J., Renger T. How proteins trigger excitation energy transfer in the FMO complex of green sulfur bacteria, *Biophys. J.*, 2006, Vol. 91, p. 2778–2797.

Albertsson P.-A., Andreasson E., Persson A., Svensson P. Organization of the thylakoid membrane with respect to the four photosystems, PSIα, PSIβ, PSIIα, PSIIβ. *Current Research in Photosynthesis*, Baltschefsky M. ed., Kluwer Academic Publishers: Dordrecht, 1990, p. 923–926

Allahverdiyeva Y., Aro E.-M. Photosynthetic responses of plants to excess light: mechanisms and conditions for photoinhibition, excess energy dissipation and repair, *Photosynthesis Plastid Biology, Energy Conversion and Carbon Assimilation*, Eaton-Rye J.J., Tripathy B.C., Sharkey T.D. eds., Springer, Dordrecht, 2012, p. 275–298.

Allakhverdiev S.I., Murata N. Environmental stress inhibits the synthesis de novo of proteins involved in the photodamage-repair cycle of photosystem II in *Synechocystis* sp. PCC 6803, *Biochim. Biophys. Acta*, 2004, Vol. 1657, p. 23–32.

Allakhverdiev S.I., Los D.A., Mohanty P., Nishiyama Y., Murata N. Glycinebetaine alleviates the inhibitory effect of moderate heat stress on the repair of photosystem II during photoinhibition, *Biochim. Biophys. Acta*, 2007, Vol. 1767, p. 1363-1371.

Alscher R.G., Erturk N., Heath L.S. Role of superoxide dismutases (SODs) in controlling oxidative stress in plants, *J. Exp. Bot.*, 2002, Vol. 53, p. 1331–1341.

Ananyev G., Wydrzynski T., Renger G., Klimov V. Transient peroxide formation by the manganese containing redox active donor side of photosystem II upon inhibition of O_2 evolution with lauroylcholine chloride, *Biochim. Biophys. Acta*, 1992, Vol. 1100, p. 303–311.

Andresen M., Wahl M.C., Stiel A.C., Gräter F., Schäfer L.V., Trowitzsch S., Weber G., Eggeling C., Grubmüller H., Hell S.W., Jakobs S. Structure and mechanism of the reversible photoswitch of a fluorescent protein, *Proc. Natl. Acad. Sci. U.S.A.*, 2005, Vol. 102, p. 13070–13074.

Andresen M., Stiel A.C., Trowitzsch S., Weber G., Eggeling C., Wahl M.C., Hell S.W., Jakobs S. Structural basis for reversible photoswitching in Dronpa, *Proc. Natl. Acad. Sci. U.S.A.*, 2007, Vol. 104, p. 13005–13009.

Apel K., Hirt H., Reactive oxygen species: metabolism, oxidative stress, and signal transduction, *Annu. Rev. Plant Biol.*, 2004, Vol. 55, p. 373–399.

Asada K., Kiso K., Yoshikawa K. Univalent reduction of molecular oxygen by spinach chloroplasts on illumination, *J. Biol. Chem.*, 1974, Vol. 249, p. 2175–2181.

Asada K., The water–water cycle in chloroplasts: scavenging of active oxygens and dissipation of excess photons, *Annu. Rev. Plant Physiol. Plant Mol. Biol.*, 1999, Vol. 50, p. 601–639.

Asada K. Production and scavenging of reactive oxygen species in chloroplasts and their functions, *Plant Physiol.*, 2006, Vol. 141, p. 391–396.

Atkins P.W. *Physical Chemistry*, Oxford University Press, Oxford, 10th edition, 2014, 1008 p.

Baake E., Shloeder J.P. Modelling the fast fluorescence rise of photosynthesis, *Bull. Math. Biol.*, 1992, Vol. 54, p. 999–1021.

Babu T.S., Jansen M.A.K., Greenberg B.M., Gaba V., Malkin S., Mattoo A.K., Edelman M. Amplified degradation of photosystem II D1 and D2 proteins under a mixture of photosynthetically active radiation and UV-B radiation: dependence on redox status of photosystem II, *Photochem. Photobiol.*, 1999, Vol. 69, p. 553–559.

Baier J., Maier M., Engl R., Landthaler M., Baumler W. Time-resolved investigations of singlet oxygen luminescence in water, in phosphatidylcholine, and in aqueous suspensions of phosphatidylcholine or HT29 cells, *J. Phys. Chem. B*, 2005, Vol. 109, p. 3041–3046.

Baier M., Dietz K.-J. Chloroplasts as source and target of cellular redox regulation: a discussion on chloroplast redox signals in the context of plant physiology, *J. Exp. Bot.*, 2005, Vol. 56, p. 1449–1462.

Barber J., Nield J., Morris E.P., Zheleva D., Hankamer B. The structure, function and dynamics of photosystem two, *Physiol. Plant.*, 1997, Vol. 100, p. 817.

Barinov A.V., Goryachev N.S., Schmitt F.-J., Renger G., Kotel'nikov A.I. Luminescent analysis of microsecond relaxation dynamics of viscous media, ISSN 0030_400X, *Optics and Spectroscopy* (Condensed Matter Spectroscopy), 2009, Vol. 107, p. 95–100.

Baruah A., Simkova K., Hincha D.K., Apel K., Laloi C. Modulation of $^1O_2$-mediated retrograde signaling by the PLEIOTROPIC RESPONSE LOCUS 1 (PRL1) protein, a central integrator of stress and energy signaling, *Plant J.*, 2009, Vol. 60, p. 22–32.

Bechtold U., Richard O., Zamboni A., Gapper C., Geisler M., Pogson B., Karpinski S., Mullineaux P.M. Impact of chloroplastic- and extracellular-sourced ROS on high light-responsive gene expression in *Arabidopsis*, *J. Exp. Bot.*, 2008, Vol. 59, 2008, p. 121–133.

Bekker A., Holland H.D., Wang P.-L., Rumble D., Stein H.J., Coetzee L., Beukes N. J. Dating the rise of atmospheric oxygen, *Nature*, 2004, Vol. 427, p. 117–120.

Belousov V.V., Fradkov A.F., Lukyanov K.A., Staroverov D.B., Shakhbazov K.S., Terskikh A.V., Lukyanov S., Genetically encoded fluorescent indicator for intracellular hydrogen peroxide, *Nat. Methods*, 2006, Vol. 3, p. 281–286.

Belyaeva N.E. *Generalized Model of Primary Photosynthetic Processes in Chloroplasts*, Ph D thesis, Moscow, 2004.

Belyaeva N.E., Lebedeva G.V., Riznichenko G.Yu. Kinetic model of primary photosynthetic processes in chloroplasts. Modeling of thylakoid membranes electric potential, *Mathematics Computer Education*, Vol. 10, Riznichenko G.Yu., ed., Moscow, Progress-Traditsiya, 2003, p. 263–276 (in Russian).

Belyaeva N.E., Schmitt F.-J., Steffen R., Paschenko V.Z., Riznichenko G.Y., Chemeris Y.K., Renger G., Rubin A.B. PS II model-based simulations of single turnover flash-induced transients of fluorescence yield monitored within the time domain of 100 ns–10 s on dark-adapted *Chlorella pyrenoidosa* cells, *Photosynth. Res.*, 2008, Vol. 98, p. 105–119.

Belyaeva N.E., Schmitt F.-J., Paschenko V.Z., Riznichenko G.Y., Rubin A.B., Renger G. PS II model based analysis of transient fluorescence yield measured on whole leaves of *Arabidopsis thaliana* after excitation with light flashes of different energies, *BioSystems*, 2011, Vol. 103, p. 188–195.

Belyaeva N.E., Schmitt F.-J., Paschenko V.Z., Riznichenko G.Y., Rubin A.B., Renger G. Model based analysis of transient fluorescence yield induced by actinic laser flashes in spinach leaves and cells of green alga *Chlorella pyrenoidosa* Chick, *Plant Physiol. Biochem.*, 2014, Vol. 77, p. 49–59.

Belyaeva N.E., Schmitt F.-J., Paschenko V.Z., Riznichenko G.Y., Rubin A.B. Modeling of the redox state dynamics in photosystem II of *Chlorella pyrenoidosa* Chick cells and leaves of spinach and *Arabidopsis thaliana* from single flash-induced fluorescence quantum yield changes on the 100 ns–10 s time scale, *Photosynth. Res.*, 2015, Vol. 125, p. 123–140.

Benson A.A. Following the path of carbon in photosynthesis: A personal story, *Photosynth. Res.*, 2002, Vol. 73, p. 29–49.

Bergantino E., Segalla A., Brunetta A., Teardo E., Rigoni F., Giacometti G.M., Szabo I. Light- and pH-dependent structural changes in the PsbS subunit of photosystem II, *Proc. Natl. Acad. Sci. U.S.A.*, 2003, Vol. 100, p. 15265–15270.

Bergmann A. *Picosekunden-Fluoreszenzspektroskopie Mithilfe Doppeltkorrelierter Einzelphotonendetek-tion zur Untersuchung der Primärprozesse im Photosystem II*, PhD thesis, Mensch und Buch Verlag, Berlin, 1999.

Bhattacharjee S. The language of reactive oxygen species signalling in plants, *J. Bot.*, 2012, Vol. 2012, 22 p.

Bibby T.S., Nield J., Chen M., Larkum A.W.D., Barber J., Structure of a photosystem II supercomplex isolated from Prochloron didemni retaining it´s chlorophyll a/b light-harvesting system, *Proc. Natl. Acad. Sci. U.S.A.*, 2003, Vol. 100, p. 9050–9054.

Biel K., Fomina I.R., Kreslavski V.D., Allakhverdiev S.I. Methods for assessment of activity and stress acclimation of photosynthetic machinery, in: *cyanobacteria and symbiotic microalgae. Protocols on algal and cyanobacterial research,* Kliner W., Nath Bogchi S., Mohanty P. eds., Narosa Publishing House., New Delhi., 2009, Chapter 13.

Bienert G.P., Møller A.L., Kristiansen K.A., Schulz A., Møller I.M., Schjoerring J.K., Jahn T.P. Specific aquaporins facilitate the diffusion of hydrogen peroxide across membranes, *J. Biol. Chem.*, 2007, Vol. 282, p. 1183–1192.

Bigler W., Schreiber U. Chlorophyll luminescence as an indicator of stress-induced damage to the photosynthetic apparatus. Effects of heat-stress in isolated chloroplasts, *Photosynth. Res.*, 1990, Vol. 25, p. 161–171.

Biswal U.C., Biswal B., Raval M.K. *Chloroplast Biogenesis. From Propastid to Gerontoplast*, Springer, Dordrecht, Kluwer Academic Publishers, 2003, 353 p.

Bizzarri R., Serresi M., Luin S., Beltram F. Green fluorescent protein based pH indicators for *in vivo* use: a review, *Anal. Bioanal. Chem.*, 2009, Vol. 393, p. 1107–1122.

Blokhina O., Fagerstedt K.V., Reactive oxygen species and nitric oxide in plant mitochondria: origin and redundant regulatory systems, *Physiol. Plant.*, 2010, Vol. 138, p. 447–462.

Boccalandro H.E., Rugnone M.L., Moreno J.E., Ploschuk E.L., Serna L., Yanovsky M.J., Casal J.J. Phytochrome B enhances photosynthesis at the expense of water-use efficiency in *Arabidopsis*, *Plant Physiol.*, 2009, Vol. 150, p.1083–1092.

Bolwell G.P., Bindschedler L.V., Blee K.A., Butt V.S., Davies D.R., Gardner S.L., Gerrish C., Minibayeva F. The apoplastic oxidative burst in response to biotic stress in plants: a three-component system, *J. Exp. Bot.*, 2002, Vol. 53, p. 1367–1376.

Boulay C., Wilson A., D'Haene S., Kirilovsky D. Identification of a protein required for recovery of full antenna capacity in OCP-related photoprotective mechanism in cyanobacteria, *Proc. Natl. Acad. Sci. U.S.A.*, 2010, Vol. 107, p. 11620–11625.

Bowler C., Van Montagu M., Inzé D., Superoxide dismutase and stress tolerance, *Annu. Rev. Plant Physiol. Plant Mol. Biol.*, 1992, Vol. 43, p. 83–116.

Bowler C., Fluhr R. The role calcium and activated oxygen as signals for controlling cross-tolerance, *Trends Plant Sci.*, 2000, Vol. 5, p. 241–246.

Brakemann T., Stiel A., Weber G., Andresen M., Testa I., Grotjohann T., Leutenegger M., Plessmann U., Urlaub H., Eggeling C., Wahl M.C., Hell S.W., Jakobs S. A reversibly photoswitchable GFP-like protein with fluorescence excitation decoupled from switching, *Nat. Biotechnol.*, 2011, Vol. 29, p. 942–947.

Buchanan B.B., Luan S. Redox regulation in the chloroplast thylakoid lumen: A new frontier in photosynthesis research, *J. Exp. Bot.*, 2005, Vol. 56, p. 1439–1447.

Buick R. The antiquity of oxygenic photosynthesis: evidence from stromatolites in sulphate-deficient Archaean lakes, *Science*, 1992, Vol. 255, p. 74–79.

Bulychev A.A., Niyazova M.M., Rubin A.B. Fluorescence changes of chloroplasts caused by the shifts of membrane-potential and their dependence on the redox state of the acceptor of photosystem II, *Biologicheskie Membrany*, 1987, Vol. 4, p. 262–269.

Bulychev A.A., Vredenberg W.J. Light-triggered electrical events in the thylakoid membrane of plant chloroplast, *Physiol. Plant.*, 1999, Vol. 105, p. 577–584.

Bulychev A.A., Vredenberg W.J. Modulation of photosystem II chlorophyll fluorescence by electrogenic events generated by photosystem I, *Bioelectrochemistry*, 2001, Vol. 54, p. 157–168.

Butz M.V., Sigaud O., Gérard P. *Anticipatory Behavior in Adaptive Learning Systems: Foundations, Theories, and Systems*, Springer, 2003.

Byrdin M. *Messungen und Modellierungen zur Dynamik angeregter Zustände in Photosystem I*, PhD thesis, Freie Universität Berlin, 1999.

Calas-Blanchard C., Catanante G.E., Noguer T. Electrochemical sensor and biosensor strategies for ROS/RNS detection in biological systems, *Electroanalysis*, 2014, Vol. 26, p. 1277–1286.

Calhoun T.R., Ginsberg N.S., G. S. Schlau-Cohen G.S., Cheng Y.-C., Ballottari M., Bassi R., Fleming G.R. Quantum coherence enabled determination of the energy landscape in light-harvesting complex II, *J. Phys. Chem. B*, 2009, Vol. 113, 16291–16295.

Campbell T.N., Choy F.Y.M. The Effect of pH on green fluorescent protein: a brief review, *Molecular Biology Today*, 2001, Vol. 2, p. 1–4.

Campbell N.A., Reece J.B. *Biologie*, 8. Auflage, Pearson Studium, München, 2009.

Carbonera D., Gerotto C., Posocco B., Giacometti G.M., Morosinotto T. NPQ activation reduces chlorophyll triplet state formation in the moss *Physcomitrella patens, Biochim. Biophys. Acta*, 2012, Vol. 1817, p. 1608–1615.

Carvalho R.F., Campos M.L., Azevedo R.A. The role of phytochrome in stress tolerance, *J. Int. Plant. Biol.*, 2011, Vol. 53, p. 920–929.

Chemeris Yu.K., Korol'kov N.S., Seifullina N.Kh., Rubin A.B. PSII complexes with destabilized primary quinone acceptor of electrons in dark-adapted chlorella, *Russ. J. Plant Physiol.*, 2004, Vol. 51, p. 9–14.

Chen Z., Silva H., Klessig D.F. Active oxygen species in the induction of plant systemic acquired resistance by salicylic acid, *Science*, 1993, Vol. 262, p. 1883–1886.

Chen M., Bibby T.S., Nield J., Larkum A.W.D., Barber J. Structure of a large Photosystem II supercomplex from acaryochloris marina, *FEBS Lett.*, 2005, Vol. 579, p. 1306.

Christen G., Seeliger A., Renger G. P680•+ reduction kinetics and redox transition probability of the water oxidizing complex as a function of pH and H/D isotope exchange in spinach thylakoids, *Biochemistry*, 1999, Vol. 38, p. 6082–6092.

Christen G., Steffen R., Renger G. Delayed fluorescence emitted from light-harvesting complex II and photosystem II of higher plants in the 100 ns–5 μs time domain, *FEBS Lett.*, 2000, Vol. 475, p. 103–106.

Cohen J., .Stewart I. *Chaos und Anti-Chaos*, dtv, 1997.

Cohen-Tannouji C., Diu B., Laloe F. *Quantenmechanik, Teil 1*, de Gruyter, Berlin, 1999a.

Cohen-Tannouji C., Diu B., Laloe F. *Quantenmechanik, Teil 2*, de Gruyter, Berlin, 1999b.

Cohn C.A., Simon S.R., Schoonen M.A. Comparison of fluorescence-based techniques for the quantification of particle-induced hydroxyl radicals, *Part Fibre Toxicol.*, 2008, Vol. 5, 9 p.

Coombes S. The Geometry and Pigmentation of Seashells, 2009, *https://www.maths.nottingham.ac.uk/personal/sc/pdfs/Seashells09.pdf*, (26.12.2016).

Corpas F.J., Barroso J.B., del Río L.A. Peroxisomes as a source of reactive oxygen species and nitric oxide signal molecules in plant cells, *Trends Plant Sci.*, Vol. 6, 2012, p. 145–150.

Crofts A.R., Wraight C.A. The electrochemical domain of photosynthesis, *Biochim. Biophys. Acta*, 1983, Vol. 726, p. 149.

Crooks G. Nonequilibrium measurements of free energy differences for microscopically reversible markovian systems, *J. Stat. Phys.*, 1998, Vol. 90, p. 1481.

Darwin C. *On the Origin of Species*, London, 1859

Dat J.F., Delgado H.L., Foyer C.H., Scott I.M. Parallel changes in H_2O_2 and catalase during termotolerance induced by salicylic acid or heat acclimation in mustard seedlings, *Plant Physiol.*, 1998, Vol. 116, p. 1351–1357.

Dat J.F., Vandenabeele S., Vranova E., Van Montagu M., Inze D., Van Breusegem F. Dual action of the active oxygen species during plant stress responses, *Cell Mol. Life Sci.*, 2000, Vol. 57, p. 779–795.

Dau H., Sauer K. Electric field effect on the picosecond fluorescence of photosystem II and its relation to the energetics and kinetics of primary charge separation, *Biochim. Biophys. Acta*, 1992, Vol. 1102, p. 91–106.

Davies M.J. Recent developments in spin trapping, *Electron Paramagnetic Resonance*, Gilbert B.C., Davies M.J., Murphy D.M. eds, The Royal Society of Chemistry, 2002, p. 47–73.

Davies M.J. Singlet oxygen-mediated damage to proteins and its consequences, *Biochem. Biophys. Res. Commun.*, 2003, Vol. 303, p. 761–770.

Davletova S., Schlauch K., Coutu J., Mittler R. The zinc-finger protein Zat12 plays a central role in reactive oxygen and abiotic stress signaling in *Arabidopsis*, *Plant Physiol.*, 2005, Vol. 139, p. 847–856.

Debreczeny M.P., Sauer K., Zhou J., Bryant D.A. Monomeric C-phycocyanin at room temperature and 77K: Resolution of the absorption and fluorescence spectra of the individual chromophores and the energy-transfer rate constants, *J. Phys. Chem.*, 1993, Vol. 97, p. 9852–9862.

Debreczeny M.P., Sauer K., Zhou J., Bryant D.A. Comparison of calculated and experimentally resolved rate constants for excitation energy transfer in C-phycocyanin 1. Monomers, *J. Phys. Chem.*, 1995a, Vol. 99, p. 8412–8419.

Debreczeny M.P., Sauer K., Zhou J., Bryant D.A. Comparison of calculated and experimentally resolved rate constants for excitation energy transfer in C-phycocyanin 2. Trimers, *J. Phys. Chem.*, 1995b, Vol. 99, p. 8420–8431.

Dedecker P., Hotta J., Flors C., Sliwa M., Uji-i H., Roeffaers M., Ando R., Mizuno H., Miyawaki A., Hofkens J. Subdiffraction imaging through the selective donut-mode depletion of thermally stable photoswitchable fluorophores: numerical analysis and application to the fluorescent protein Dronpa. *J. Am. Chem. Soc.*, 2007, Vol. 129, p. 16132–16141.

Delaunay A., Isnard A.D., Toledano M.B. H_2O_2 sensing through oxidation of the Yap1 transcription factor, *EMBO J.*, 2000, Vol. 19, p. 5157–5166.

Del Rio L.A., Puppo A. *Reactive Oxygen Species in Plant Signaling*, Series: Signaling and Communication in Plants, Springer, 2009.

Demidchik V., Shabala S.N., Koutts K.B., Tester M.A., Davies J.M. Free oxygen radicals regulate plasma membrane Ca^{2+}- and K^{+}-permeable channels in plant root cells, *J. Cell Sci.*, 2003, Vol. 116, p. 81–88.

Demmig-Adams B., Gilmore A.M., Adams W.W. The role of the xanthophyll cycle carotenoids in the protection of photosynthesis, *Trends Plant Sci.*, 1996, Vol. 1, p. 21–26.

Demmig-Adams B., Garab G., Adams W., Govindjee *Non-Photochemical Quenching and Energy Dissipation in Plants, Algae and Cyanobacteria (Advances in Photosynthesis and Respiration)*, Springer, Dodrecht, 2014, 649 p.

Demtröder W. *Molekülphysik: Theoretische Grundlagen und experimentelle Methoden*, Oldenbourg, München. 2003

Des Marais D.J. When did photosynthesis emerge on Earth? *Science,* 2000, Vol. 289, p. 1703–1705.

Desikan R., Mackerness S.A.-H., Hancock J.T., Neill S.J., Regulation of the *Arabidopsis* transcriptome by oxidative stress, *Plant Physiol.*, 2001, Vol. 127, p. 159–172.

Desikan R., Hancock J.T., Neill S.J., Oxidative stress signaling, *Plant Responses to Abiotic Stress: Topic in Current Genetics,* Hirt H., Shinozaki K. eds., Springer-Verlag, New-York, 2003, p.121–148.

Desikan R., Hancock J.T., Bright J., Harrison J., Weir I., Hooley R., Neil S.J. A role for ETR1 in hydrogen peroxide signaling in stomatal guard cells, *Plant Physiol.*, 2005, Vol. 137, p. 831–834.

Dickinson B.C., Chang C.J. Chemistry and biology of reactive oxygen species in signaling or stress responses, *Nat. Chem. Biol.*, 2011, Vol. 7, p. 504–511.

Dixit R., Cyr R. Cell damage and reactive oxygen species production induced by fluorescence microscopy: effect on mitosis and guidelines for non-invasive fluorescence microscopy, *Plant J.*, 2003, Vol. 36, p. 280–290.

Dixon D.P., Skipsey M., Grundy N.M., Edwards R. Stress induced protein S-glutathionylation in *Arabidopsis thaliana, Plant Physiol.*, 2005, Vol. 138, p. 2233–2244.

Dmitriev A.P. Signal plant molecules for activation of protective reactions in response to biotic stress, *Russ. J. Plant Physiol.*, 2003, Vol. 50, p. 465–474.

Dyson E. A meeting with Enrico Fermi, *Nature,* 2004, Vol. 427, p. 297.

Eggeling C., Hilbert M., Bock H., Ringemann C., Hofmann M., Stiel A., Andresen M., Jakobs S., Egner A., Schönle A., Hell S.W. Reversible photoswitching enables single-molecule fluorescence fluctuation spectroscopy at high molecular concentration, *Microsc. Res. Tech.*, 2007, Vol. 70, p. 1003–1009.

Egorov S.Y., Kamalov V.F., Koroteev N.I., Krasnovsky (Jr) A.A., Toleutaev B.N., Zinukov S.V. Rise and decay kinetics of photosensitized singlet oxygen luminescence in water. Measurements with nanosecond time-correlated single photon counting technique, *Chem. Phys. Lett.*, 1989, Vol. 163 , p. 421–424.

Eichler J., Eichler H.J. *Laser, Grundlagen, systeme, Anwendungen*, 2. Auflage, Springer Verlag, Berlin, 1991.

Enomoto J., Matharu Z., Revzin A. Electrochemical biosensors for on-chip detection of oxidative stress from cells, *Methods in Enzymology: Hydrogen Peroxide and cell Signaling: Part A*, 2013, Vol. 526, p. 107–121.

Feussner I., Wasternack C. The lipoxygenase pathway, *Ann. Rev. Plant Biol.*, 2002, Vol. 53, p. 275–297.

Fey V., Wagner R., Bräutigam K., Pfannschmidt T. Photosynthetic redox control of nuclear gene expression, *J. Exp. Bot.*, 2005a, Vol. 56, p. 1491–1498.

Fey V., Wagner R., Braütigam K., Wirtz M., Hell R., Dietzmann A., Leister D., Oelmüller R., Pfannschmidt T. Retrograde plastid redox signals in the expression of nuclear genes for chloroplast proteins of *Arabidopsis thaliana*, *J. Biol. Chem.*, 2005b, Vol. 280, p. 5318–5328.

Fischer B.B., Krieger-Liszkay A., Hideg E., Snyrychová I., Wiesendanger M., Eggen R.I. Role of singlet oxygen in chloroplast to nucleus retrograde signaling in *Chlamydomonas reinhardtii*, *FEBS Lett.*, 2007, Vol. 581, p. 5555–5560.

Fischer B.B., Ledford H.K., Wakao S., Huang S.Y.G., Casero D., Pellegrini M., Merchant S.S., Koller A., Eggen R.I.L., Niyogi K.K. Singlet oxygen resistant 1 links reactive electrophile signaling to singlet oxygen acclimation in *Chlamydomonas reinhardtii*, *Proc. Nat. Acad. Sci. U.S.A.*, 2012, Vol. 109, p. 1302–1311.

Flors C., Fryer M.J., Waring J., Reeder B., Bechtold U., Mullineaux P.M., Nonell S., Wilson M.T., Baker N.R., Imaging the production of singlet oxygen *in vivo* using a new fluorescent sensor, singlet oxygen sensor green, *J. Exp. Bot.*, 2006, Vol. 57, p. 1725–1734.

Förster T. Intramolecular energy transfer and fluorescene, *Ann. Phys.*, 1948, Vol. 2, p. 55–75.

Förster T. *Fluoreszenz Organischer Verbindungen*, Vandenhoeck & Ruprecht, Göttingen, 1982.

Foyer C.H., Noctor G. Redox homeostis and antioxidant signaling: a metabolic interface between stress perception and physiological responses, *Plant Cell.*, 2005, Vol. 17, p. 866–1875.

Foyer C.H., Noctor G.D. Redox regulation in photosynthetic organisms: Signaling, acclimation, and practical implications, *Antiox. Redox Signal.*, 2009, Vol. 11, p. 861–905.

Foyer C.H., Shigeoka S., Understanding oxidative stress and antioxidant functions to enhance photosynthesis, *Plant Physiol.*, 2011, Vol. 155, p. 93–100.

Frank H.A., Cogdell R.J. The photochemistry and function of carotenoids in photosynthesis, *Carotenoids in Photosynthesis*, Young A., Britton G. eds., Chapmen & Hall, London, 1993, p. 252–326.

Freer A., Prince S., Sauer K., Papiz M., Lawless A.H., McDermott G., Cogdell R., Isaacs N.W. Pigment-pigment interactions and energy transfer in the antenna complex of the photosynthetic bacterium *Rhodopseudomonas acidophila*, *Structure*, 1996, Vol. 4, p. 449.

Fridovich I. Superoxide radical: an endogenous toxicant, *Annu. Rev. Pharmacol. Toxicol.*, 1983, Vol. 23, p. 239–257.

Fryer M.J., Oxborough K., Mullineaux P.M., Baker N.R. Imaging of photo-oxidative stress responses in leaves, *J. Exp. Bot.*, 2002, Vol. 53, p. 1249–1254.

Gadjev I., Vanderauwera S., Gechev T.S., Laloi C., Minkov I.N., Shulaev V., Apel K., Inze D., Mittler R., van Breusegem F. Transcriptomic footprints disclose specificity of reactive oxygen species signaling in *Arabidopsis, Plant Physiol.*, 2006, Vol. 141, p. 436–445.

Galetskiy D., Lohscheider J.N., Kononikhin A.S., Popov I.A., Nikolaev E.N., Adamska I., Mass spectrometric characterization of photooxidativeprotein modifications in *Arabidopsis thaliana* thylakoids membranes, *Rapid Commun. Mass Spectrom.*, 2011, Vol. 25, p. 184–190.

Galvez-Valdivieso G., Fryer M.J., Lawson T., Slattery K., Truman W., Smirnoff N., Asami T., Davies W.J., Jones A.M., Baker N.R., Mullineaux P.M. The high light response in *Arabidopsis* involves ABA signaling between vascular and bundle sheath cells, *Plant Cell*, 2009, Vol. 21, p. 2143–2162.

Galvez-Valdivieso G., Mullineaux P.M., The role of reactive oxygen species in signaling from chloroplasts to the nucleus, *Physiol. Plant.*, 2010, Vol. 138, p. 430–439.

Garab G., Mustárdy L. Role of LHCII-containing macrodomains in the structure, function and dynamics of grana, *Aust. J. Plant Physiol.*, 1999, Vol. 26, p. 649.

Gechev T.S., Hille J. Hydrogen peroxide as a signal controlling plant programmed cell death, *J. Cell Biol.*, 2005, Vol. 168, p. 17–20.

Gechev T.S., Van Breusegem F., Stone J.M., Denev I., Laloi C. Reactive oxygen species as signals that modulate plant stress responses and programmed cell death, *Bioessays*, 2006, Vol. 28, p. 1091–1101.

Gell-Mann M. *The Quark and the Jaguar*, W.H. Freeman, New York, 1994.

Genty B., Briantais J.M., Baker N.R. The relationship between the quantum yield of photosynthetic electron transport and quenching of chlorophyll fluorescence, *Biochim. Biophys. Acta*, 1989, Vol. 990, p. 87–92.

Gershwin L.-a. *Stung! On Jellyfish Blooms and the Future of the Ocean*, University of Chicago Press, 2013, 456 p.

Gibasiewicz K., Dobek A., Breton J., Leibl W. Modulation of primary radical pair kinetics and energetics in photosystem II by the redox state of the quinone electron acceptor QA, *Biophys. J.*, 2001, Vol. 80, p. 1617–1630.

Gidrol X., Sabelli P.A., Fern Y.S., Kush A.K., Annexin-like protein from *Arabidopsis thaliana* rescues Δ*oxyR* mutant of *Escherichia coli* from H_2O_2 stress, *Proc. Natl. Acad. Sci. U.S.A.*, 1996, Vol. 93, p. 11268–11273.

Gillbro T., Sandström A., Sundström V., Wendler J., Holzwarth A.R. Picosecond Study of energy transfer kinetics in phycobilisomes of *Synechococcus* 6301 and the mutant AN 112, *Biochim. Biophys. Acta*, 1985, Vol. 808, p. 52–65.

Glazer A.N. Light harvesting by phycobilisomes, *Ann. Rev. Biophys. Chem.*, 1985, Vol. 14, p. 47.

Godrant A., Rose A.L., Sarthou G., Waite D. New method for the determination of extracellular production of superoxide by marine phytoplankton using the chemiluminescence probes MCLA and red-CLA, *Limnol. Oceanogr. Methods*, 2009, Vol. 7, p. 682–692.

Goh C.-H., Ko S.-M., Koh S., Kim Y.-J., Bae H.-J. Photosynthesis and environments: Photoinhibition and repair mechanisms in plants, *J. Plant Biol.*, 2012, Vol. 55, p. 93–101.

Goh C.-H., Schreiber U., Hedrich R. New approach of monitoring changes in chlorophyll a fluorescence of single guard cells and protoplasts in response to physiological stimuli, *Plant Cell Environ.*, 1999, Vol. 22, p. 1057–1070.

Golbeck J., Radmer R. Is the rate of oxygen uptake by reduced ferredoxin sufficient to account for photosystem I – mediated O_2 reduction?, *Advances in Photosynthesis Research*, Sybesma C. ed., M. Nijhoff/Dr W. Junk Publ, Lancaster, 1984, p. 561–564.

Goltsev V., Zaharieva I., Chernev P., Strasser R.J. Delayed fluorescence in photosynthesis, *Photosynth. Res.*, 2009, Vol. 101, p. 217–232.

Gomes A., Fernandes E., Lima J.L. Fluorescence probes used for detection of reactive oxygen species, *J. Biochem. Biophys. Methods*, 2005, Vol. 65, p. 45–80.

Goral T.K., Johnson M.P., Duffy C.D., Brain A.P., Ruban A.V., Mullineaux C.W. Light-harvesting antenna composition controls the macrostructure and dynamics of thylakoid membranes in *Arabidopsis*, *Plant J.*, 2012, Vol.69, p. 289–301.

Gorman A.A., Rodgers M.A. Current perspectives of singlet oxygen detection in biological environments, *J. Photochem. Photobiol. B: Biology*, 1992, Vol. 14, p. 159–176.

Govindjee, Jursinic P.A. Photosynthesis and fast changes in light emission by green plants, *Photochem. Photobiol. Reviews*, Vol. 4, Smith, K.C. ed., The Plenum Press, NY, 1979, p. 125–205.

Gracanin M., Hawkins C.L., Pattison D.I., Davies M.J. Singlet oxygen-mediated amino acid and protein oxidation: formation of tryptophan peroxides and decomposition products, *Free Radic. Biol. Med.*, 2009, Vol. 47, p. 92–102.

Gruissem W., Lee C.-H., Oliver M., Pogson B., The global plant council: Increasing the impact of plant research to meet global challenges, *J. Plant Biol.*, 2012, Vol. 55, p. 343–348.

Gudkov N.D. Some thermodynamics of chemiluminescence, *Journal of Luminescence*, 1998, Vol. 79, p. 85.

Gupta R., Luan S. Redox control of protein tyrosine phosphatases and mitogen-activated protein kinases in plants, *Plant Physiol.*, 2003, Vol. 132, p. 1149–1152.

Haber F., Weiss J. The catalytic decomposition of hydrogen peroxide by iron salts, *Proc. R. Soc. London, Ser. A.*, 1934, Vol. 147, p. 332–351.

Häder D-P. *Photosynthese*, Thieme Verlag, Stuttgart, 1999.

Haken H. *Light: Waves, Photons, and Atoms*, North Holland, Amsterdam, 1981.

Haken H. *Synergetik. Eine Einführung*, Springer, Berlin, 1990.

Halliwell B., Generation of hydrogen peroxide, superoxide and hydroxyl radicals during the oxidation of dihydroxyfumaric acid by peroxidase, *Biochem. J.*, 1977, Vol. 163, p. 441–448.

Halliwell B., Gutteridge J.M. *Free radicals in biology and medicine*, Clarendon Press, Oxford, 1985, 266 p.

Halliwell B. Reactive species and antioxidants. Redox biology is a fundamental theme of aerobic life, *Plant Physiol.*, 2006, Vol. 141, p. 312–322.

Hanson G.T., McAnaney T.B., Park E.S., Rendell M.E., Yarbrough D.K., Chu S., Xi L., Boxer S.G., Montrose M.H., Remington S.J. Green fluorescent protein variants as ratiometric dual emission pH sensors. 1. Structural characterization and preliminary application, *Biochemistry*, 2002, Vol. 41, p. 15477–15488.

Hanson G.T., Aggeler R., Oglesbee D., Cannon M., Capaldi R.A., Tsien R.Y., Remington S.J. Investigating mitochondrial redox potential with redox-sensitive green fluorescent protein indicators, *J. Biol. Chem.*, 2004, Vol. 279, p. 13044–13053.

Härtel H., Lokstein H., Grimm B., Rank B. Kinetic studies on the xanthophyll cycle in barley leaves (Influence of antenna size and relations to nonphotochemical chlorophyll fluorescence quenching), *Plant Physiol.*, 1996, Vol. 110, p. 471–482.

Heim R., Cubitt A.B., Tsien R.Y. Improved green fluorescence, *Nature*, 1995, Vol. 373, p. 663–664.

Heisenberg W. *Der Teil und das Ganze, Gespräche im Umkreis der Atomphysik*, Piper, München, 6. Auflage, 1986.

Henzler T., Steudle E. Transport and metabolic degradation of hydrogen peroxide in *Chara corallina*: model calculations and measurements with the pressure probe suggest transport of H_2O_2 across water channels, *J. Exp. Bot.*, 2000, Vol. 51, p. 2053–2066.

Herbig A.F., Helmann J.D. Roles of metal ions and hydrogen peroxide in modulating the interaction of the *Bacillus subtilis* PerR peroxide regulon repressor with operator DNA, *Mol. Microbiol.*, 2001, Vol. 41, p. 849–859.

Hideg É., Spetea C., Vass I. Singlet oxygen and free radical production during acceptor- and donor-side-induced photoinhibition: Studies with spin trapping EPR spectroscopy, *Biochim. Biophys. Acta*, 1994, Vol. 186, p. 143–152.

Hideg E., Kalai T., Hideg K., Vass I. Photoinhibition of photosynthesis *in vivo* results in singlet oxygen production detection via nitroxide-induced fluorescence quenching in broad bean leaves, *Biochemistry*, 1998, Vol. 37, p. 11405–11411.

Hideg E., Vass I., Kalai T., Hideg K. Singlet oxygen detection with sterically hindered amine derivatives in plants under light stress, *Methods Enzymol.*, 2000, Vol. 319, p. 77–85.

Hideg E., Ogawa K., Kalai T., Hideg K. Singlet oxygen imaging in *Arabidopsis thaliana* leaves under photoinhibition by excess photosynthetically active radiation, *Physiol. Plant.*, 2001, Vol. 112, p. 10–14.

Hideg E., Kos P.B., Vass I. Photosystem II damage induced by chemically generated singlet oxygen in tobacco leaves, *Physiol. Plant.*, 2007, Vol. 131, p. 33–40.

Hideg E., Kálai T., Hideg K. Direct detection of free radicals and reactive oxygen species in thylakoids, *Methods Mol. Biol.*, 2011, Vol. 684, p. 187–200.

Hintze K.J., Theil E.C. Cellular regulation and molecular interactions of the ferritins, *Cell Mol. Life Sci.*, 2006, Vol. 63, p. 591–600.

Hoeberichts F.A., Woltering E.J. Multiple mediators of plant programmed cell death: Interplay of conserved cell death mechanisms and plant-specific regulators, *BioEssays*, 2003, Vol. 25, p. 47–57.

Hofmann M., Eggeling C., Jakobs S., Hell S.W. Breaking the diffraction barrier in fluorescence microscopy at low light intensities by using reversibly photoswitchable proteins, *Proc. Natl. Acad. Sci. U.S.A.*, 2005, Vol. 102, p. 17565–17569.

Holzwarth A.R. Structure-function relationships and energy transfer in phycobiliprotein antennae, *Physiol. Plant.*, 1991, Vol. 83, p. 518–528.

Horváth E., Janda T., Szalai G., Páldi E. *In vitro* salicylic acid inhibition of catalase activity in maize: differences between the isozymes and a possible role in the induction of chilling tolerance, *Plant Sci.*, 2002, Vol. 163, p. 1129–1135.

Hsiao H.-Y., He Q., van Waasbergen L.G., Grossman A.R. Control of photosynthetic and high-light-responsive genes by the histidine kinase DspA: negative and positive regulation and interactions between signal transduction pathways, *J. Bacteriol.*, 2004, Vol. 186, p. 3882–3888.

Hung S.-H., Yu C.-W., Lin C.H. Hydrogen peroxide functions as a stress signal in plants, *Bot. Bull. Acad. Sinica.*, 2005, Vol. 46, p. 1–10.

Hung Y.P., Albeck J.G., Tantama M., Yellen G. Imaging cytosolic NADH-NAD(+) redox state with a genetically encoded fluorescent biosensor, *Cell Metab.*, 2011, Vol. 14, p. 545–554.

Imlay J.A. Pathways of oxidative damage, *Annu. Rev. Microbiol.*, 2003, Vol. 57, p. 395–418.

Ivanov B.N., Mubarakshina M.M., Khorobrykh S.A. Kinetics of the plastoquinone pool oxidation following illumination. Oxygen incorporation into photosynthetic electron transport chain, *FEBS Lett.*, 2007, Vol. 581, p. 1342–1346.

Iwai M., Yokono M., Inada N., Minagawa J. Live-cell imaging of photosystem II antenna dissociation during state transitions, *Proc.Natl. Acad. Sci. U.S.A.*, 2010, Vol. 107, p. 2337–2342.

Jahns P., Wehner A., Paulsen W., Hobe S. De-epoxidation of violaxanthin after reconstitution into different carotenoid binding sites of light-harvesting complex II, *Biol. Chem.*, 2001, Vol. 276, p. 22154.

Jaspers P., Kangasjärvi J. Reactive oxygen species in abiotic stress signaling, *Physiol. Plant.*, 2010, Vol. 138, p. 405–413.

Jennings R.C., Engelmann E., Garlaschi F., Casazza A.P., Zucchelli G. Photosynthesis and negative entropy production, *Biochim. Biophys. Acta*, 2005, Vol. 1709, p. 251.

Jennings R.C., Casazza A.P., Belgio E., Garlaschi F.M., Zucchelli G. Reply to commentary on: Photosynthesis and negative entropy production, *Biochim. Biophys. Acta*, 2006, Vol. 1757, p. 1460.

Jennings R.C., Belgio E., Casazza A.P.,Garlaschi F.M., Zucchelli G. Entropy consumption in primary photosynthesis, *Biochim. Biophys. Acta*, 2007, Vol. 1767, p. 1194.

Jensen R.L., Arnbjerg J., Ogilby P.R. Reaction of singlet oxygen with tryptophan in proteins: A pronounced effect of the local environment on the reaction rate, *J. Am. Chem. Soc.*, 2012, Vol. 134, p. 9820–9826.

Jimenez-Banzo A., Nonell S., Hofkens J., Flors C. Singlet oxygen photosensitization by EGFP and its chromophore HBDI, *Biophys. J.*, 2008, Vol. 94, p. 168–172.

Joliot P., Joliot A. Quantification of cyclic and linear flows in plants, *Proc. Nat. Acad. Sci. U.S.A.*, 2005, Vol. 102, p. 4913–4918.

Joo J.H., Wang S., Chen J.G., Jones A.M., Fedoroff N.V. Different signaling and cell death roles of heterotrimeric G-protein α and β subunits in the *Arabidopsis* oxidative stress response to ozone, *Plant Cell*, 2005, Vol. 17, p. 957–970.

Jung J.-Y., Shin R., Schachtaman D.P. Ethylene mediates response and tolerance to potassium deprivation in *Arabidopsis, Plant Cell*, 2009, Vol. 21, p. 607–621.

Junge W., Evolution of Photosynthesis, *Primary Processes of Photosynthesis: Priniciples and Apparatus, Part II Reaction Centers/Photosystems, Electron Transport Chains, Photophosphorylation and Evolution*, Renger G. ed., Royal Society Chemistry, Cambridge, 2008, p. 447–487.

Junghans C., Schmitt F.-J., Vukojević V., Friedrich T. Monitoring the diffusion behavior of Na,K-ATPase by fluorescence correlation spectroscopy (FCS) upon fluorescence labelling with eGFP or Dreiklang, *Optofluid., Microfluid. Nanofluid.*, 2016, Vol. 2, p. 1–14.

Kanesaki Y., Suzuki I., Allakhverdiev S.I., Mikami K., Murata N. Salt stress and hyperosmotic stress regulate the expression of different sets of genes in *Synechocystis* sp. PCC 6803, *Biochem. Biophys. Res. Commun.*, 2002, Vol. 290, p. 339–348.

Kanesaki Y., Yamamoto H., Paithoonrangsarid K., Shoumskaya M., Suzuki I., Hayashi H., Murata N. Histidine kinases play important roles in the perception and signal transduction of hydrogen peroxide in the cyanobacterium *Synechocystis* sp. PCC 6803, *Plant J.*, 2007, Vol. 49, p. 313–324.

Kanesaki Y., Los D.A., Suzuki I., Murata N. Sensors and signal transducers of environmental stress in cyanobacteria, *Abiotic Stress Adaptation in Plants: Physiological, Molecular and Genomic Foundation*, Pareek A., Sopory S.K., Bohnert H.J., Govindjee, eds., Springer, Dordrecht, 2010, chap. 2.

Karpinski S., Gabrys H., Mateo A., Karpinska B., Mullineaux P.M. Light perception in plant disease defence signaling, *Curr. Opin. Plant Biol.*, 2003, Vol. 6, p. 390–396.

Karpinski S., Reynolds H., Karpinska B., Wingsle G., Creissen G., Mullineaux P.M. Systemic signaling and acclimation in response to excess excitation energy in *Arabidopsis, Science*, 1999, Vol. 284, p. 654–657.

Kasson T.M., Barry B.A. Reactive oxygen and oxidative stress: N-formyl kynurenine in photosystem II and non-photosynthetic proteins, *Photosynth. Res.*, 2012, Vol. 114, p. 97–110.

Kasting J.F., Siefert J.L. Life and the evolution of Earth's Atmosphere, *Science*, 2002, Vol. 296, p. 1066–1068.

Kaur N., Gupta A.K. Signal transduction pathways under abiotic stresses in plants, *Current Sci.*, 2005, Vol. 88, p. 1771–1780.

Kehrer J.P. The Haber-Weiss reaction and mechanisms of toxicity, *Toxicology*, 2000, Vol. 149, p. 43–50.

Kern J., Renger G. Photosystem II: structure and mechanism of the water: plastoquinone oxidoreductase, *Photosynth. Res.*, 2007, Vol. 94, p. 183–202.

Ketelaars M., van Oijen A. M., Matsushita M., Köhler J., Schmidt J., Aartsma T. J. Spectroscopy on the B850 band of individual light-harvesting 2 complexes of *Rhodopseudomonas acidophila* I. Experiments and Monte Carlo simulations, *Biophys. J.*, 2001, Vol. 80, p. 1591.

Khorobrykh S.A., Karonen M., Tyystjärvi E. Experimental evidence suggesting that H_2O_2 is produced within the thylakoid membrane in a reaction between plastoquinol and singlet oxygen, *FEBS Lett.*, 2015, Vol. 589, p. 779–786.

Kim M.C., Chung W.S., Yun D., Cho M.J. Calcium and calmodulin-mediated regulation of gene expression in plants, *Mol. Plant.*, 2009, Vol. 2, p. 13–21.

Kirilovsky D., Kerfeld C.A., the orange carotenoid protein: a blue-green light photoactive protein, *Photochem Photobiol. Sci.*, 2013, Vol. 12, p. 1135–1143.

Kirwan A.D. Intrinsic photon entropy? The darkside of light, *Int. J. Eng Sci.*, 2004, Vol. 42, p. 725 (2004).

Klar T.A., Jakobs S., Dyba M., Egner A. and Hell S.W. Fluorescence microscopy with diffraction resolution limit broken by stimulated emission, *Proc. Natl. Acad. Sci. U.S.A.*, 2000, Vol. 97, p. 8206.

Klimov V.V., Allakhverdiev S. I., Paschenko V. Z. Measurement of energy activation and life-time of the fluorescence of photosystem II chlorophyll, *Dokl. Acad. Nauk SSSR*, 1978, Vol. 242, p. 1204–1207.

Klimov V.V., Ananyev G., Zastryzhnaya O., Wydrzynski T., Renger G. Photoproduction of hydrogen peroxide in photosystem II particles, *Photosynth. Res.*, 1993, Vol. 38, p. 409–416.

Klotz L.-O., Kröncke K.-D., Sies H. Singlet oxygen-induced signaling effects in mammalian cells, *Photochem. Photobiol. Sci.*, 2003, Vol. 2, p. 88–94.

Kobayashi M., Ishizuka T., Katayama M., Kanehisa M., Bhattacharyya-Pakrasi M., Pakrasi H.B., Ikeuchi M. Response to oxidative stress involves a novel peroxiredoxin gene in the unicellular cyanobacterium *Synechocystis* sp. PCC 6803, *Plant Cell Physiol.*, 2004, Vol. 45, p. 290–299.

Kojima K., Oshita M., Nanjo Y., Kasai K., Tozawa Y., Hayashi H., Nishiyama Y. Oxidation of elongation factor G inhibits the synthesis of the D1 protein of photosystem II, *Mol. Microbiol.*, 2007, Vol. 65, p. 936–947.

Kojima K., Motohashi K., Morota T., Oshita M., Hisabori T., Hayashi H., Nishiyama Y. Regulation of translation by the redox state of elongation factor G in the cyanobacterium *Synechocystis* sp. PCC 6803, *J. Biol. Chem.*, 2009, Vol. 284, p. 18685–18691.

Kovtun Y., Chiu W.L., Tena G., Sheen J. Functional analysis of oxidative stress-activated mitogen-activated protein kinase cascade in plants, *Proc. Natl. Acad. Sci. U.S.A.*, 2000, Vol. 97, p. 2940–2945.

Kozuleva M.A., Ivanov B.N. Evaluation of the participation of ferredoxin in oxygen reduction in the photosynthetic electron transport chain of isolated pea thylakoids, *Photosynth. Res.*, 2010, Vol. 105, p. 51–61.

Kozuleva M.A., Petrova A.A., Mamedov M.D., Semenov A.Y., Ivanov B.N. O_2 reduction by photosystem I involves phylloquinone under steady-state illumination, *FEBS Lett.*, 2014, Vol. 588, p. 4364–4368.

Kramer D.M., Crofts A.R. Control of photosynthesis and measurement of photosynthetic reactions in intact plants, *Photosynthesis and the Environment*, N.R. Baker ed., Series: Advances in Photosynthesis, Kluiver Academic Press, Dordrecht, 1996, p. 25–66.

Kremers G.J., Gilbert S.G., Cranfill P.J., Davidson M.W., Piston D.W. Fluorescent proteins at a glance, *J. Cell. Sci.*, 2011, Vol. 124, p. 157–160.

Kreslavski V.D., Ivanov A.A., Kosobrukhov A.A. Low energy light in the 620–660 nm range reduces the UV-B-induced damage to photosystem II in spinach leaves, *Biophysics*, 2004, Vol. 49, p. 767–771.

Kreslavski V.D., Carpentier R., Klimov V.V., Murata N., Allakhverdiev S.I., Molecular mechanisms of stress resistance of the photosynthetic apparatus, *Membr. Cell Biol.*, 2007, Vol. 1, p. 185–205.

Kreslavski V.D., Carpentier R., Klimov V.V., Allakhverdiev S.I. Transduction mechanisms of photoreceptor signals in plant cells, *J. Photochem. Photobiol. C. Photochem. Rev.*, 2009, Vol. 10, p. 63–80.

Kreslavski V.D., Luybimov V.Y., Kotova L.M., Kotov A.A., Effect of pretreatment with chlorocholinechloride on resistance of bean PSII to UV-radiation, phytohormone and H_2O_2 contents, *Russ. J. Plant Physiol.*, 2011, Vol. 58, p. 324–329.

Kreslavski V.D., Fomina I.R., Los D.A., Carpentier R., Kuznetsov V.V., Allakhverdiev S.I. Red and near infra-red signaling: Hypothesis and perspectives, *J. Photochem. Photobiol. C. Photochem. Rev.*, 2012a, Vol. 13, p. 190–203.

Kreslavski V.D., Los D.A., Allakhverdiev S.I., Kuznetsov V.V. Signaling role of reactive oxygen species in plants under stress, *Russ. J. Plant Physiol.*, 2012b, Vol. 59, p. 141–154.

Kreslavski V.D., Khristin M.S., Shabnova N.I., Lyubimov V.Y. Preillumination of excised spinach leaves with red light increases the resistance of photosynthetic apparatus to UV radiation, *Russ. J. Plant Physiol.*, 2012c, Vol. 59, p. 717–723.

Kreslavski V.D. Zorina A.A., Los D.A., Fomina I.R., Allakhverdiev S.I. Molecular mechanisms of stress resistance of photosynthetic machinery, *Molecular Stress Physiology of Plants*, Rout G.R., Das A.B. eds., Springer India, 2013a, p. 21–51.

Kreslavski V.D., Shirshikova G.N., Lyubimov V.Y., Shmarev A.N., Boutanaev A., Kosobryukhov A.A., Schmitt F.-J., Friedrich T., Allakhverdiev S.I. Effect of pre-illumination with red light on photosynthetic parameters and oxidant-/antioxidant balance in *Arabidopsis thaliana* in response to UV-A, *J. Photochem. Photobiol. B:Biol.*, 2013b, Vol. 127, p. 229–236.

Kreslavski V.D., Lyubimov V.Y., Shirshikova G.N., Shmarev A.N., Kosobryukhov A.A., Schmitt F.-J., Friedrich T., Allakhverdiev S.I. Preillumination of lettuce seedlings with red light enhances the resistance of photosynthetic apparatus to UV-A, *J. Photochem. Photobiol. B.*, 2013c, Vol. 122, p. 1–6.

Kreslavski V.D., Lyubimov V.Y., Shabnova N.I., Shirshikova G.N., Shmarev A.N., Kosobryukhov A.A. Growth in the UV-A irradiation resistance of the photosynthetic apparatus of lettuce seedlings as a result of activation of phytochrome B, *Russian Agricultural Sciences*, 2014a, Vol. 40, p. 100–103.

Kreslavski V.D., Lankin A.V., Vasilyeva G.K., Luybimov V.Yu., Semenova G.N., Schmitt F.-J., Friedrich T., Allakhverdiev S.I. Effects of polyaromatic hydrocarbons on photosystem II activity in pea leaves, *Plant Physiol. Biochem.*, 2014b, Vol. 81, p. 135–142.

Krieger-Liszkay A. Singlet oxygen production in photosynthesis, *J. Exp. Bot.*, 2005, Vol. 56, p. 337–346.

Kruk J., Szymańska R. Singlet oxygen and non-photochemical quenching contribute to oxidation of the plastoquinone-pool under high light stress in *Arabidopsis, Biochim. Biophys. Acta*, 2012, Vol. 1817, p. 705–710.

Krumova K., Friedland S., Cosa G. How lipid unsaturation, peroxyl radical partitioning, and chromanol lipophilic tail affect the antioxidant activity of α-tocopherol: direct visualization via high-throughput fluorescence studies conducted with fluorogenic α-tocopherol analogues, *J. Am. Chem. Soc.*, 2012, Vol. 134, p. 10102–10113.

Kuznetsov V.V., Shevyakova N.I. Proline under stress: biological role, metabolism, and regulation, *Russ. J. Plant Physiol.*, 1999, Vol. 46, p. 274–288.

Lai A.G., Doherty C.J., Mueller-Roeber B., Kay S.A., Schippers J.H.M., Dijkwel P.P. *CIRCADIAN CLOCK-ASSOCIATED 1* regulates ROS homeostasis and oxidative stress responses, *Proc.Nat. Acad. Sci. U.S.A.*, 2012, Vol. 109, p. 17129–17134.

Lakowicz J.R. *Principles of Fluorescence Spectroscopy*, Springer, New York, 2006.

Laloi C., Apel K., Danon A. Reactive oxygen signaling: the latest news, *Curr. Opin. Plant Biol.*, 2004, Vol. 7, p. 323–328.

Laloi C., Stachowiak M., Pers-Kamczyc E., Warzych E., Murgia I., Apel K. Cross-talk between singlet oxygen- and hydrogen peroxide-dependent signaling of stress responses in *Arabidopsis thaliana*, *Proc. Natl. Acad. Sci. U.S.A.*, 2007, Vol. 104, p. 672–677.

Lambrev P.H., Schmitt F.-J., Kussin S., Schoengen M., Várkonyi Z., Eichler H.J., Garab G., Renger G. Functional domain size in aggregates of light-harvesting complex II and thylakoid membranes, *Biochim. Biophys. Acta (BBA)-Bioenergetics*, 2011, Vol. 1807, p. 1022–1031.

Lane N. *Oxygen – the Molecule that Made the World*, Oxford University Press, Oxford, 2002, 384 p.

Lavergne J. Commentary on: Photosynthesis and negative entropy production by Jennings and coworkers, *Biochim. Biophys. Acta*, 2006, Vol. 1757, p. 1453.

Law C.J., Cogdell R.J. The light-harvesting system of purple anoxygenic photosynthetic bacteria, *Primary Processes of Photosynthesis: Basic Principles and Apparatus*, Vol. I, Renger G. ed., RSC Publ, Cambridge, U.K., 2007, p. 205–259.

Lazár D. The polyphasic chlorophyll *a* fluorescence rise measured under high intensity of exciting light, *Funct. Plant Biol.*, 2006, Vol. 33, p. 9–30.

Lebedeva G.V., Belyaeva N.E., Demin O.V., Riznichenko G.Yu., Rubin A.B. Kinetic model of primary photosynthetic processes in chloroplasts. Description of the fast phase of chlorophyll fluorescence induction under different light intensities, *Biophysics*, 2002, Vol. 47, p. 968–980.

Lee K.P., Kim C., Langraf F., Apel K., EXECUTER1- and EXECUTER2-dependent transfer of stress-related signals from the plastid to the nucleus of *Arabidopsis thaliana*, *Proc. Natl. Acad. Sci. U.S.A.*, 2007, Vol. 104, p. 10270–10275.

Leibl W., Breton J., Deprez J., Trissl H.W. Photoelectric study on the kinetics of trapping and charge stabilization in oriented PS II membranes, *Photosynth. Res.*, 1989, Vol. 22, p. 257–275.

Leverenz R.L., Jallet D., Li M.D., Mathies R.A., Kirilovsky D., Kerfeld C.A. Structural and functional modularity of the orange carotenoid protein: Distinct roles for the N- and C-terminal domains in cyanobacterial photoprotection, *Plant Cell*, 2014, Vol. 26, p.426–437.

Levinthal, C. How to Fold Graciously Mossbauer Spectroscopy in Biological Systems: *Proceedings of a meeting held at Allerton House*, Monticello, Illinois, 1969, p. 22–24.

Levinthal, C. Are there pathways for protein folding? *Journal de Chimie Physique et de Physico-Chimie Biologique*, 1968, Vol. 65, p. 44–45.

Li H., Singh A.K., McIntyre L.M., Sherman L.A. Differential gene expression in response to hydrogen peroxide and the putative PerR regulon of *Synechocystis* sp. strain PCC 6803, *J. Bacteriol.*, 2004, Vol, 186, p. 3331–3345.

Li H., Melø T.B., Arellano J.B., Razi Naqvi K.R. Temporal profile of the singlet oxygen emission endogenously produced by photosystem II reaction centre in an aqueous buffer, *Photosynth. Res.*, 2012, Vol. 112, p. 75–79.

Li Z., Wakao S., Fischer B.B., Niyogi K.K. Sensing and responding to excess light, *Ann. Rev. Plant Biol.*, 2009, Vol. 60, p. 239–260.

Lingakumar K., Kulandaivelu G. Regulatory role of phytochrome on ultraviolet-B (280–315 nm) induced changes in growth and photosynthetic activities of *Vigna sinensis* L., *Photosynthetica*, 1993, Vol. 29, p. 341–351.

Liou G.-Y., Storz P. Reactive oxygen species in cancer, *Free Radic Res.*, 2010, Vol. 44, p. 479–496.

Liu H., Colavitti R., Rovira I.I., Finkel T. Redox-dependent transcriptional regulation, *Circ. Res.*, 2005, Vol. 97, p. 967–974.

Liu Y., Ren D., Pike S., Pallardy S., Gassmann W., Zhang S. Chloroplast-generated reactive oxygen species are involved in hypersensitive response-like cell death mediated by a mitogen-activated protein kinase cascade, *Plant J.*, 2007, Vol. 51, p. 941–954.

Liu Z., Yan H., Wang K., Kuang T., Zhang J., Gui L., An X., Chang W. Crystal structure of spinach major light-harvesting complex at 2.72 Å resolution, *Nature*, 2004, Vol. 428, p. 287.

Löffler G., Petrides P.E., Heinrich P.C., *Biochemie und Pathobiochemie*, Springer, Heidelberg, 8. Auflage, 2007, 1266 p.

Loll B., Broser M., Kos P.B., Kern J., Biesiadka J., Vass I., Saenger W., Zouni A. Modeling of variant copies of subunit D1 in the structure of photosystem II from *Thermosynechococcus elongatus*, *Biol. Chem.*, 2008, Vol. 389, p. 609–617.

Los D.A., Zorina A., Sinetova M., Kryazhov S., Mironov K., Zinchenko V.V. Stress sensors and signal transducers in cyanobacteria, *Sensors*, 2010, Vol. 10, p. 2386–2415.

Los D.A., Mironov K.S., Allakhverdiev S.I. Regulatory role of membrane fluidity in gene expression and physiological functions, *Photosynth. Res.*, 2013, Vol. 116, p. 489–509.

McCord J.M., Crapo J.D., Fridovich I. *Superoxide and Superoxide Dismutases*, Michelson A.M., McCord J.M., Fridovich I., ed., Academic Press, London, 1977, p. 11–17.

Maksimov E.G., Schmitt F.-J., Hätti P., Klementiev K.E., Paschenko V.Z., Renger G., Rubin A.B. Anomalous temperature dependence of the fluorescence lifetime of phycobiliproteins, *Laser Phys. Lett.*, 2013, Vol. 10, p. 055602.

Maksimov E.G., Schmitt F.-J., Shirshin E.A., Svirin M.D., Elanskaya I.V., Friedrich T., Fadeev V.V., Paschenko V.Z., Rubin A.B. The time course of non-photochemical quenching in phycobilisomes of *Synechocystis* sp. PCC6803 as revealed by picosecond time-resolved fluorimetry, *Biochim. Biophys. Acta*, 2014a, Vol. 1837, p. 1540–1547.

Maksimov E.G., Schmitt F.-J., Tsoraev G.V., Ryabova A.V., Friedrich T., Paschenko V.Z. Fluorescence quenching in the lichen *Peltigera aphthosa* due to desiccation, *Plant Phys. Biochem.*, 2014b, Vol. 81, p. 67–73.

Maksimov E.G., Shirshin E.A., Sluchanko N.N., Zlenko D.V., Parshina E.Y., Tsoraev G.V., Klementiev K.E., Budylin G.S., Schmitt F.-J., Friedrich T., Fadeev V.V., Paschenko V.Z., Rubin A.B. The signaling state of orange carotenoid protein, *Biophys. J.*, 2015, Vol. 109, p. 595–607.

Maksimov E.G., Moldenhauer M., Shirshin E.A., Parshina E.A., Sluchanko N.N., Klementiev K.E., Tsoraev G.V., Tavraz N.N., Willoweit M., Schmitt F.-J., Breitenbach J., Sandmann G., Paschenko V.Z., Friedrich T., Rubin A.B. A comparative study of three signaling forms of the orange carotenoid protein. *Photosynth. Res.*, 2016, Vol, 130, p. 389–401.

Maly F.E., Nakamura M., Gauchat J.F., Urwyler A., Walker C., Dahinden C.A., Cross A.R., Jones O.T., de Weck A.L. Superoxide-dependent nitroblue tetrazolium reduction and expression of cytochrome b-245 components by human tonsillar B lymphocytes and B cell lines, *J. Immunol.*, 1989, Vol. 142, p. 1260–1267.

Mamedov M., Govindjee, Nadtochenko V., Semenov A., Primary electron transfer processes in photosynthetic reaction centers from oxygenic organisms, *Photosynth. Res.*, 2015, Vol. 125, p. 51–63; DOI 10.1007/s11120-015-0088-y.

Märk J., Theiss C ., Schmitt F.-J., Laufer J. Experimental validation of a theoretical model of dual wavelength photoacoustic (PA) excitation in fluorophores, *SPIE BiOS*, 2015a, 93233G-93233G-10.

Märk J., Schmitt F.-J., Theiss C ., Dortay H., Friedrich T., Laufer J. Photoacoustic imaging of fluorophores using pump-probe excitation, *Biomedical Optics Express*, 2015b, Vol. 6, p. 2522–2535.

Marquardt J., Senger H., Miyashita H., Miyachi S.,Mörschel E. Isolation and characterization of biliprotein aggregates from *Acaryochloris marina*, a Prochloron-like prokaryote containing mainly chlorophyll *d*, *FEBS Lett.*, 1997, Vol. 410, p. 428.

Maruta T., Noshi M., Tanouchi A., Tamoi M., Yabuta Y., Yoshimura K., Ishikawa T., Shigeoka S. H_2O_2-triggered retrograde signaling from chloroplasts to nucleus plays specific role in response to stress, *J. Biol. Chem.*, 2012, Vol. 287, p. 11717–11729.

Mateo A., Funck D., Mühlenbock P., Kular B., Mullineaux P.M., Karpinski S. Controlled levels of salicylic acid are required for optimal photosynthesis and redox homeostasis, *J. Exp. Bot.*, 2006, Vol. 57, p. 1795–1807.

Mattila H., Khorobrykh S., Havurinne V., Tyystjärvi E, Reactive oxygen species: Reactions and detection from photosynthetic tissues, *J. Photochem. Photobiol. B*, 2015, Vol. 152, p. 176–214; DOI: 10.1016/j.jphotobiol.2015.10.001.

Maulucci G., Labate V., Mele M., Panieri E., Arcovito G., Galeotti T., Østergaard H., Winther J.R., De Spirito M., Pani G., High-resolution imaging of redox signaling in live cells through an oxidation-sensitive yellow fluorescent protein, *Sci. Signal.*, 2008, Vol. 1, p. l3.

Mertenskötter A., Keshet A., Gerke P., Paul R.J. The p38 MAPK PMK-1 shows heat-induced nuclear translocation, supports chaperone expression, and affects the heat tolerance of *Caenorhabditis elegans*, *Cell Stress Chaperones*, 2013, Vol. 18, p. 293-306; DOI: 10.1007/s12192-012-0382-y.

Messiah A. *Quantenmechanik I*, 2. Auflage, de Gruyter, Berlin, 1991a.

Messiah A. *Quantenmechanik II*, 2. Auflage, de Gruyter, Berlin, 1991b.

Meyer A.J., Dick T.P. Fluorescent protein-based redox probes, *Antioxid. Redox Signal.*, 2010, Vol. 13, p. 621–650.

Michelet L., Krieger-Liszkay A. Reactive oxygen intermediates produced by photosynthetic electron transport are enhanced in short-day grown plants, *Biochim. Biophys. Acta*, 2012, Vol. 1817, p. 1306–1313.

Miesenböck G., De Angelis D.A., Rothman J.E. Visualizing secretion and synaptic transmission with pH-sensitive green fluorescent proteins, *Nature*, 1998, Vol. 394, p. 192–195.

Miller G., Nobuhiro S., Rizhsky L., Hegie A., Koussevitzky S., Mittler R. Double mutants deficient in cytosolic and thylakoid ascorbate peroxidase reveal a complex mode of interaction between reactive oxygen species, plant development, and response to abiotic stresses, *Plant Physiol.*, 2007, Vol. 144, p. 1777–1785.

Miller G., Schlauch K., Tam R., Cortes D., Torres M.A., Shulaev V., Dangl J.L., Mittler R. The plant NADPH oxidase RBOHD mediates rapid systemic signaling in response to diverse stimuli, *Sci. Signal*, 2009, Vol. 2, ra45.

Mimuro M., Füglistaller P., Rümbeli R., Zuber H. Functional assignment of chromophores and energy transfer in C phycocyanin isolated from the thermophilic cyanobacterium *Mastigocladus laminosus, Biochim. Biophys. Acta*, 1986, Vol. 848, p. 155.

Mimuro M., Kobayashi M., Murakami A., Tsuchiya T., Miyashita H. Oxygen-evolving cyanobacteria, *Primary Processes of Photosynthesis: Basic Principles and Apparatus*, Vol. I, Renger G. ed., RSC Publ., Cambridge, U.K., 2008, p. 261–300.

Minagawa J. Live-cell imaging of photosystem II antenna dissociation during state transitions, *Seminar Talk in the Seminar for Optics and Photonics*, 14.5.2010, Institute of Optics and Atomic Physics, TU Berlin, 2010.

Minibayeva F., Kolesnikov O.P., Gordon L.K. Contribution of a plasma membrane redox system to the superoxide production by wheat root cells, *Protoplasma*, 1998, Vol. 205, p. 101–106.

Minibayeva F., Gordon L.K. Superoxide production and activity of outcellular peroxidase under stress, *Russ. J. Plant Physiol.*, 2003, Vol. 50, p. 459–464.

Minibayeva F., Kolesnikov O., Chasov A., Beckett R.P., Luthje S., Vylegzhanina N., Buck F., Bottger M. Wound-induced apoplastic peroxidase activities: their roles in the production and detoxification of reactive oxygen species, *Plant Cell. Environ.*, 2009, Vol. 32, p. 497–508.

Mishra N.P., Francke C., van Gorkom H.J., Ghanotakis D.F., Destructive role of singlet oxygen during aerobic illumination of the photosystem II core complex, *Biochim. Biophys. Acta*, 1994, Vol. 1186, p. 81–90.

Mittler R., Vanderauwera S., Suzuki N., Miller G., Tognetti V.B., Vandepoele K., Gollery M., Shulaev V., Van Breusegem F. ROS signaling: the new wave?, *Trends in Plant Science*, 2011, Vol. 16, p. 300–309.

Miyashita H., Ikemoto H., Kurano N., Adachi K., Chihara M., Miyachi S. Chlorophyll *d* as a major pigment, *Nature (London)*, 1996, Vol. 383, p. 402.

Miyashita H., Adachi K., Kurano N., Ikemoto H., Chihara M., Miyachi S. Pigment composition of a novel oxygenic photosynthetic prokaryote containing chlorophyll *d* as the major chlorophyll, *Plant Cell Physiol.*, 1997, Vol. 38, p. 274.

Moan J., On the diffusion length of singlet oxygen in cells and tissues, *J. Photochem. Photobiol. B*, 1990, Vol. 6, p. 343–344.

Mongkolsuk S., Helmann J.D., Regulation of inducible peroxide stress responses, *Mol. Microbiol.*, 2002, Vol. 45, p. 9–15.

Mori I.C., Schroeder J.I., Reactive oxygen species activation of plant Ca^{2+} channels: a signaling mechanism in polar growth, hormone transduction, stress signaling, and hypothetically mechanotransduction, *Plant Physiol.*, 2004, Vol. 135, p. 702–708.

Morosinotto T., Baronio R., Bassi R. Dynamics of chromophore binding to Lhc proteins *in vivo* and *in vitro* during operation of the xantophyll cycle, *Biol.Chem.*, 2002, Vol. 277, p. 36913.

Mubarakshina M.M., Ivanov B.N., Naidov I.A., Hillier W., Badger M.R., Krieger-Liszkay A. Production and diffusion of chloroplastic H_2O_2 and its implication to signaling, *J. Exp. Bot.*, 2010, Vol. 61, p. 3577–3587.

Müh F., El-Amine Madjet M., Adolphs J., Abdurahman A., Rabenstein B., Ishikita H., Knapp E.-W., Renger T. α-Helices direct excitation energy flow in the Fenna–Matthews–Olson protein, *Proc. Natl. Acad. Sci. U.S.A.*, 2007, Vol. 104, p. 16862.

Mukamel S. *Principles of nonlinear optical spectroscopy*, Oxford University Press, New York, 1995

Müller I. *Entropy and Energy*, Springer, Berlin, 2005.

Müller P., Li X.P., Niyogi K.K., Non-photochemical quenching. A response to excess light energy, *Plant Physiol.*, 2001, Vol. 125, p. 1558–1566.

Mullineaux C.W., Holzwarth A.R. Kinetics of excitation energy transfer in the cyanobacterial phycobilisome-photosystem II complex, *Biochim. Biophys. Acta*, 1991, Vol. 1098, p. 68–78.

Mullineaux P.M., Ball L., Escobar C., Karpinska B., Creissen G., Karpinski S. Are diverse signaling pathways integrated in the regulation of arabidopsis antioxidant defence gene expression in response to excess excitation energy?, *Phil. Transact. R. Soc. London*, 2000, Vol. 355, p. 1531–1540.

Mullineaux P.M., Rausch T. Glutathione, photosynthesis and the redox regulation of stress-responsive gene expression, *Photosynth. Res.*, 2005, Vol. 86, p. 459–474.

Mullineaux P.M., Karpinski S., Baker N.R.. Spatial dependence for hydrogen peroxide-directed signaling in light-stressed plants, *Plant Physiol.*, 2006, Vol. 141, p. 346–350.

Mullineaux P.M., Lawson T. Measuring redox changes *in vivo* in leaves: prospects and technical challenges, *Methods Mol. Biol.*, 2009, Vol. 476, p. 65–75.

Mullineaux P.M., Baker N.R., Oxidative stress: antagonistic signaling for acclimation or cell death?, *Plant Physiol.*, 2010, Vol. 154, p. 2521–2525.

Munnik T., Irvine R.F., Musgrave A. Phospholipid signaling in plants, *Biochim. Biophys. Acta*, 1998, Vol. 1389, p. 222–272.

Murata N., Takahashi S., Nishiyama Y., Allakhverdiev S.I., Photoinhibition of photosystem II under environmental stress, *Biochim. Biophys. Acta*, 2007, Vol. 1767, p. 414–421.

Murata N., Allakhverdiev S.I., Nishiyama Y. The mechanism of photoinhibition *in vivo*: Re-evaluation of the roles of catalase, α-tocopherol, non-photochemical quenching, and electron transport, *Biochim. Biophys. Acta*, 2012, Vol. 1817, p. 1127–1133.

Murray J.W., Maghlaoui K., Barber J. The structure of allophycocyanin from *Thermosynechococcus* elongatus at 3.5 A resolution, *Acta Crystallogr., Sect.* F, 2007, Vol. 63, p. 998.

Najafpour M.M., Pashaei B., Zand Z. Photodamage of the manganese-calcium oxide: a model for UV-induced photodamage of the water oxidizing complex in photosystem II, *Dalton Trans.*, 2013, Vol. 42, p. 4772–4776.

Najafpour M.M., Ghobadi M.Z., Larkum A.W., Shen J-R., Allakhverdiev S.I. The biological water-oxidizing complex at the nano-bio interface, *Trends in Plant Science,* 2015, Vol. 20(9), p. 559–568.

Najafpour M.M., Renger G., Hołyńska M., Moghaddam A.N., Aro E.M., Carpentier R., Nishihara H., Eaton-Rye J.J., Shen J-R., Allakhverdiev S.I. Manganese compounds as water-oxidizing catalysts: from the natural water-oxidizing complex to nanosized manganese oxide structures, *Chem Rev.,* 2016, Vol. 116(5), p. 2886–2936.

Nazarenko L.V., Andreev I.M., Lyukevich A.A., Pisareva T.V., Los D.A. Calcium release from *Synechocystis* cells induced by depolarization of the plasma membrane: MscL as an outward Ca^{2+} channel, *Microbiology,* 2003, Vol. 149, p. 1147–1153.

Neill S.J., Desikan R., Clarke A., Hurst R.D., Hancock J.T. Hydrogen peroxide and nitric oxide as signaling molecules in plants, *J. Exp. Bot.*, 2002, Vol. 53, p. 1237–1247.

Neubauer C., Schreiber U. The polyphasic rise of chlorophyll fluorescence upon onset of strong continuous illumination: I. Saturation characteristics and partial control by the photosystem II acceptor side, *Z. Naturforsch. C*, 1987, Vol. 42, p. 1246–1254.

Nieuwenburg P., Clarke R.J., Cai Z.L., Chen M., Larkum A.W., Cabral N.M., Ghiggino K.P., Reimers J.R. Examination of the photophysical processes of chlorophyll d leading to a clarification of proposed uphill energy transfer processes in Cells of *Acaryochloris marina, Photochem. Photobiol.*, 2003, Vol. 77, p. 628.

Nicholls D.G., Ferguson S.J., *Bioenergetics,* Academic Press, London, 2013, 434 p.

Nield J., Rizkallah P.J., Barber J., Chayen N.E. The 1.45A three-dimensional structure of C-phycocyanin from the thermophilic cyanobacterium *Synechococcus elongates, J. Struct. Biol.*, 2003, Vol. 141, p. 149.

Niethammer P., Grabher C., Look A.T., Mitchison T.J. A tissue-scale gradient of hydrogen peroxide mediates rapid wound detection in zebrafish, *Nature,* 2009, Vol. 459, p. 996–999.

Nishimura M.T., Dangl J.L. *Arabidopsis* and the plant immune system, *Plant J.*, 2010, Vol. 61, p. 1053–1066.

Nishiyama Y., Yamamoto H., Allakhverdiev S.I., Inaba M., Yokota A., Murata N. Oxidative stress inhibits the repair of photodamage to the photosynthetic machinery, *EMBO J.*, 2001, Vol. 20, p. 5587–5594.

Nishiyama Y., Allakhverdiev S.I., Yamamoto H., Hayashi H., Murata N. Singlet oxygen inhibits the repair of photosystemII by suppressing the translation elongation of the D1 protein in *Synechocystis* sp. PCC 6803, *Biochemistry,* 2004, Vol. 43, p. 11321–11330.

Nishiyama Y., Allakhverdiev S.I., Murata N. A new paradigm for the action of reactive oxygen species in the photoinhibition of photosystem II, *Biochim. Biophys. Acta,* 2006, Vol. 1757, p. 742–749.

Nishiyama Y., Allakhverdiev S.I., Murata N. Protein synthesis is the primary target of reactive oxygen species in the photoinhibition of photosystem II, *Physiol. Plant.*, 2011, Vol. 142, p. 35–46.

Nixon P.J., Michoux F., Yu J., Boehm M., Komenda J. Recent advances in understanding the assembly and repair of photosystem II, *Ann Bot.*, 2010, Vol. 106, p. 1–16.

Niyogi K.K., Truong T.B. Evolution of flexible non-photochemical quenching mechanisms that regulate light-harvesting in oxygenic photosynthesis, *Curr. Opin. Plant Biol.*, 2013, Vol. 16, p. 307–314.

Noctor G., Arisi A-C., Jouanin L., Kunert K.J., Rennenberg H., Foyer C.H. Glutathione: biosynthesis, metabolism and relationship to stress tolerance explored in transformed plants, *J. Exp. Bot.*, 1998, Vol. 49, p. 623–647.

Noctor G., Foyer C.H., Ascorbate and glutathione: Keeping active oxygen under control, *Annu. Rev. Plant. Physiol. Plant. Mol. Biol.*, 1998, Vol. 49, p. 249–279.

Noctor G., Gomez L., Vanacker H., Foyer C.H. Interactions between biosynthesis, compartmentation and transport in the control of glutathione homeostasis and signalling, *J. Exp. Bot.*, 2002, Vol. 53, p. 1283–1304.

Ochsenbein C., Przybyla D., Danon A., Landgraf F., Göbel C., Imboden A., Feussner I., Apel K. The role of EDS1 (enhanced disease susceptibility) during oxygen-mediated stress responses of *Arabidopsis*, *Plant J.*, 2006, Vol. 47, p. 445–456.

Oelze M.-L., Kandlbinder A., Dietz K.-J. Redox regulation and overreduction control in the photosynthesizing cell: Complexity in redox regulatory networks, *Biochim. Biophys. Acta*, 2008, Vol. 1780, p. 1261–1272.

Ogilby P.R., Foote C.S. Chemistry of singlet oxygen. Effect of solvent, solvent isotopic substitution and temperature on the lifetime of singlet molecular oxygen ($^1\Delta g$), *J. Am. Chem. Soc.*, 1983, Vol. 105, p. 3423–3430.

Ohnishi N., Allakhverdiev S.I., Takahashi S., Higashi S., Watanabe M., Nishiyama Y., Murata N. Two-step mechanism of photodamage to photosystem II: Step 1 occurs at the oxygen-evolving complex and Step 2 occurs at the photochemical reaction center, *Biochemistry*, 2005, Vol. 44, p. 8494–8499.

Op den Camp R., Przybyla D., Ochsenbein C., Laloi C., Kim C., Danon A., Wagner D., Hideg E., Göbel C., Feussner I., Nater M., Apel K. Rapid induction of distinct stress responses after the release of singlet oxygen in *Arabidopsis*, *Plant Cell*, 2003, Vol. 15, p. 2320–2332.

Ostergaard H., Henriksen A., Flemming H., Hansen G., Winther J.R. Shedding light on disulfide bond formation: engineering a redox switch in green fluorescent protein, *EMBO J.*, 2001, Vol. 20, p. 5853–5862.

Overmyer K., Brosché M., Kangasjärvi J. Reactive oxygen species and hormonal control of cell death, *Trends Plant Cell Sci.*, 2003, Vol. 8, p. 335–342.

Pap E.H., Drummen G.P., Winter V.J., Kooij T.W., Rijken P., Wirtz K.W., Op den Kamp J.A., Hage W.J., Post J.A. Ratio-fluorescence microscopy of lipid oxidation in living cells using C11-BODIPY(581/591), *FEBS Lett.*, 1999, Vol. 453, p. 278–282.

Papageorgiou G.C., Govindjee, *Chlorophyll a Fluorescence: A Signature of Photosynthesis*, Series: Advances in Photosynthesis and Respiration, Springer, Dordrecht, 2005, 820 p.

Parks B.M., Quail P.H. Phytochrome-deficient hy1 and hy2 long hypocotyl mutants of *Arabidopsis* are defective in phytochrome chromophore biosynthesis, *Plant Cell*, 1991, Vol. 3, p. 1177–1186.

Parson W. Functional patterns of reaction centers in anoxygenic photosynthetic bacteria, *Primary Processes of Photosynthesis: Basic Principles and Apparatus*, Vol. II, Renger G. ed., RSC Publ., Cambridge, U.K., 2008, p. 57–109.

Peschek G.A. Electron transport in oxygenic cyanobacteria, *Primary Processes of Photosynthesis: Principles and Applications*, Vol. II, Renger G. ed., RSC Publishing, Cambridge, 2008, p. 383–415.

Petrášek Z., Schmitt F.-J., Theiss C., Huyer J., Chen M., Larkum A., Eichler H.J., Kemnitz K., Eckert H.-J. Excitation energy transfer from phycobiliprotein to chlorophyll *d* in intact cells of *Acaryochloris marina* studied by time- and wavelength resolved fluorescence spectroscopy, *Photochem. Photobiol. Sci.*, 2005, Vol. 4, p. 1016–1022.

Pfannschmidt T., Schütze K., Fey V., Sherameti I., Oelmüller R. Chloroplast redox control of nuclear gene expression. A new class of plastid signals in interorganellar communication, *Antioxid. Redox Signal.*, 2003, Vol. 5, p. 95–101.

Pfannschmidt T., Bräutigam K., Wagner R., Dietzel L., Schröter Y., Steiner S., Nykytenko A. Potential regulation of gene expression in photosynthetic cells by redox and energy state: approaches towards better understanding, *Ann. Bot.*, 2009, Vol. 103, p. 599–607.

Pieper J., Ratsep M., Trostmann I., Schmitt F.-J., Theiss C., Paulsen H., Freiberg A., Renger G., Eichler H.J. Excitonic energy level structure and pigment-protein interactions in the recombinant water-soluble chlorophyll protein (WSCP). Part II: Spectral hole-burning experiments, *J Phys Chem B.*, 2011, Vol. 115, p. 4053–4065.

Pitzschke P., Hirt H. Mitogen-activated protein kinases and reactive oxygen species signaling in plants, *Plant Physiol.*, 2006, Vol. 141, p. 351–356.

Pogson B.J., Rissler H.M., Frank H.A., The role of carotenoids in energy quenching, *The Light-Driven Water: Plastoquinone Oxidoreductase*, Wyrdzynski T., Satoh K. eds., Springer, Dodrecht, 2005, p. 515–537.

Pogson B.J., Woo, B. Förster, I.D. Small, Plastid signaling to the nucleus and beyond, *Trends Plant Sci.*, 2008, Vol. 13, p. 602–609.

Pohanka M., Skladai P. Electrochemical biosensors – principles and applications, *J. Appl. Biomed.*, 2008, Vol. 6, p. 57–64.

Polívka T., Sundström V., Ultrafast dynamics of carotenoid excited states – from solution to natural and artificial systems, *Chem. Rev.*, 2004, Vol. 104, p. 2021–2072.

Pospisil P. Production of reactive oxygen species by photosystem II, *Biochim. Biophys. Acta*, 2009, Vol. 1787, p. 1151–1160.

Pradedova E.V., Isheeva O.D., Salyaev R.K. Classification of the antioxidant defense system as the ground for reasonable organization of experimental studies of the oxidative stress in plants, *Russ. J. Plant Physiol.*, 2011, Vol. 58, p. 210–217.

Prasad A., Kumar A., Suzuki M., Kikuchi H., Sugai T., Kobayashi M., Pospíšil P., Tada M., Kasai S. Detection of hydrogen peroxide in photosystem II (PSII) using catalytic amperometric biosensor, *Front. Plant Sci. (Methods)*, 2015, Vol. 6, Article 862; http://dx.doi.org/10.3389/fpls.2015.00862.

Przybyla D., Göbel C., Imboden A., Hamberg M., Feussner I., Apel K. Enzymatic, but not non-enzymatic, $^1O_2$-mediated peroxidation of polyunsaturated fatty acids forms part of the EXECUTER1-dependent stress response program in the *flu* mutant of *Arabidopsis thaliana, Plant J.*, 2008, Vol. 54, p. 236–248.

Qi Z., Yue M., Wang X.L. Laser pretreatment protects cells of broad bean from UV-B radiation damage, *J. Photochem. Photobiol. B:Biol*, 2000, Vol. 59, p. 33–37.

Qi Z., Yue M., Han R., Wang X.L. The damage repair role of He-Ne laser on plants exposed to different intensities of ultraviolet-B radiation, *Photochem. Photobiol.*, 2002, Vol. 75, p. 680–686.

Ramel F., Birtic S., Ginies C., Soubigou-Taconnat L., Triantaphylidès C., Havaux M. Carotenoid oxidation products are stress signals that mediate gene responses to singlet oxygen in plants, *Proc. Nat. Acad. Sci. U.S.A.*, 2012, Vol. 109, p. 5535–5540.

Rao A.Q., Irfan M., Saleem Z., Nasir I.A., Riazuddin S., Husnain T. Overexpression of the phytochrome B gene from *Arabidopsis thaliana* increases plant growth and yield of cotton (*Gossypium hirsutum*), *J. Zhejiang Univ. Sci. B*, 2011, Vol. 12, p. 326–334.

Rappaport F., Guergova-Kuras M., Nixon P. J., Diner B. A., Lavergne J. Kinetics and pathways of charge recombination in photosystem II, *Biochemistry*, 2002, Vol. 41, p. 8518–8527.

Rehman A.U., Cser K., Sass L., Vass I. Characterization of singlet oxygen production and its involvement in photodamage of Photosystem II in the cyanobacterium *Synechocystis* PCC 6803 by histidine-mediated chemical trapping, *Biochim. Biophys. Acta*, 2013, Vol. 1827, p. 689–698.

Renger G. Biological energy conservation, *Biophysics*, Lohmann W., Markl H., Ziegler H., Hoppe W. eds., Springer, Berlin, 1983, p. 347–371.

Renger G. Energy transfer and trapping in photosystem II, *Topics in Photosynthesis, The Photosystems: Structure, Function and Molecular Biology*, Barber J. ed., Elsevier Science Publishers B.V., 1992, p. 45–92.

Renger G. Photosynthetic water oxidation to molecular oxygen: apparatus and mechanism, *Biochim. Biophys. Acta*, 2001, Vol. 1503, p. 210–228.

Renger G. Oxidative photosynthetic water splitting: energetics, kinetics and mechanism, *Photosynth. Res.*, 2007, Vol. 92, p. 407–425.

Renger G. Functional pattern of photosystem II, *Primary Processes of Photosynthesis: Principles and Apparatus, Part II: Reaction Centers/Photosystems, Electron Transport Chains, Photophosphorylation and Evolution*, Renger G. ed., Royal Society Chemistry, Cambridge, 2008, p. 237–290.

Renger G. *Primary Processes of Photosynthesis – Part I: Principles and Apparatus*, RSC Publ., Cambridge, U.K., 2008b.

Renger G. Photosynthetic water splitting: apparatus and mechanism, *Photosynthesis: Plastid Biology, Energy Conversion and Carbon Assimilation*, Eaton-Rye J.J., Tripathy B.C., Sharkey T.D. eds., Springer, Dordrecht, 2012, p. 359–414.

Renger G., Eckert H.J., Bergmann A., Bernarding J., Liu B., Napiwotzki A., Reifarth F., Eichler H.J. Fluorescence and spectroscopic studies on exciton trapping and electron transfer in photosystem II of higher plants, *Aust. J. Plant Physiol.*, 1995, Vol. 22, p. 167–181.

Renger G., Holzwarth A.R. Primary electron transfer, *Photosystem II: The Water:Plastoquinone Oxido-Reductase in Photosynthesis*, Wydrzynski T., Satoh K. eds., Springer, Dordrecht, 2005, p. 139–175.

Renger T., Holzwarth A.R. Theory of excitation energy transfer and optical spectra of photosynthetic systems, *Biophysical Techniques in Photosynthesis*, Matysik J., Aartsma T., ed., Springer Dodrecht, 2007

Renger T., Kühn O. *Molekulare Energietrichter*, Wissenschaftsmagazin fundiert, Energie, 1/2007, FU Berlin ISSN: 1616-524, 2007.

Renger G., Ludwig B. Mechanism of photosynthetic production and respiratory reduction of molecular dioxygen: A biophysical and biochemical comparison, *The Bioenergetic Processes of Cyanobacteria-from Evolutionary Singularity to Ecological Diversity*, Peschek G., Obinger C., Renger G. eds., Springer, Dordrecht, The Netherlands, 2011, p. 337–394.

Renger T., Madjet M.E., Müh F., Trostmann I., Schmitt F.-J., Theiss C., Paulsen H., Eichler H.J., Knorr A., Renger G. Thermally activated superradiance and intersystem crossing in the water-soluble chlorophyll binding protein, *J. Phys. Chem. B*, 2009, Vol. 113, p. 9948–9957.

Renger G., Pieper J., Theiss C., Trostmann I., Paulsen H., Renger T., Eichler H.J., Schmitt F.-J. Water soluble chlorophyll binding protein of higher plants: a most suitable model system for basic analyses of pigment-pigment and pigment-protein interactions in chlorophyll protein complexes, *J. Plant Physiol.*, 2011, Vol. 168, p. 1462–1472.

Renger G., Renger T. Photosystem II: The machinery of photosynthetic water splitting, *Photosynth. Res.*, 2008, Vol. 98, p. 53–80.

Renger G., Schulze A. Quantitative analysis of fluorescence induction curves in isolated spinach chloroplasts, *Photobiochemistry and Photobiophysics*, 1985, Vol. 9, p. 79–87.

Renger T., Schlodder E. Modeling of optical spectra and light-harvesting in photosystem I, *Photosystem I: The Light Driven Plastocyanin: Ferredoxin Oxidoreductase*, Series: Advances in Photosynthesis and Respiration, Kluwer Academic Publishers, 2005, p. 5595–5610.

Richter M., Renger T., Knorr A. A Bloch equation approach to intensity dependent optical spectra of light-harvesting complex II: excitation dependence of light-harvesting complex II pump-probe spectra, *Photosynth. Res.*, 2008, Vol. 95, p. 119.

Rigo A., Stevanato R., Finazzi-Agro A., Rotilio G. An attempt to evaluate the rate of the Haber-Weiss reaction by using OH radical scavengers, *FEBS Lett.*, 1977, Vol. 80, p. 130–132.

Ritz T., Park S., Schulten K. Kinetics of excitation migration and trapping in the photosynthetic unit of purple bacteria, *J. Phys. Chem. B*, 2001, Vol. 105, p. 8259.

Rizhsky L., Liang H., Mittler R., The water-water cycle is essential for chloroplast protection in the absence of stress, *J. Biol. Chem.*, 2003, Vol. 278, p. 38921–38925.

Roelofs T.A., Lee C.H., Holzwarth A.R. Global target analysis of picosecond chlorophyll fluorescence kinetic from pea chloroplasts, *Biophys. J.*, 1992, Vol. 61, p. 1147–1163.

Roháček K. Chlorophyll fluorescence parameters: the definitions, photosynthetic meaning, and mutual relationships, *Photosynthetica*, 2002, Vol. 40, p. 13–29.

Rossel J.B., Wilson P.B., Hussain D., Woo N.S., Gordon M.J., Mewett O.P., Howell K.A., Whelan J., Kazan K., Pogson B.J. Systemic and intracellular responses to photooxidative stress in *Arabidopsis, Plant Cell*, 2007, Vol. 19, p. 4091–4110.

Ruban A.V., Johnson M.P., Duffy C.D.P. The photoprotective molecular switch in the photosystem II antenna, *Biochim. Biophys. Acta*, 2012, Vol. 1817, p. 167–181.

Rutherford A.W., Osyczka A., Rappaport F., Back-reactions, short-circuits, leaks and other energy wasteful reactions in biological electron transfer: Redox tuning to survive life in O_2, *FEBS Lett.*, 2012, Vol. 586, p. 603–616.

Sagi M., Fluhr R. Production of reactive oxygen species by plant NADPH oxidases, *Plant Physiol.*, 2006, Vol. 141, p. 336–340.

Sairam R.K., Srivastava G.C. Increased antioxidant activity under elevated temperatures: a mechanism of heat stress tolerance in wheat genotypes, *Biol. Plant.*, 2000, Vol. 43, p. 381–386.

Sang Y., Cui D., Wang X. Phospholipase D and phosphatidic acid-mediated generation of superoxide in *Arabidopsis, Plant Physiol.*, 2001, Vol. 126, p. 1449–1458.

Sauer K., Scheer H. Excitation transfer in C-phycocyanin. Förster transfer rate and exciton calculations based on new crystal struture data for C-phycocyanins from *Agmenellum quadruplicatum* and *Mastigocladus laminosus*, Biochim. Biophys. Acta, 1988, Vol, 936, p. 157.

Scarpeci T.E., Zanor M.I., Carrillo N., Mueller-Roeber B., Valle E.M. Generation of superoxide anion in chloroplasts of *Arabidopsis thaliana* during active photosynthesis: A focus on rapidly induced genes, *Plant Mol. Biol.*, 2008, Vol. 66, p. 361–378.

Schatz G., Brock H., Holzwarth A.R. Kinetic and energetic model for the primary processes in photosystem II, *Biophys. J.*, 1988, Vol. 54, p. 397–405.

Scheibe R., Backhausen J.E., Emmerlich V., Holtgrefe S. Strategies to maintain redox homeostasis during photosynthesis under changing conditions, *J. Exp. Bot.*, 2005, Vol. 56, p. 1481–1489.

Scheuring S., Reiss-Husson F., Engel A., Rigaud J.-L., Ranck J.-L. High-resolution AFM topographs of *Rubrivivax gelatinosus* light-harvesting complex LH2, *The EMBO J.*, 2001, Vol. 20, p. 3029.

Schlodder E., cetin M., Eckert H.-J., Schmitt F.-J., Barber J., Telfer A. Both chlorophylls *a* and *d* are essential for the photochemistry in photosystem II of the cyanobacterium *Acaryochloris marina*, *Biochim. Biophys. Acta*, 2007, Vol. 1767, p. 589–595.

Schmitt F.-J. Temperature induced conformational changes in hybrid complexes formed from CdSe/ZnS nanocrystals and the phycobiliprotein antenna of *Acaryochloris marina*, *J. Opt.*, 2010, Vol. 12, 084008; doi: 10.1088/2040-8978/12/8/084008.

Schmitt F.-J., Junghans C., Sturm M., Csongor K., Eichler H.J., Friedrich T. Laser switching contrast microscopy to monitor free and restricted diffusion inside the cell nucleus, *Optofluidics, Microfluidics and Nanofluidics*, 2, 2016 (accepted for publication).

Schmitt F.-J., Maksimov E., Junghans C., Weißenborn J., Hätti P., Allakhverdiev S.I., Friedrich T. Structural organization and dynamic processes in protein complexes determined by multiparameter imaging, *Signpost Open Access J. NanoPhotoBioSciences*, 2013, Vol. 1, p. 1–45.

Schmitt F.-J., Maksimov E.G., Hätti P., Weißenborn J., Jeyasangar V., Razjivin A.P., Paschenko V.Z., Friedrich T., Renger G. Coupling of different isolated photosynthetic light-harvesting complexes and CdSe/ZnS nanocrystals via Förster resonance energy transfer, *Biochim. Biophys. Acta*, 2012 Vol. 1817, p. 1461–70.

Schmitt F.-J., Maksimov E.G., Suedmeyer H., Jeyasangar V., Theiss C., Paschenko V.Z., Eichler H.J., Renger G. Time resolved temperature switchable excitation energy transfer processes between CdSe/ZnS nanocrystals and phycobiliprotein antenna from *Acaryochloris marina*, *Photon. Nanostruct.: Fundam. Appl.*, 2011, Vol. 9, p. 190–195.

Schmitt F.-J., *Picobiophotonics for the Investigation of Pigment-Pigment and Pigment-Protein Interactions in Photosynthetic Complexes*, PhD thesis, Berlin, 2011, 230 p.

Schmitt F.-J., Renger G., Friedrich T., Kreslavski V.D., Zharmukhamedov S.K., Los D.A., Kuznetsov V.V., Allakhverdiev S.I. Reactive oxygen species: Re-evaluation of generation, monitoring and role in stress-signaling in phototrophic organisms, *Biochim. Biophys. Acta*, 2014a, Vol. 1837, p. 835–848.

Schmitt F.-J., Thaa B., Junghans C., Vitali M., Veit M., Friedrich T., eGFP-pHsens as a highly sensitive fluorophore for cellular pH determination by fluorescence lifetime imaging microscopy (FLIM), *Biochim. Biophys. Acta*, 2014b, Vol. 1837, p. 1581–1593.

Schmitt F.-J.Theiss C., Wache K., Fuesers J., Andree S., Eichler H.J., Eckert H.-J. Investigation of metabolic changes in living cells of the Chl *d*- containing cyanobacterium *Acaryochloris marina* by time- and wavelength correlated single-photon counting, *ZiG-Print*, Zentrum für innovative Gesundheitstechnologien der TU Berlin, 2007, 6 p.

Schmitt F.-J., Theiss C., Wache K., Fuesers J., Andree S., Handojo A., Karradt A., Kiekebusch D., Eichler H.J., Eckert H.-J. Investigation of the excited states dynamics in the Chl *d*- containing cyanobacterium *Acaryochloris marina* by time- and wavelength correlated single-photon counting, *Proc. SPIE*, 2006, Vol. 6386, 6 p; doi:10.1117/12.689127.

Schmitt F.-J., Trostmann I., Theiss C., Pieper J., Renger T., Fuesers J., Hubrich E.H., Paulsen H., Eichler H.J., Renger G. Excited state dynamics in recombinant water-soluble chlorophyll proteins (WSCP) from cauliflower investigated by transient fluorescence spectroscopy, *J. Phys. Chem. B*, 2008, Vol. 112, p. 13951–13961.

Schreiber U., Schliwa U., Bilger W. Continuous recording of photochemical and nonphotochemical fluorescence quenching with a new type of modulation fluorometre, *Photosynth. Res.*, 1986, Vol. 10, p. 51–62.

Schreiber U., Krieger A. Two fundamentally different types of variable chlorophyll fluorescence *in vivo*, *FEBS Lett.*, 1996, Vol. 397, p. 131–135.

Schulten K. From simplicity to complexity and back: Function, architecture and mechanism of light-harvesting systems in photosynthetic bacteria, *Simplicity and Complexity in Proteins and Nucleic Acids*, H. Frauenfelder, J. Deisenhofer, P.G. Wolynes eds., Dahlem University Press, Berlin, 1999, p. 227–253..

Schürmann P., Buchanan B.B. The ferredoxin/thioredoxin system of oxygenic photosynthesis, *Antioxid. Redox Signal.*, 2008, Vol. 10, p. 1235–1274.

Schwabl F. *Quantenmechanik (QM I): Eine Einführung*, Springer, Berlin, 2007.

Schwabl F. *Quantenmechanik für Fortgeschrittene (QM II)*, Springer, Berlin, 2008.

Schwarzländer M., Fricker M.D., Müller C., Marty L., Brach T., Novak J., Sweetlove L.J., Hell R., Meyer, A.J. Confocal imaging of glutathione redox potential in living plant cells, *J. Microsc.*, 2008, Vol. 279, p. 299–316.

Schwarzländer M., Fricker M.D., Sweetlove L.J. Monitoring the *in vivo* redox state of plant mitochondria: Effect of respiratory inhibitors, abiotic stress and assessment of recovery from oxidative challenge, *Biochim. Biophys. Acta*, 2009, Vol. 1787, p. 468–475.

Schweitzer C., Schmidt R. Physical mechanisms of generation and deactivation of singlet oxygen. *Chem. Rev.*, 2003, Vol. 103, p. 1685–1758.

Setif P., Leibl W. Functional pattern of photosystem I in oxygen evolving organisms, *Primary Processes of Photosynthesis: Basic Principles and Apparatus*, Vol. II, Renger G. ed., RSC Publ., Cambridge, U.K., 2008, p 147–191.

Shaikhali J., Heiber I., Seidel T., Ströher E., Hiltscher H., Birkmann S., Dietz K.-J., Baier M. The redox-sensitive transcription factor Rap2.4a controls nuclear expression of 2-Cys peroxiredoxin A and other chloroplast antioxidant enzymes, *BMC Plant Biol.*, 2008, p. 8–48.

Shao N., Beck C.F., Lemaire S.D., Krieger-Liszkay A. Photosynthetic electron flow affects H_2O_2 signaling by inactivation of catalase in *Chlamydomonas reinhardti, Planta*, 2008, Vol. 228, p. 1055–1066.

Shao N., Krieger-Liszkay A., Schroda M., Beck C.F. A reporter system for the individual detection of hydrogen peroxide and singlet oxygen: its use for the assay of reactive oxygen species produced *in vivo, Plant J.*, 2007, Vol. 50, p. 475–487.

Sharkov A.V., Gulbinas V., Gottschalk L., Scheer H., Gillbro T. Dipole-dipole interaction in phycobiliprotein trimers. Femtosecond dynamics of allophycocyanian. Excited state absorption, *Braz. J. Phys.*, 1996, Vol. 26, p. 553–559.

Sharkov A.V., Kryukov I., Khoroshilov E.V., Kryukov P.G., Fischer R., Scheer H., Gillbro T. Femtosecond energy transfer between chromophores in allophycocyanin trimers, *Chem. Phys. Lett.*, 1992, Vol. 191, p. 633–638.

Sharkov A.V., Kryukov I., Khoroshilov E.V., Kryukov P.G., Fischer R., Scheer H., Gillbro T. Femtosecond spectral and anisotropy study of excitation energy transfer between neighbouring α-80 and β-81 chromophores of allophycocyanin trimers, *Biochim. Biophys. Acta*, 1994, Vol. 1188, p. 349–356.

Sharma P., Jha A.B., Dubey R.S., Pessarakli M. Reactive oxygen species, oxidative damage, and antioxidative defense mechanism in plants under stressful conditions, *J. Bot.*, 2012, Vol. 2012, 26 p.

Sheldrake R. *Das Schöpferische Universum: Die Theorie der Morphogenetischen Felder und der Morphischen Resonanz*, Nymphenburger, 2008.

Shim S.-H., Xia C., Zhong G., Babcock H.P., Vaughan J.C., Huang B., Wang X., Xu C., Bi J.C., Zhuang X. Super-resolution fluorescence imaging of organelles in live cells with photoswitchable membrane probes, *Proc. Nat. Acad. Sci. U.S.A.*, 2012, Vol. 109, p. 13978–13983.

Shinkarev V.P., Govindjee Insight into the relationship of chlorophyll *a* fluorescence yield to the concentration of its natural quenchers in oxygenic photosynthesis, *Proc. Natl. Acad. Sci. U.S.A.*, 1993, Vol. 90, p. 7466–7469.

Shoumskaya M.A., Paithoonrangsarid K., Kanesaki Y., Los D.A., Zinchenko V.V., Tanticharoen M., Suzuki I., Murata N. Identical Hik-Rre systems are involved in perception and transduction of salt signals and hyperosmotic signals but regulate the expression of individual genes to different extents in *Synechocystis, J. Biol. Chem.*, 2005, Vol. 280, p. 21531–21538.

Sicora C., Máté Z., Vass I. The interaction of visible and UV-B light during photodamage and repair of photosystem II, *Photosynth. Res.*, 2003, Vol. 75, p. 127–137.

Sies H., Menck C.F.,Singlet oxygen induced DNA damage, *Mutation Res.*, 1992, Vol. 275, p. 367–375.

Skovsen E., Snyder J.W., Lambert J.D.C., Ogilby P.R., Lifetime and diffusion of singlet oxygen in a cell, *J. Phys. Chem. B.*, 2005, Vol. 109, p. 8570–8573.

Ślesak I., Libik M., Karpinska B., Karpinski S., Miszalski Z. The role of hydrogen peroxide in regulation of plant metabolism and cellular signaling in response to environmental stresses, *Acta Biochimica Polonica*, 2007, Vol. 54, p. 39–50.

Snyder J.W., Zebger I., Gao Z., Poulsen L., Frederiksen P.K., Skovsen E., McIlroy S.P., Klinger M., Andersen L.K., Ogilby P.R. Singlet oxygen microscope: from phase-separated polymers to single biological cells, *Acc. Chem. Res.*, 2004, Vol. 37, p. 894–901.

Spadaro D., Yun B.-W., Spoel S.H., Chu C., Wang Y.-Q., Loake G.J., The redox switch: dynamic regulation of protein function by cysteine modifications, *Physiol. Plant.*, 2010, Vol. 138, p. 360–371.

Stadnichuk I.N., Yanyushin M.F., Bernát G., Zlenko D.V., Krasilnikov P.M., Lukashev E.P., Maksimov E.G., Paschenko V.Z. Fluorescence quenching of the phycobilisome terminal emitter LCM from the cyanobacterium *Synechocystis* sp. PCC 6803 detected *in vivo* and *in vitro*, *J. Photochem. Photobiol. B.*, 2013, Vol. 125, p. 137–145.

Standfuss J., Terwisscha van Scheltinga A.C., Lamborghini M., Kühlbrandt W. Mechanisms of photoprotection and nonphotochemical quenching in pea light-harvesting complex at 2.5 Å resolution, *The EMBO J.*, 2005, Vol. 24, p. 919.

Steffen R., Christen G., Renger G. Time-resolved monitoring of flash-induced changes of fluorescence quantum yield and decay of delayed light emission in oxygen-evolving photosynthetic organisms, *Biochemistry*, 2001, Vol. 40, p. 173–180.

Steffen R., Eckert H.J., Kelly A.A., Dormann P., Renger G. Investigations on the reaction pattern of photosystem II in leaves from *Arabidopsis thaliana* by time-resolved fluorometric analysis, *Biochemistry*, 2005a, Vol. 44, p. 3123–3133.

Steffen R., Kelly A.A., Huyer J., Doermann P., Renger G. Investigations on the reaction pattern of photosystem II in leaves from *Arabidopsis thaliana* wild type plants and mutants with genetically modified lipid content, *Biochemistry*, 2005b, Vol. 44, p. 3134–3142.

Stirbet A., Govindjee. On the relation between the Kautsky effect (chlorophyll *a* fluorescence induction) and photosystem II: basics and applications of the OJIP fluorescence transient, *J. Photochem. Photobiol. B*, 2011, Vol. 104, p. 236–257.

Strasser A., Srivastava A., Tsimilli-Michael M. The fluorescence transient as a tool to characterize and screen photosynthetic samples, in: *Probing Photosynthesis: Mechanisms, Regulation and Adaptation*, Yunus M., Pathre U., Mohanty P. eds., Taylor and Francis, London, 2000, p. 445–483.

Strasser B., Sánchez-Lamas M., Yanovsky M.J., Casal J.J., Cerdán P.D. *Arabidopsis thaliana* life without phytochromes, *Proc. Natl. Acad. Sci.*, 2010, Vol. 107, p. 4776–4781.

Strasser R.J., Srivastava A., Govindgee. Polyphasic chlorophyll a fluorescence transient in plants and cyanobacteria, *Photochem. Photobiol.*, 1995, Vol. 61, p. 32–42.

Strasser R.J., Tsimilli-Michael M., Srivastava A. Analysis of the chlorophyll *a* fluorescence transient, *Chlorophyll Fluorescence: A Signature of Photosynthesis*, Papageorgiou G.C., Govindjee eds., Series: Advances in Photosynthesis and Respiration, Vol. 19, Kluwer Academic Publishers, the Netherlands, 2004, p. 321–362

Straus M.R., Rietz S., van Themaat E.V.L., Bartsch M., Parker J.E. Salicylic acid antagonism of EDS1-driven cell death is important for immune and oxidative stress responses in *Arabidopsis*, *Plant J.*, 2010, Vol. 62, p. 628–640.

Strid A.W., Chow S., Anderson J.M. UV-B damage and protection at the molecular level in plants, *Photosynth. Res.*, 1994, Vol. 39, p. 475–489.

Stuart Y.E., Campbell T.S., Hohenlohe P.A., Reynolds R.G., Revell L.J., Losos J.B., Rapid evolution of a native species following invasion by a congener, *Science*, Vol. 346, p. 463–466.

Subach F.V., Verkhusha V.V., Chromophore transformations in red fluorescent proteins, *Chem. Rev.*, 2012, Vol. 112, p. 4308–4327.

Suter G.W., Holzwarth A.R. A kinetic model for the energy transfer in phycobilisomes, *Biophys.*, 1987, Vol. 52, p. 673–683.

Suzuki I., Kanesaki Y., Hayashi H., Hall J.J., Simon W.J., Slabas A.R., Murata N. The histidine kinase Hik34 is involved in thermotolerance by regulating the expression of heat shock genes in *Synechocystis, Plant Physiol.*, 2005, Vol. 138, p. 1409–1421.

Suzuki N., Mittler R. Reactive oxygen species and temperature stresses: A delicate balance between signaling and destruction, *Physiol. Plant.*, 2006, Vol. 126, p. 45–51.

Svedruzic D., Jonsson S., Toyota C.G., Reinhardt L.A., Ricagno S., Lindqvist Y., Richards N.G. The enzymes of oxalate metabolism: unexpected structures and mechanisms, *Arch. Biochem. Biophys.*, 2005, Vol. 433, p. 176–192.

Swanson S., Gilroy S., ROS in plant development, *Physiol. Plant.*, 2010, Vol. 138, p. 384–392.

Swanson S.J., Choi W.-G., Chanoca A., Gilroy S. *In vivo* imaging of Ca^{2+}, pH, and reactive oxygen species using fluorescent probes in plants, *Ann. Rev. Plant Biol.*, 2011, Vol. 62, p. 273–297.

Szabo I., Bergantino E., Giacometti G.M., Light and oxygenic photosynthesis: energy dissipation as a protection mechanism against photo-oxidation, *EMBO Rep.*, 2005, Vol. 6, p. 629–634.

Tejwani V., Schmitt F.-J., Wilkening S., Zebger I., Horch M., Lenz O., Friedrich T. Investigation of the NADH/NAD^+ ratio in *Ralstoniaeutropha* using the fluorescence reporter protein Peredox, *BBA – Bioenergetics*, 2017, Vol. 1858, p. 86–94.

Theiss C. *Transiente Femtosekunden-Absorptionsspektroskopie des Anregungs-Energietransfers in Isolierten Pigment-Proteinkomplexen des Photosyntehse-apparates*, PhD thesis, Technische Universität Berlin, Mensch & Buch Verlag, Berlin, 2006; ISBN: 3-86664-188-5.

Theiss C., Andree S., Schmitt F.-J., Renger T., Trostmann I., Eichler H.J., Paulsen H.,Renger G. Pigment-pigment and pigment-protein interactions in recombinant water-soluble chlorophyll proteins (WSCP) from cauliflower, *J. Phys. Chem. B*, 2007b 111, Vol. 46, p. 13325–13335,

Theiss C., Schmitt F.-J., Andree S., Cardenas-Chavez C., Wache K., Fuesers J., Vitali M., Wess M., Kussin S., Eichler H.J., Ecker H.-J. Excitation energy transfer in the phycobiliprotein antenna of *Acaryochloris marina* studied by transient sub-ps absorption and fluorescence spectroscopy, *Photosynth. Res.*, 2007a, Vol. 91, p. 165–166.

Theiss C., Schmitt F.-J., Andree S., Cardenas-Chavez C., Wache K., Fuesers J., Vitali M., Wess M., Kussin S., Eichler H.J., Eckert H.-J. Excitation energy transfer in the phycobiliprotein antenna of *Acaryochloris marina* studied by transient fs absorption and fluorescence spectroscopy, *Photosynthesis. Energy from the Sun*, Allen J.F., Gantt E., Golbeck J.H., Osmond B. eds., Springer, Dordrecht, 2008, p. 339–342.

Theiss C., Schmitt F.-J., Pieper J., Nganou C., Grehn M., Vitali M., Olliges R., Eichler H.J., Eckert H.-J. Excitation energy transfer in intact cells and in the phycobiliprotein antennae of the chlorophyll *d* containing cyanobacterium *Acaryochloris marina*, *J. Plant Physiol.*, 2011, Vol. 168, p. 1473–1487.

Thiele A., Herold M., Lenk I., Quail P.H., Gatz C. Heterologous expression of *Arabidopsis* phytochrome B in transgenic potato influences photosynthetic performance and tuber, *Plant Physiol.*, 1999, Vol. 120, p. 73–81.

Thimm O., Bläsing O., Gibon Y., Nagel A., Meyer S., Krüger P., Selbig J., Müller L.A., Rhee S., Stitt M. MAPMAN: a user-driven tool to display genomics data sets onto diagrams of metabolic pathways and other biological processes, *Plant J.*, 2004, Vol. 37, p. 914–939.

Thordal-Christensen H., Zhang Z., Wei Y., Collinge D.B. Subcellular localization of H_2O_2 in plants. H_2O_2 accumulation in papillae and hypersensitive response during the barley - powdery mildew interaction, *The Plant Journal*, 1997, Vol. 11, p. 1187–1194.

Tognetti V.B., Mühlenbock P., van Breusegem F. Stress homeostasis-the redox and auxin perspective, *Plant Cell Environ.*, 2012, Vol. 35, p. 321–333.

Triantaphylidès C., Krischke M., Hoeberichts F. A., Ksas B., Gresser G., Havaux M., Van Breusegem F., Mueller M.J. Singlet oxygen is the major reactive oxygen species involved in photooxidative damage to plants, *Plant Physiol.*, 2008, Vol. 148, p. 960–968.

Triantaphylidès C., Havaux M. Singlet oxygen in plants: production, detoxification and signaling, *Trends Plant Sci.*, 2009, Vol. 14, p. 219–228.

Trissl H.-W. Modeling the excitation capture in thylakoid membranes, *Photosynthesis in Algae*, T.W. Larkum, S.E. Douglas, J.A. Raven eds., Kluwer Academic Publishers, S., 2003, p. 245–276.

Tronrud D.E., Wen J., Gay L., Blankenship R.E The structural basis for the difference in absorbance spectra for the FMO antenna protein from various green sulfur bacteria, *Photosynth.Res.* 2009, Vol. 100, p. 79–87.

Trouillard M., Shahbazi M., Moyet L., Rappaport F., Joliot P., Kuntz M., Finazzi G. Kinetic properties and physiological role of the plastoquinone terminal oxidase (PTOX) in a vascular plant, *Biochim. Biophys. Acta*, 2012, Vol. 1817, p. 2140–2148.

Tsien R.Y. The green fluorescent protein, *Annu. Rev. Biochem.*, 2008, Vol. 67, p. 509–544.

Umena Y., Kawakami K., Shen J.R., Kamiya N. Crystal structure of oxygen-evolving photosystem II at a resolution of 1.9 A°, *Nature*, 2011, Vol. 473, p. 55–60.

Vallad G.E., Goodman R.M. Systemic acquired resistance and induced systemic resistance in conventional agriculture, *Crop. Sci.*, 2004, Vol. 44, p. 1920–1934.

Van Amerongen H., Croce R. Structure and function of photosystem II Light harvesting proteins (Lhcb) of higher plants, *Primary Processes of Photosynthesis: Basic Principles and Apparatus*, Vol. I, Renger G. ed., RSC Publ., Cambridge, U.K., 2008, p 329–368.

Van Kooten O., Snel J.F.H., Vredenberg W.J. Free energy transduction related to the electric potential changes across the thylakoid membrane, *Photosynth. Res.*, 1986, Vol. 9, p. 211–227.

Vandenabeele S., Vanderauwera S., Vuylsteke M., Rombauts S., Langebartels C., Seidlitz H.K., Zabeau M., van Montagu M., Inzé D., van Breusegem F. Catalase deficiency drastically affects gene expression induced by high light in *Arabidopsis thaliana, Plant J.*, 2004, Vol. 39, p. 45–58.

Vass I., Aro E.-M., Photoinhibition of photosynthetic electron transport, *Primary Processes of Photosynthesis: Basic Principles and Apparatus, Part 1: Photophysical Principles, Pigments Light Harvesting/Adaptation/Stress*, Renger G. ed., Royal Society Chemistry, Cambridge, 2008, p. 393–425.

Vass I., Cser K., Janus-faced charge recombinations in photosystem II photoinhibition, *Trends Plant. Sci.*, 2009, Vol. 14, p. 200–205.

Vellosillo T., Vicente J., Kulasekaran S., Hamberg M., Castresana C. Emerging complexity in reactive oxygen species production and signaling during the response of plants to pathogens, *Plant Physiol.*, 2010, Vol. 154, p. 444–448.

Vener A.V., Ohad I., Andersson B. Protein phosphoorylation and redox sensing in chloroplast thylakoids, *Curr. Opin. Plant Biol.*, 1998, Vol. 1, p. 217–223.

Vitali M. *Long-Term Observation of Living Cells by Wide-Field Fluorescence Lifetime Imaging Microscopy*, PhD thesis, Technical University of Berlin, 2011, 126 p.

Vranova E., Inze D., van Breusegem F. Signal transduction during oxidative stress, *J. Exp. Bot.*, 2002, Vol. 53, p. 1227–1236.

Vredenberg W.J., Bulychev A.A. Photoelectric effects on chlorophyll fluorescence of photosystem II *in vivo*. Kinetic in the absence and presence of valinomycin, *Bioelectrochemistry*, 2003, Vol. 60, p. 87–95.

Wagner D., Przybyla D., Op den Camp R., Kim C., Landgraf F., Lee K.P., Würsch M., Laloi C., Nater M., Hideg E. The genetic basis of singlet oxygen-induced stress responses of *Arabidopsis thaliana*, *Science*, 2004, Vol. 306, p. 1183–1185.

Wen J., Zhang H., Gross M.L., Blankenship R. Native electrospray mass spectrometry reveals the nature and stoichiometry of pigments in the FMO photosynthetic antenna protein, *Biochemistry*, 2011, Vol. 50, p. 3502–3511; dx.doi.org/10.1021/bi200239k

Wessels J.M., Rodgers M.A.J. Detection of the $O_2(^1\Delta_g)$ – $O_2(^3\Sigma g^-)$ transition in aqueous environments: A Fourier-transform near-infrared luminescence study, *J. Phys. Chem.*, 1995, Vol. 99, p. 15725–15727.

Westphal V., Seeger J., Salditt T. and Hell S.W. Stimulated Emission Depletion Microscopy on Lithographic Nanostructures, *J. Phys. B: At. Mol. Opt. Phys.*, 2005, Vol. 38, p. 695.

Wilkening S., Schmitt F.-J., Horch M., Zebger I., Lenz O., Friedrich T. Characterization of Frex as an NADH sensor for in vivo applications in the presence of NAD+ and at various pH values, *Photosynth. Research*, 2017, in press, doi:10.1007/s11120-017-0348-0.

Wilkinson F., Helman W., Ross A.B. Rate constants for the decay and reactions of the lowest electronically excited singlet state of molecular oxygen in solution. An expanded and revised compilation, *J. Phys. Chem. Ref. Data.*, 1995, Vol. 24, p. 663–1021.

Wilson A., Punginelli C., Gall A., Bonetti C., Alexandre M., Routaboul J.M., Kerfeld C.A., van Grondelle R., Robert B., Kennis J.T.M., Kirilovsky D. A photoactive carotenoid protein acting as light intensity sensor, *Proc. Natl. Acad. Sci. U S A*, 2008, Vol. 105, p. 12075–12080.

Wilson A., Kinney J.N., Zwart P.H., Punginelli C., D' Haene S., Perreau F., Klein M.G., Kirilovsky D., Kerfeld C.A., Structural determinants underlying photoprotection in the photoactive orange carotenoid protein of cyanobacteria, *J. Biol. Chem.*, 2010, Vol. 285, p. 18364–18375.

Wilson I.D., Neil S. J., Hancock J. T., Nitric oxide synthesis and signaling in plants, *Plant Cell Environ.*, 2008, Vol. 31, p. 622–631.

Wolfram S. *A New Kind of Science*, Wolfram Media Inc., 2002.

Worrest R.C., Caldwell M.M. eds., *Stratospheric Ozone Reduction, Solar Ultraviolet Radiation and Plant Life*, Springer, Berlin, 1986, p. 171–184.

Xiong J., Bauer C.E. Complex evolution of photosynthesis, *Annu. Rev. Plant Biol.,* 2002, Vol. 53, p. 503–521.

Yabuta Y., Maruta T., Yoshimura K., Ishikawa T., Shigeoka S. Two distinct redox signaling pathways for cytosolic APX induction under photooxidative stress, *Plant Cell Physiol.*, 2004, Vol. 45, p. 1586–1594.

Yamamoto N., Koga N., Nagaoka M. Ferryl–Oxo species produced from Fenton's Reagent via a two-step pathway: minimum free-energy path analysis, *J. Phys. Chem. B*, 2012, Vol. 116, p. 14178–14182.

Yang D.-H., Andersson B., Aro E.-M., Ohad I. The Redox state of the plastoquinone pool controls the level of the light-harvesting chlorophyll *a/b* binding protein complex II (LHC II) during photoacclimation, *Photosynth. Res.*, 2001, Vol. 68, p. 163–174.

Yang F., Moss L.G., Phillips G.N.Jr. The molecular structure of green fluorescent protein., *Nat. Biotechnol.*, 1996, Vol. 14, p. 1246–1251.

Yin D., Kuczera K., Squier T.C. The Sensitivity of carboxyl-terminal methionines in calmodulin isoforms to oxidation by H_2O_2 modulates the ability to activate the plasma membrane Ca-ATPase, *Chem. Res. Toxicol.*, 2000, Vol. 13, p. 103–110.

Yu J.W., Rubio V., Lee N.Y., Bai S., Lee S.Y., Kim S.S., Liu L., Zhang Y., Irigoyen M.L., Sullivan J.A., Zhang Y., Lee I., Xie Q., Paek N.C., Deng X.W. COP1 and ELF3 control circadian function and photoperiodic flowering by regulating GI stability, *Molecular Cell*, 2008, Vol. 32, p. 617–630.

Zamaraev K.I., Parmon V.N. Potential methods and perspectives of solar energy conversion via photocatalytic processes, *Rev. Sci. Eng.*, 1980, Vol. 22, p. 261–324.

Zhang L., Paakkarinen V., van Wijk K.J., Aro E.-M. Biogenesis of the chloroplast-encoded D1 protein: regulation of mechanisms of D1 protein translation elongation, insertion and assembly, *Plant Cell.*, 2000, Vol. 12, p. 1769–1782.

Zhang W.-B., *Synergetic economics*, Springer, Berlin, 1991.

Zhao J., Zhou J.J., Wang Y.Y., Gu J.W., Xie X.Z. Positive regulation of phytochrome B on chlorophyll biosynthesis and chloroplast development in rice. *Rice Science*, 2013, Vol. 20, p. 243–248.

Zheng M., Aslund F., Storz G. Activation of OxyR transcription factor by reversible disulfide bond formation, *Science*, 1998, Vol. 279, p. 1718–1721.

Zhou L., Aon M.A., Almas T., Cortassa S., Winslow R.L., O`Rourke B. A reaction-diffusion model of ROS-induced ROS release in a mitochondrial network, *PLoS Comput. Biol.*, 2010, Vol. 6, e1000657.

Zhou L., O'Rourke B. Cardiac mitochondrial network excitability: insights from computational analysis, *Am. J. Physiol. Heart Circ. Physiol.*, 2012, Vol. 302, p. H2178–H2189.

Zoia L., Argyropoulos D.S. Characterization of free radical spin adducts of 5-diisopropyloxy-phosphoryl-5-methyl-1-pyrroline-N-oxide using mass spectrometry and P-31 nuclear magnetic resonance, *Eur. J. Mass. Spectrom.*, 2010, Vol. 16, p. 175–185.

Zorina A.A., Mironov K.S., Stepanchenko N.S., Sinetova M.A., Koroban N.V., Zinchenko V.V., Kupriyanova E.V., Allakhverdiev S.I., Los D.A. Regulation systems for stress responses in cyanobacteria, *Russ. J. Plant. Physiol.*, 2011, Vol. 58, p. 749–767.

Zulfugarov I., Tovuu A., Kim J.-H., Lee C.-H. Detection of reactive oxygen species in higher plants, *J. Plant Biol.*, 2011, Vol. 54, p. 351–357.

Appendix

i Light Matter Interaction in Classical Electrodynamics

In classical electrodynamics the interaction of light described as an electric field or electromagnetic field with a certain material is given by Maxwell's equations in matter:

$$\nabla \times \overline{E} + \dot{\overline{B}} = 0, \tag{74}$$

$$\nabla \cdot \overline{B} = 0, \tag{75}$$

$$\nabla \cdot \overline{D} = \rho, \tag{76}$$

$$\nabla \times H - \dot{\overline{D}} = \overline{j}. \tag{77}$$

The material parameters are found in the displacement field

$$\overline{D}(\overline{r},t) = \varepsilon_0 \overline{E}(\overline{r},t) + \overline{P}(\overline{r},t) = \varepsilon_0 \overline{E}(\overline{r},t) + \varepsilon_0 \chi_e \overline{E}(\overline{r},t) = \varepsilon_0 \varepsilon_r \overline{E}(\overline{r},t) \tag{78}$$

which is formed by the time and space dependent vacuum electric field $\overline{E}(\overline{r},t)$ and the polarisation $\overline{P}(\overline{r},t)$. The polarisation depends on the electric susceptibility χ_e which determines the relative dielectric constant $\varepsilon_r = (\chi_e + 1)$.

In general the electric field $\overline{E}(\overline{r},t)$ is not necessarily parallel to the polarization $\overline{P}(\overline{r},t)$ as given by eq. 78 but there might be an anisotropy of the polarisation leading to an angle between $\overline{E}(\overline{r},t)$ and $\overline{P}(\overline{r},t)$. In that case the susceptibility has to be a mathematical entity which can turn the electric field vector, i.e. $\overline{\overline{\chi}}_e$ is a tensor and not a scalar. In that general case the polarisation calculates to

$$\overline{P}(\overline{r},t) = \varepsilon_0 \overline{\overline{\chi}}_e \overline{E}. \tag{79}$$

For strong electric fields the polarization $\overline{P}(\overline{r},t)$ as given by eq. 78 is not necessarily proportional to the electric field $\overline{E}(\overline{r},t)$. The expansion as a Taylor series generally denotes to

$$\overline{P}(\overline{r},t)=\varepsilon_0\left(\overline{\overline{\chi}}_e^{(1)}\overline{E}+\overline{\overline{\chi}}_e^{(2)}\overline{E}^2+\overline{\overline{\chi}}_e^{(3)}\overline{E}^3+\ldots\right). \tag{80}$$

Eq. 80 contains anisotropy effects due to the tensor character of $\overline{\overline{\chi}}_e$ but also the nonlinear relation between $\overline{E}(\overline{r},t)$ and $\overline{P}(\overline{r},t)$. According to eq. 80 the polarisation is treated to depend linearly on the electric field as 1st order approximation $\overline{P}(\overline{r},t)\approx\varepsilon_0\overline{\overline{\chi}}_e^{(1)}\overline{E}$ with corrections of second order $\overline{\overline{\chi}}_e^{(2)}\overline{E}^2$ and higher orders.

In full analogy to eq. 78 the magnetic response of the material denotes to

$$\overline{B}(\overline{r},t)=\mu_0\left(\overline{H}(\overline{r},t)+\overline{M}(\overline{r},t)\right)=\mu_0\left(\overline{H}(\overline{r},t)+\chi\overline{H}(\overline{r},t)\right)=\mu_0\mu_r\overline{H}(\overline{r},t)$$

with the magnetic field $\overline{B}(\overline{r},t)$, the magnetizing field $\overline{H}(\overline{r},t)$, and the magnetization $\overline{M}(\overline{r},t)$ which is determined by the relative magnetic permeability $\mu_r=(1+\chi_M)$ or magnetic susceptibility χ_M. For non-ferromagnetic materials which are only slightly diamagnetic or paramagnetic $\mu_r\approx 1$, i.e. the magnetic susceptibility vanishes and the material is transparent for magnetic fields. This is the case for all pigment-protein complexes studied in this thesis. Therefore we neglect the magnetic properties of our materials and focus on the electric properties only. In the following $\mu_r\approx 1$ is assumed.

The relative dielectric constant ε_r contains effects of absorption and refraction. Therefore ε_r must be a complex tensor. For the simple case of isotropic media and linear response ε_r is a complex number $\tilde{\varepsilon}_r=\varepsilon_1+i\varepsilon_2$ with the real part ε_1 of the electric field denoting the refraction and reflection of the electric field amplitude and the imaginary part ε_2 which denotes the damping of the electric field amplitude.

The dielectric constant is the square of the complex refraction index $\tilde{n}$: $\tilde{\varepsilon}_r=\tilde{n}^2=(n+i\kappa)^2$ with the real part n of the complex refraction index and the imaginary part κ. From that one can derive $\varepsilon_1=n^2-\kappa^2$ and $\varepsilon_2=2n\kappa$.

Generally the Maxwell's equations in materia are macroscopic equations, i.e. the material parameters are assumed to be macroscopic parameters which are common mathematical entities for the whole solid state (see eq. 79).

ii The Solid State in Quantum Mechanics

In the following the quantum mechanics of the solid state is briefly introduced and it is shown how Bloch waves are suitable solutions for the eigenfunctions describing these systems assuming periodic potentials and the Born Oppenheimer approximation.

Due to the fact that the potential of Pigment-Protein complexes is nonperiodic it is not possible to use Bloch waves. The theory has to be improved and was widely improved for the description of organic supermolecules like photosynthetic pigment-protein conplexes. Today a good agreement between the calculated and measured time dependent spectra of coupled biological pigment-protein complexes is achieved with new theoretical approaches (Renger and Schlodder, 2005; Renmger and Holzwarth, 2007; Renger et al., 2009). The quantum formalism presented in this chapter is beyond the scope of this book.

The possible energy values measurable on a quantum system characterized by a Hamiltonian $\hat{H}$ are described by the time independent Schrödinger equation:

$$\hat{H}\left|\Psi\right\rangle = E\left|\Psi\right\rangle. \quad (81)$$

The quantum dynamics of states, e.g. time dependent polarisation and population follows the time dependent Schrödinger equation:

$$i\hbar\left|\dot{\Psi}\right\rangle = \hat{H}\left|\Psi\right\rangle. \quad (82)$$

In general we will focus on the temporal dynamics of the population of excited states. The most simple form of the Hamiltonian of a single electron bound to a potential $V(\overline{r})$ denotes to

$$\hat{H} = \frac{\hat{p}_{el}^{\,2}}{2m} + V(\overline{r}). \quad (83)$$

The potential $V(\overline{r})$ has to contain all electromagnetic contributions that influence the electron determining the quantum dynamics of the electronic state $\left|\Psi\right\rangle$.

The Hamiltonian denoted in eq. 83 has to contain all possible energetic contributions as operators that contribute to the total energy of the electronic ground or excited state ${}^{0}S$ or ${}^{1}S$, respectively. These are the energy of the free electron $\frac{\hat{p}_{el}^{\,2}}{2m}$ and all terms of the potential $V(\overline{r})$ contributing to the energy. Generally the potential $V(\overline{r})$ is composed by the Coulomb-potentials of the nuclei $\hat{H}_C$, the electronic repulsion of other electrons $\hat{H}_{el-el}$, the interaction with vibrations $\hat{H}_{el-phon}$ or any other interaction $\hat{H}_{WW}$:

$$\hat{H} = \hat{H}_0 + \hat{H}_C + \hat{H}_{el-el} + \hat{H}_{el-phon} + \hat{H}_{WW} \cdots \tag{84}$$

From the stationary Schrödinger equation (eq. 81) one can achieve a rough estimation for the absorption spectrum of a molecule.

If the potential of a single molecule is modeled e.g. with a rectangular quantum well one finds that the difference between ground and first excited state shrinks with the size of the quantum well.

In fact as bigger the delocalized Pi-electron system of a molecule is as lower the energy difference of the transition ${}^0S \rightarrow {}^1S$.

As mentioned $\hat{H}$ is not a simple function for the coupled pigments in a protein environment.

The Hamiltonian generally looks like (Cohen-Tannouji et al., 1999a; 1999b; Schwabl, 2007; 2008)

$$\begin{aligned}
\hat{H} &= \sum_i \frac{\hat{p}_i^2}{2m_e} + \sum_j \frac{\hat{P}_j^2}{2M_j} \\
&+ \frac{1}{4\pi\varepsilon_0}\left(\sum_{j>j'} \frac{e^2 Z_j Z_{j'}}{|\bar{R}_j - \bar{R}_{j'}|} - \sum_{i,j} \frac{e^2 Z_j}{|\bar{r}_i - \bar{R}_j|} + \sum_{i>i'} \frac{e^2}{|\bar{r}_i - \bar{r}_{i'}|}\right) \\
&= \frac{-\hbar^2}{2m_e} \sum_i \Delta_{r_i} + \frac{-\hbar^2}{2} \sum_j \frac{\Delta_{R_j}}{M_j} \\
&+ \frac{1}{4\pi\varepsilon_0}\left(\sum_{j>j'} \frac{e^2 Z_j Z_{j'}}{|\bar{R}_j - \bar{R}_{j'}|} - \sum_{i,j} \frac{e^2 Z_j}{|\bar{r}_i - \bar{R}_j|} + \sum_{i>i'} \frac{e^2}{|\bar{r}_i - \bar{r}_{i'}|}\right)
\end{aligned} \tag{85}$$

where the kinetic terms of electrons with mass m_e and the kinetic terms of nuclei with variable mass M_j for the nucleus j with charge eZ_j are described as well as the coulombic interaction of the nuclei, between the electrons and the nuclei and between the electrons themselves. Principally all neighbouring charges have to be summarized. Simplifications can be done e.g. assuming only the next neighbouring charges in the lattice.

In eq. 55 coupling between the spin momentum of the electrons and cores are neglected which have to be included into the Schroedinger equation separately or which are calculated from relativistic formulations of the electron dynamics. As rate equations are used for the description of optical spectra in this thesis these aspects are not elaborated here. For an elaborated discussion of the quantum mechanical description of electrons see (Cohen-Tannouji et al., 1999a; 1999b; Schwabl, 2007; 2008; Messiah, 1991a; 1991b) for large molecules see (Demtröder, 2003).

According to the Born–Oppenheimer approximation the wave function $\left|\Psi_k\left(r_i,R_j\right)\right\rangle$ solving the Hamiltonian of eq. 85 factorises into the wave function of the electrons and the nuclei:

$$\left|\Psi_k\left(r_i,R_j\right)\right\rangle=\left|\Psi_k\left(r_i\right)\right\rangle\left|\Psi_k\left(R_j\right)\right\rangle. \tag{86}$$

The wave functions solving eq. 86 could be found employing the concept of Bloch waves if the potential given in the Hamiltonian is periodic. These Bloch waves comprise a term expressing the structural lattice, multiplied with a plane wave part.

$$\left|\Psi_k\left(r,R_j\right)\right\rangle=e^{ikr}\varphi_k\left(R_j\right). \tag{87}$$

Every solution for a given time point can be expressed by a superposition of spatial Bloch waves forming a wave packet

$$\left|\Psi_{k0}\left(r_i,R_j\right)\right\rangle=\int a(k)e^{i\bar{k}\bar{r}}dk. \tag{88}$$

The electromagnetic interactions between the electrons are different for electrons in the ground state or in the excited state (see Figure 81). In addition we have modulations of the ground and excited states due to the coupling of the electrons and the phonons in the system.

The spectrum of possible interactions of electrons in a crystal lattice or protein environment covers electron-electron interaction (see Figure 81, 1st row), electron-phonon interaction (see Figure 81, second row), light coupling and other couplings (see Figure 81, 3rd row). For simplicity the electron is often treated as 2-level system.

Coupled 2-level states can undergo several transitions, e.g. by spontaneous emission of light or phonons or by inter system crossing (ISC). These transitions are discussed later and evaluated at the experimental data. Schematic illustrations of excitation energy transfer transitions also including electron delocalisation and electron transfer are shown in Figure 81, 4th row.

The electron-phonon interaction as pointed out in Figure 81 shifts lifetimes (i.e. the transition probabilities) and the energy of the states.

In that case the Born–Oppenheimer approximation is not longer valid and the electronic wave function does not factorise into electronic part and vibrational part (of the core lattice) as shown in eq. 86 (Cohen-Tannouji et al., 1999a; 1999b; Schwabl, 2007; 2008; Messiah, 1991a; 1991b; Demtröder, 2003).

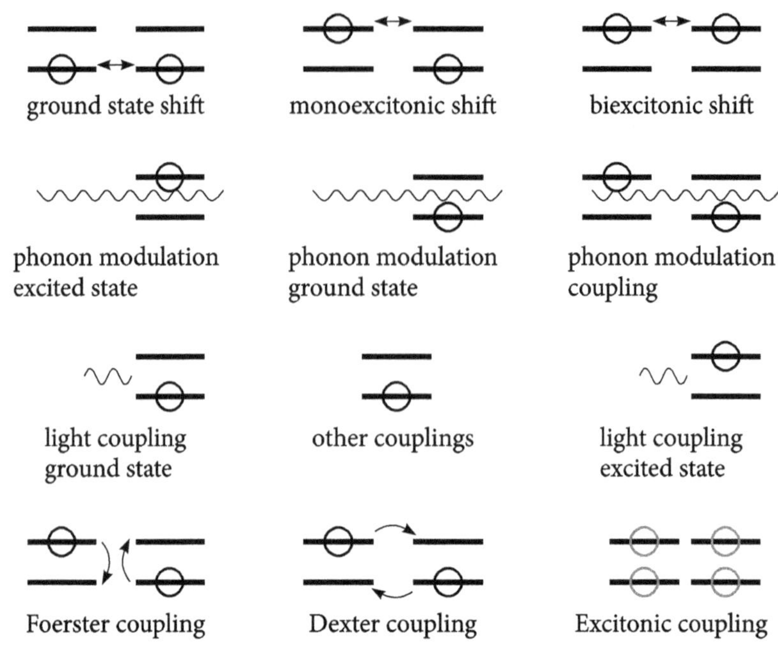

Figure 81. Interactions and EET/ET processes of coupled 2-Niveau systems bound to a protein matrix.

If the electron is assumed to move in a periodic potential one finds a description of the electron where it is treated perturbatively like a free particle with an effective mass describing the inertia of the electronic mass:

$$\frac{1}{m_{eff.}} = \frac{1}{\hbar^2} \frac{\partial^2 E}{\partial k_i \partial k_j}. \tag{89}$$

In the case of pigment-protein complexes the lattice is not periodic. Therefore the Born–Oppenheimer approximation and eq. 86–89 are not valid. Especially $\left|\Psi_k\left(r,R_j\right)\right\rangle = e^{ikr}\varphi_k\left(R_j\right)$ in eq. 87 can not be expressed as a periodic wave function. Therefore the electron can not be treated like a free particle with an effective mass as given in eq. 89.

In first order perturbation theory the probability for the transition from a state i to a state j is proportional to the square of the matrix element V_{ji} which is known as Fermi´s Golden rule. This approximation calculates the Einstein coefficient for the spontaneous emission to $A_{01} = \left|\left\langle \Psi_0 \middle| \hat{p} \middle| \Psi_1 \right\rangle\right|^2$, where $\left|\Psi_1\right\rangle$ and $\left|\Psi_0\right\rangle$ denote the excited and the ground state, respectively.

iii Full Quantum Mechanical Description

The harmonic oscillator can be described by the creation and annihilation operators. It had been shown that this formalism is suitable for a description of pigment-protein complexes (Richter et al., 2008). For a complete elaboration of this formalism see (Mukamel 1995).

The Hamiltonian of the 1-dim harmonic oscillator denotes to

$$\hat{H} = \hbar\omega\left(a^{+}a + \frac{1}{2}\right) \tag{90}$$

where a^{+} denotes the creation operator and a the annihilation operator which are defined according to

$$\begin{aligned} a^{+} &= \frac{1}{\sqrt{2}}\left(-\frac{d}{dq} + q\right), \\ a &= \frac{1}{\sqrt{2}}\left(\frac{d}{dq} + q\right) \end{aligned} \tag{91}$$

with $q = \sqrt{\frac{m\omega}{\hbar}}x$.

With eq. 91 the Schroedinger equation 90 reads as

$$\hat{H} = \frac{\hbar\omega}{2}\left(-\frac{d^2}{dq^2} + q^2\right) = -\frac{\hbar^2\Delta_x}{2m} + \frac{1}{2}m\omega^2x^2 = \frac{\hat{p}_{el}^{\,2}}{2m} + V(\overline{r}) \tag{92}$$

which is the well known form of a free electron in the harmonic potential $V(\overline{r}) = \frac{1}{2}m\omega^2x^2$ according to eq. 83. In that notation one can write the uncoupled part of the electron and phonon Hamiltonian of a big pigment-protein complex containing N chlorophyll molecules as

$$\hat{H}_0 = \hat{H}_{0,el.} + \hat{H}_{0,phon.} = \sum_{i=1}^{N}\left(\varepsilon_{gi}a_{gi}^{\ +}a_{gi} + \varepsilon_{ei}a_{ei}^{\ +}a_{ei}\right) + \sum_{j=1}^{M}\hbar\omega_j b_j^{\ +}b_j \tag{93}$$

where the energies ε_{gi} of the ground state of pigment number i with the corresponding operators $a_{gi}^{\ +}$ and a_{gi} and the energies of the excited states ε_{ei} with the corresponding operators $a_{ei}^{\ +}$ and a_{ei} are summarized for the electrons. The Chlorophyll molecules are treated as 2-level systems of single electrons which can exist in the ground state or excited states and the term $\hat{H}_{0,phon.} = \sum_{i=1}^{M}\hbar\omega_j b_j^{\ +}b_j$ is the operator for the phonon energy $\hbar\omega_j$ of the jth phonon mode in the system. b_j^+ and b_j describe the creation and the annihilation operator, respectively, for phonons in the jth eigenmode.

The eigenvalue of $\sum_{i=1}^{N}\left(a_{gi}^{+}a_{gi}+a_{ei}^{+}a_{ei}\right)$ is $N=N_g+N_e$ denoting the numbers N_g and N_e of Chl molecules in the ground state and excited state, respectively. The eigenvalue of $b_j^+b_j$ is the number of phonons in the jth eigenmode. Therefore the eigenvalue of eq. 64 is the overall energy in the uncoupled system summarizing all energies of ground state electrons, excited state electrons and phonons.

As pointed out in Figure 81 there exist many couplings between electrons and between electrons and phonons in the pigment-protein-complex that have to be incorporated in the Hamiltonian of eq. 93.

With electric dipole approximation the light coupling of the ground state electrons is described as

$$\hat{H}_{el-h\nu}=\sum_{i=1}^{N}\bar{E}(t)\cdot\hat{\bar{d}}_i a_{ei}^{+}a_{gi}+h.c.$$

denoting the fact that the incoming electric light field $\bar{E}(t)$ couples to the electric dipole operator of the ith Chl molecule $\hat{\bar{d}}_i$ and leads to an annihilation of a ground state while an excited state is created. The hermitian conjugated expression (h.c.) describes the induced emission annihilating an excited state while the ground state is populated.

According to (Richter et al., 2008) all the couplings shown in Figure 81 are incorporated into the coulomb interaction-Hamiltonian. The coulomb coupling is containing all relevant electronic interactions of electrons bound to ground or excited states:

$$\hat{H}_C=\sum_{i>j}V_{ij}^{gg}a_{gi}^{+}a_{gj}^{+}a_{gj}a_{gi}+\sum_{i\neq j}V_{ij}^{eg}a_{ei}^{+}a_{gj}^{+}a_{gj}a_{ei}$$
$$+\sum_{i>j}V_{ij}^{ee}a_{ei}^{+}a_{ej}^{+}a_{ej}a_{ei}+\sum_{i\neq j}V_{ij}^{F}a_{ei}^{+}a_{gj}^{+}a_{ej}a_{gi}.$$

$\hat{H}_C$ is containing all terms for ground state shift due to the potential V_{ij}^{gg} of the neighbouring electrons in ground states, monoexcitonic shifts due to the potential V_{ij}^{eg} and biexcitonic shift due to V_{ij}^{ee} (see Figure 81).

The term V_{ij}^{F} denotes the dipole coupling between neighbouring Chl molecules, i.e. the Förster Coupling. If Förster Coupling occurs between the ith pigment in the ground state and the jth pigment in the excited state. Then the Förster Resonance Energy Transfer leads to an annihilation of the ground state of pigment i and the excited state of pigment j while a ground state of pigment j and an excited state of pigment i is created.

The electron–phonon coupling for the electrons in the excited state can be formulated as following:

$$\hat{H}_{el-phon} = \sum_i \sum_j \left(g_{ij}^{(c)} b_j + g_{ij}^{(c)*} b_j^+ \right) a_{ei}^+ a_{ei}$$

where the electron-phonon coupling is described by the matrix elements $g_{ij}^{(c)}$ expressing the probability for phonon absorption or phonon emission. $g_{ij}^{(c)}$ can be determined through the spectral density function which can be determined by experiments or molecular mechanic calculations (Richter et al., 2008).

Index

Abiotic stress, 157, 172
Anoxygenic photosynthetic bacteria, 67, 68, 71
Antenna complexes, 23, 43, 67, 68, 70–91, 92, 95, 101, 113, 115, 178
Antioxidant enzymes, 123, 136, 168, 177
Apoplast, 130, 154, 166, 178

Bacteriochlorophyll, 71, 75
Big bang of the biosphere, 201, 207
Biotic stress, 157, 158, 164
Bloch equation, 20
Bottom up, 4, 6, 9, 11, 14, 24–26, 124, 168, 174, 176–179, 211

Calcium signaling, 191
Calvin–Benson cycle, 131, 135
Cascade, 125, 130, 151, 165, 166, 170, 171, 173
Cell apoptosis, 165
Cellular automata, 5–13, 15, 200–202
Charge separation, 38, 51, 68–70, 92, 105, 109, 111, 112, 114, 129, 130
Charge stabilization, 92, 105
Charge transfer, 14, 64, 127
Chloroplast, 21, 67, 119–121, 124, 130–132, 143, 152–154, 156, 159, 160, 163–167, 169–172, 174–175, 192, 196
Closed reaction centers, 105, 111–113
Color intensity plot, 40, 94
Compartment, 22–24, 42, 43, 45, 47, 83, 95, 96, 125, 135, 147, 153, 154, 176, 196
Complex networks, 5, 15, 174
Control parameter, 12, 26, 211
Core antenna, 23, 43, 68, 98, 118
Core complex, 66–69, 86

D1-repair cycle, 129
Darwin, Charles, 199, 206, 207
Decay associated spectra (DAS), 39–41, 47–52, 95, 96

Decay time, 45, 46, 48, 62, 92, 99, 100, 137
Detailed balance, 18
Dexter transfer, 54, 58, 75, 116
Dichlorofluorescein (DCF), 140–144, 195
Differential equation, 15, 32, 33, 42, 44, 45, 109, 110
Dipole moment, 25, 56, 57, 77

Ecologic niche, 201, 206, 210
Ecosphere, 199, 200, 209
Einstein coefficient, 33–35, 254
Electrochemical biosensors, 150–151
Electromagnetic waves, 27, 29
Electron paramagnetic resonance (EPR), 131, 138, 145–147
Electron transfer (ET), 3, 13, 15, 19–20, 22, 38, 42, 52, 54, 67, 68, 86, 93–94, 99, 105, 107, 109, 110, 112, 114, 131, 134, 149, 193
Electron transfer chain (ETC), 86, 131, 134, 154, 169, 175, 179, 193
Elongation factor G (EF-G), 135, 159, 174
Emergence, 6, 9, 11, 25, 119
Entropy, 19, 95, 96, 100, 179–190, 200
Entropy dynamics, 186
Equilibrium, 18, 34, 44, 46, 179, 180, 182–185, 188–190, 211
Evolution, 4, 11–13, 15–17, 31, 68, 115, 149, 151, 153, 161, 199–207, 209, 211
Excitation energy transfer (EET), 13, 15, 19, 20, 22–24, 28, 38, 42, 43, 48, 51, 54, 59–60, 70, 72, 79, 83–84, 86, 91–105, 115, 117, 118, 254
Exogenous dyes, 138
Exogenous fluorescence probes, 138, 139
Exponential decay, 38, 45, 47, 54, 96, 112
Extrinsic antenna complex, 22

Fenton reaction, 132, 135
Fluorescence dynamics, 19, 35–39, 46, 64, 94–96, 100
Fluorescence induction, 72, 92, 114
Fluorescence lifetime, 38, 39, 55, 61, 62
Fluorescence quantum yield, 39, 73, 92, 93, 105, 107, 117
Fluorescence rise kinetics, 41, 64, 72, 94
Fluorescence spectroscopy, 48, 99
Fluorescent proteins, 138, 147, 149
Förster Resonance Energy Transfer (FRET), 19, 51, 52, 55–64, 75, 77, 89
Franck Condon principle, 35–37

Gene activation, 4
Gene expression, 118, 123, 124, 154, 160, 162, 163, 167–174, 177, 178, 191
Genetic diversity, 11, 199, 200, 205–207
Genetic signaling, 124, 152
Genetically encoded ROS sensors, 137, 146–150
Green fluorescent protein (GFP), 125, 138, 146–150

Haber–Weiss reaction, 136
Haken, Hermann, 12–14, 25, 211
Hierarchical architecture, 118–121
Higher plants, 19, 66–69, 71–75, 86, 93, 105–114, 116, 121, 138, 154, 159, 164, 173, 192, 196
Histidine kinase, 153, 161, 162, 167, 168, 173, 176, 177
Hydrogen peroxide, 125, 139, 152, 153, 158–167
Hydroxide, 123
Hydroxyl radical, 125, 139, 145, 147

Induced emission, 33–35, 256
Infinite monkey theorem, 203–205
Intrinsic antenna complex, 22, 86, 95

Jablonski Diagram, 35, 36

Laser switching contrast microscopy (LSCM), 147
Levinthal-paradox, 201, 203
Lifetime of fluorescence, 38, 39
Lifetime of ROS, 123
Light harvesting complex, 13, 23, 65, 66, 69, 71–75, 115, 117, 169, 178, 193
Light reaction, 3, 21, 103, 121

MAPK cascade, 151, 165, 166, 172, 173, 178
Marcus theory, 19
Master equation, 17, 18, 32, 33
Maxwell equations, 28, 31–33, 249, 250
Membrane potential, 178
Membrane protein, 21, 119, 120
Metastructure, 2
Midpoint potential, 136
Mitochondria, 119, 130, 147, 149, 150, 152, 158, 160, 191, 192
Mitogen-activated protein kinase (MAPK), 26, 118, 151, 165, 166, 172–175, 178
Multiexponential, 41, 45, 55, 62, 94
Multiscale hierarchical processes, 1–26
Multisensory, 161, 167, 176, 177

Networking, 11, 13, 174, 178, 191–192
Nonequilibrium, 19, 25, 61, 180, 182–184, 186, 190
Nonphotochemical quenching, 12, 14, 92, 110, 114–118
Nucleus, 119, 139, 148, 160, 163, 164, 169–172, 175, 192

Order parameter, 25–26
Oxidants, 123, 125
Oxidative burst, 125
Oxidative degradation, 123, 136
Oxygenic photosynthesis, 67, 71, 199–201, 207

Pathogen infection, 126, 136, 153, 165
Peroxidase, 123, 130, 132, 133, 135, 136, 138, 139, 151, 154, 162, 164–166, 168, 170, 177
Peroxide, 125, 128, 139, 152, 153, 158, 163
Peroxisomes, 130, 152, 160, 175, 192
Phosphorescence, 37, 127, 137

Photobiology, 14, 27
Photobleaching, 142, 150
Photochemical quenching, 110
Photoinhibition, 129, 159, 179
Photosynthetic apparatus, 129, 151, 191
Photosynthetic complexes, 20, 53, 59, 64, 65, 184
Photosystem I (PS I), 131, 170
Photosystem II (PS II), 19, 66, 92, 94, 107, 108, 114, 126, 130, 143, 155, 164, 169, 181
Point-dipole approximation, 54–57
Prey-predator model, 25, 192–195
Probability per time unit, 20, 35, 53, 58
Programmed cell death, 126, 136, 157, 163–165, 191
Prokaryotic cyanobacteria, 159, 161
Prompt fluorescence, 72, 92, 114
Protective mechanism, 115, 135, 155
Protein translation, 135
Protochlorophyllide, 164
Proton transfer, 3, 19
Pseudocyclic electron transport, 133, 134

Rate equations, 11–13, 15–23, 27, 28, 31–33, 41–64, 67, 93–114, 118, 184, 186, 193, 194, 252
Reaction center, 23, 38, 43, 51, 66–68, 70, 92, 105, 114, 115, 126, 165
Reactive nitrogen species, 125, 129
Reactive oxygen species (ROS), 12–14, 123–156, 199
Redox-regulated, 159, 178
Reduction potential, 136
Repair mechanisms, 129, 136
Respiratory burst, 124
ROS scavenging, 14, 115, 135, 136, 174, 194, 199
ROS signaling, 20, 152, 157–197
ROS-sensitive fluorescent proteins, 149
ROS-waves, 25, 134, 143, 192–195

Second law of thermodynamics, 179–182, 184, 185, 189, 190
Second messenger, 164, 168, 170, 171, 178, 191–192, 196
Secondary ROS, 14, 130
Selection pressure, 199, 201, 205–207
Self-organization, 25, 26, 211
Signal transduction, 123, 161, 169–171, 173–175
Signaling chains, 191
Signaling role of ROS, 20, 123, 151–156
Singlet oxygen, 37, 38, 65, 70, 116, 118, 125, 126, 128, 137, 139, 145, 146, 158, 163, 193
Slaving principle, 25, 26, 119
Special pair, 67, 69
Spin traps, 131, 137–139, 145–147
Stoichiometry of photosynthesis, 3
Stokes shift, 36
Stress signals, 152, 153, 161, 164, 167, 176
Stress-resistance, 154, 192
Superoxide dismutase (SOD), 123, 132, 133, 136, 160, 169
Superoxide radicals, 125, 132, 145, 147

Thylakoid membrane, 13, 20, 22, 43, 65, 68, 71, 74, 85, 86, 89, 93, 101, 115, 116, 119–121, 133, 155, 164–166, 169, 175
Time-correlated single photon counting (TCSPC), 39, 48, 54, 70, 92–95, 97, 100, 114
Time dependency, 18, 33
Time resolved fluorescence spectra, 40, 51, 95
Top down, 9–11, 14, 24–26, 118, 119, 124, 168, 174, 176–179, 211
Transcription factor, 150, 151, 154, 162, 163, 166, 170, 172–175, 178, 196
Transfer matrix, 16, 17, 32, 33, 44–46
Translation, 135, 159, 168, 170, 174, 176, 177
Transmembrane electrochemical potential, 22, 67, 112, 121, 130
Triplet oxygen, 118, 126
Triplet-triplet interaction, 38, 126

Uncertainty principle, 30

Water-oxidizing complex (WOC), 69, 109, 129, 130, 134
Water-water cycle, 128, 130, 133

CPSIA information can be obtained
at www.ICGtesting.com
Printed in the USA
BVHW03*1931110518
515556BV00010B/27/P